FRAMINGHAM STATE COLLEGE
D1035144

KEYWORDS IN EVOLUTIONARY BIOLOGY

KEYWORDS IN EVOLUTIONARY BIOLOGY

EDITED BY

Evelyn Fox Keller and Elisabeth A. Lloyd

HARVARD UNIVERSITY PRESS
CAMBRIDGE, MASSACHUSETTS
LONDON, ENGLAND
1992

Printed in the United States of America
10 9 8 7 6 5 4 3 2 1

This book is printed on acid-free paper, and its binding
materials have been chosen for strength and durability.

Library of Congress Cataloging-in-Publication Data

Keywords in evolutionary biology /
edited by Evelyn Fox Keller and Elisabeth A. Lloyd.
p. cm.
Includes bibliographical references and index.
ISBN 0-674-50312-0
1. Evolution (Biology)—Terminology.
I. Keller, Evelyn Fox, 1936– . II. Lloyd, Elisabeth Anne.
QH360.6.K49 1992
575′.0014—dc20 92-8283
CIP

Designed by Gwen Frankfeldt

CONTENTS

ACKNOWLEDGMENTS

BY ITS very nature, the realization of this project has depended entirely on the good will and cooperation of our colleagues in evolutionary biology and the history and philosophy of science who agreed to write essays for it. Our first and primary debt is therefore to the contributors, especially for their patience with and responsiveness to our many requests. We are also indebted to Renee Courey for her good-humored, diligent, and vast help in compiling the bibliography; to Howard Boyer, Lindsay Waters, and Elizabeth Gretz at Harvard University Press for their unfailingly generous support and assistance in organizing and editing this manuscript; and to Jennifer Culbert for her help in the final stages of editing.

Permission has been granted by the University of California Press to reprint in "Competition: Current Usages" and "Fitness: Reproductive Ambiguities" portions of Evelyn Fox Keller, "Language and Ideology in Evolutionary Biology," in *The Boundaries of Humanity: Humans, Animals, and Machines,* ed. James Sheehan and Morton Sosna, pp. 90–95, 97–99, copyright © 1991 The Regents of the University of California; by Princeton University Press to reprint in "Natural Selection: Current Usages" material from chapters 1 and 2 of John A. Endler, *Natural Selection in the Wild,* copyright © 1986 Princeton University Press; by the American Statistical Association to reproduce in "Heritability: Some Theoretical Ambiguities" Figure 1 from Sewall Wright, "Statistical Methods in Biology," *Journal of the American Statistical Association* 26:155–163 (1931); by W. H. Freeman to reproduce in "Genotype and Phenotype" Figure 1-7 from D. T. Suzuki, A. J. F. Griffiths, J. H. Miller, and R. C. Lewontin, *An Introduction to Genetic Analysis* (New York: W. H. Freeman, 1981); and by the director of the Bancroft Library to quote in "Macromutation" from Sewall Wright's letter to Richard Goldschmidt, August 12, 1949, Goldschmidt Papers, Bancroft Library, University of California, Berkeley. Michael J. Donoghue acknowledges the grant support of the NSF (BSR-8822658) in the preparation of "Homology."

CONTRIBUTORS

PETER ABRAMS
Department of Ecology and
Behavioral Biology
University of Minnesota
Minneapolis, Minnesota

JOHN BEATTY
Department of Ecology and
Behavioral Biology
University of Minnesota
Minneapolis, Minnesota

DOUGLAS H. BOUCHER
Appalachian Environmental
Laboratory
University of Maryland
Frostburg, Maryland

PETER J. BOWLER
Department of Social Anthropology
The Queen's University of Belfast
Belfast, Northern Ireland

ROBERT N. BRANDON
Department of Philosophy
Duke University
Durham, North Carolina

RICHARD M. BURIAN
Department of Philosophy
Virginia Polytechnic and
State University
Blacksburg, Virginia

ROBERT K. COLWELL
Department of Ecology and
Evolutionary Biology
University of Connecticut
Storrs, Connecticut

HELENA CRONIN
Department of Philosophy
London School of Economics
London, England

JAMES F. CROW
Genetics Department
University of Wisconsin
Madison, Wisconsin

JOHN DAMUTH
Department of Biological Sciences
University of California, Santa Barbara
Santa Barbara, California

LINDLEY DARDEN
Department of Philosophy
University of Maryland
College Park, Maryland

RICHARD DAWKINS
Department of Zoology
University of Oxford
Oxford, England

MICHAEL R. DIETRICH
Department of Philosophy, History and Philosophy of Science Program
University of California, Davis
Davis, California

MICHAEL J. DONOGHUE
Department of Ecology and Evolutionary Biology
University of Arizona
Tucson, Arizona

LEE A. DUGATKIN
Department of Biology
Mt. Allison University
Sackville, New Brunswick
Canada

JOHN DUPRÉ
Department of Philosophy
Stanford University
Stanford, California

JOHN A. ENDLER
Department of Biological Sciences
University of California, Santa Barbara
Santa Barbara, California

MARCUS W. FELDMAN
Department of Biological Sciences
Stanford University
Stanford, California

KURT FRISTRUP
Biology Department
Woods Hole Oceanographic Institution
Woods Hole, Massachusetts

DEBORAH M. GORDON
Department of Biological Sciences
Stanford University
Stanford, California

STEPHEN JAY GOULD
Museum of Comparative Zoology
Harvard University
Cambridge, Massachusetts

JAMES R. GRIESEMER
Department of Philosophy and Center for Population Biology
University of California, Davis
Davis, California

M. J. S. HODGE
Philosophy Department
Leeds University
Leeds, England

DAVID L. HULL
Department of Philosophy
Northwestern University
Evanston, Illinois

EVELYN FOX KELLER
Program in Science, Technology, and Society
Massachusetts Institute of Technology
Cambridge, Massachusetts

DANIEL J. KEVLES
Department of History
California Institute of Technology
Pasadena, California

MOTOO KIMURA
National Institute of Genetics
Mishima, Japan

PHILIP KITCHER
Philosophy Department
University of California, San Diego
La Jolla, California

JAMES G. LENNOX
Department of History and Philosophy of Science
University of Pittsburgh
Pittsburgh, Pennsylvania

RICHARD C. LEWONTIN
Museum of Comparative Zoology
Harvard University
Cambridge, Massachusetts

ELISABETH A. LLOYD
Department of Philosophy
University of California, Berkeley
Berkeley, California

JANE MAIENSCHEIN
Department of Philosophy
Arizona State University
Tempe, Arizona

JUDITH C. MASTERS
College of Science
University of the Witwatersrand
Wits, South Africa

ROBERT McINTOSH
Biology Department
University of Notre Dame
Notre Dame, Indiana

DIANE PAUL
Department of Political Science
University of Massachusetts
Boston, Massachusetts

ROBERT J. RICHARDS
Committee on the Conceptual
Foundations of Science
University of Chicago
Chicago, Illinois

ALEXANDER ROSENBERG
Department of Philosophy
University of California, Riverside
Riverside, California

MICHAEL RUSE
Departments of History
and Philosophy
University of Guelph
Guelph, Ontario
Canada

ELLIOTT SOBER
Department of Philosophy
University of Wisconsin
Madison, Wisconsin

HAMISH G. SPENCER
Department of Zoology
University of Otago
Dunedin, New Zealand

PETER F. STEVENS
The Gray Herbarium of
Harvard University
Harvard University
Cambridge, Massachusetts

PETER TAYLOR
Program on Science, Technology
and Society
Cornell University
Ithaca, New York

MARCY K. UYENOYAMA
Department of Zoology
Duke University
Durham, North Carolina

MICHAEL J. WADE
Department of Biology
University of Chicago
Chicago, Illinois

MARY JANE WEST-EBERHARD
Smithsonian Tropical Research
Institute
Balboa, Panama

MARY B. WILLIAMS
Center for Science and Culture
University of Delaware
Newark, Delaware

DAVID SLOAN WILSON
Department of Biological Sciences
State University of New York
Binghamton, New York

KEYWORDS IN EVOLUTIONARY BIOLOGY

Introduction

Evelyn Fox Keller and Elisabeth A. Lloyd

UNLIKE poets, and even unlike most speakers of ordinary prose, scientists expect and indeed generally assume that their language is (or at least ought to be) both precise and clear. Scientific terms are intended to mean neither more nor less than what they say, and to say neither more nor less than what they mean. In the traditional model for scientific language, at least since Leibniz, Condillac, and Pascal, terminological ambiguity, uncertainty, and double entendre are generally seen as evidence of scientific inadequacy—as impediments simultaneously to progress and to truth and, accordingly, as impurities requiring removal. In the writings of the early positivists of this century, insistence on the univocality and unireferentiality of scientific language reached a new height.[1] It might even be said that escape from the vagaries, opacity, and imprecision of ordinary language has become one of the primary functions of technical vocabulary.

The reality, of course, is somewhat different. It would be difficult to find or even to construct a sentence composed strictly of technical terms; in practice, scientific discourse is entirely suffused with ordinary language, with terms that bring with them all varieties of the imprecision scientists seek to avoid. More distressing yet, even technical terms turn out, far more often than we had hoped, to be plagued by the unruliness of ordinary language. By virtue of their dependence on ordinary language counterparts, technical terms carry, along with their ties to the natural world of inanimate and animate objects, indissoluble ties to the social world of ordinary language speakers. In this way, even carefully delineated technical

1. Gillian Beer cites the work of Leonard Bloomfield's *Linguistic Aspects of Science* (1939) as an instance of the linguistic positivism of the earlier part of this century. Bloomfield wrote, "It is our task to discover which of our terms are undefined or partially defined or draggled with fringes of connotation, and to catch our hypotheses and exhibit them by clear statements, instead of letting them haunt us in the dark" (quoted in Beer, 1987, p. 44).

terms are bedeviled by semantic shadows that insistently blur their borders. Words, even technical terms, have insidious ways of traversing the boundaries of particular theories, of historical periods, and of disciplines—in the process contaminating the very notion of a pure culture. They serve as conduits for unacknowledged, unbidden, and often unwelcome traffic between worlds. Words also have memories; they can insinuate a theoretical or cultural past into the present. Finally, they have force. Upon examination, their multiple shadows and memories can be seen to perform real conceptual work, in science as in ordinary language.[2] They help to hold worldviews together, to bridge disparate (even contradictory) concepts, to insulate us from problems we cannot solve. They work to help make arguments persuasive, even to turn arguments into "proofs." It is words that take us from the logic of a predicate calculus to the logos of scientific reasoning.

Over the past thirty years, the traditional model of scientific language so hopefully aspired to by working scientists has come under a barrage of criticism. Not only is the practice remote from the ideal, but significant challenges to even the possibility of such an ideal language have recently been posed by scholars in the history, philosophy, and sociology of science (see, e.g., Kuhn, 1962, 1979; Black, 1962, 1979; Hesse, 1966, 1980, 1985; Rorty, 1985; Beer, 1983). Early on, Thomas Kuhn focused attention on the importance of (generally unconscious) changes in the meaning of scientific terms and showed how such changes can signal the profound shifts in worldviews that we associate with scientific revolutions. For Kuhn, as for others, this recognition provided a starting point for more intensive subsequent investigations into the complex (even tortured) relations between language and "nature." The possibility of the traditional goal of univocality and precision for scientific language recedes yet further if one believes that meanings do not simply change, but, in a certain sense, accumulate—"carry[ing] the mark of the historical (sedimented) circumstances of their origin and use in ever new ways" (Edie, 1976, pp. 154–158; see also Carlisle, 1980).

In parallel (and virtual synchrony) with Kuhn, Max Black made an important contribution from a somewhat different perspective to the view of scientific language as "open rather than closed." By calling philosophi-

2. See, e.g., the discussion of the concept of "normal" in Ian Hacking, *The Taming of Chance* (1990). Hacking writes: "The word [normal] is also like a faithful retainer, a voice from the past. It uses a power as old as Aristotle to bridge the fact/value distinction, whispering in your ear that what is normal is also right. But also . . . , it has become a soothsayer, teller of the future, of progress and ends . . ." (pp. 160–161). And finally, "The normal stands indifferently for what is typical, the unenthusiastic objective average, but it also stands for what has been, good health, and for what shall be, our chosen destiny. That is why the benign and sterile-sounding word 'normal' has become one of the most powerful ideological tools of the twentieth century" (p. 169).

cal attention to the similarities between models and metaphors, Black (1962) provided a basis for regarding the use of metaphor in the construction of scientific theories as beneficial. This initial argument for the scientific value of linguistic "open-endedness" (such as that found in the use of metaphors) has since been considerably extended by others, especially by Mary Hesse (1966, 1980, 1985). Finally, and most recently, the influence of critical theory (or deconstruction) has become detectable in discussions of language and science as authors such as Gillian Beer and Richard Rorty have tentatively begun to argue for the same kind of conceptual productivity for ambiguity (or semantic polysemy)[3] in scientific texts as was earlier argued for in literary texts.

Yet throughout all such efforts to undermine our traditional conception of a clear demarcation between scientific and ordinary or literary language, one crucial distinction remains relatively intact. Although it may not be possible, or even wholly desirable, to achieve a fixed meaning for scientific terms, the effort to "control and curtail the power of language" remains a significant feature of scientific activity. The very extent to which scientists (far more than speakers of ordinary language) *aim* at a language of fixed and unambiguous meanings constitutes, in itself, one of the most distinctive features of their enterprise. And even though never quite realizable, this effort to control the vicissitudes of language, like the commitment to objectivity, reaps distinctive cognitive benefits. The same effort also reaps distinctive social benefits, on which at least some of the cognitive benefits depend. It especially serves to delineate a disciplinary and theoretical community, a community whose participants can be identified by their tacit agreement to abide by local conventions that restrict the range of possible meanings and, hence, stabilize the discourse. Because of the abiding commitment of working scientists to precision and clarity, to fixed meaning, the elaboration of prevailing instabilities (or multiplicities) of meaning attempted here will be of value to scientists themselves. We have chosen to concentrate on evolutionary biology for the simple reason that the borders between subdisciplines in this field are less well drawn than in many other disciplines, and the conventions necessary to stabilize meaning are correspondingly less clearly established. It is because of their commitment to restabilizing their own discourse that scientists working in this field need to be able to identify the domains where meanings are unstable.

Accordingly, our goal in this book has been to identify and explicate those terms in evolutionary biology that, though commonly used, are plagued in their usage by multiple concurrent and historically varying

3. Arguments had earlier been extended for the political and/or rhetorical (rather than conceptual) productivity of ambiguity in scientific language (a noteworthy example being Robert Young's important paper, "Darwin's Metaphor: Does Nature Select?" in Young, 1985), but just how sharply the distinction between rhetorical and conceptual can be maintained remains a question for consideration.

meanings. Our choice of the term "keywords" is thus indebted to Raymond Williams, for it was he who first used it in this particular sense and who first alerted us to the social, political, and intellectual value of exploring the multiple and shifting meanings of familiar terms. Williams' *Keywords* (1976) was intended primarily for social and intellectual historians; this book, by contrast, is intended as much for scientists and philosophers actually working in the field of evolutionary biology as it is for historians and sociologists of science. These two groups of readers, however, will surely use the book in different ways.

The relevance of this project to historians of science interested in the cognitive evolution of scientific theories (that is, in the history of ideas) will be evident. But it is as a resource for the social history of science that this project bears its closest resemblance to Williams' own work on "keywords." Williams' project grew out of what he saw as a problem of *vocabulary:* "the available and developing meanings of known words, which needed to be set down; and the explicit but as often implicit connections which people were making, in what seemed to me . . . particular formations of meaning." He wrote:

> Keywords are significant, binding words in certain activities and their interpretation; they are significant, indicative words in certain forms of thought. Certain uses bound together certain ways of seeing culture and society, not least in these two most general words [i.e., culture and society]. Certain other uses seemed to me to open up issues and problems, in the same general area, of which we all needed to be very much more conscious.

Williams is a cultural historian; accordingly, he is primarily interested in "keywords" as fruitful indicators of social patterns and patterns of social change. As historians and philosophers of science, however, our interest in "keywords" is primarily as indicators of patterns of scientific meaning and of changes over time in the ways that particular scientific meanings have been structured. Attending to the multiple meanings of key terms provides a lens through which it is possible not only to understand better what is at issue in particular scientific debates but also to scrutinize the very structure of the arguments under debate. Such a lens enables an exploration of the historically evolving field of meanings from which these arguments draw and on which they depend. Gould's essay on "heterochrony," Damuth's on "extinction," Donoghue's on "homology," and Stevens' on "species" all provide good examples of such analyses.

In no case, however, and especially not in evolutionary biology, is the field of meanings on which scientific representations of nature draw strictly scientific. Indeed, it is precisely because of the large overlap between forms of scientific thought and forms of social thought that "keywords"—terms whose meanings chronically and insistently traverse the boundaries between ordinary and technical discourse—can serve not simply as indicators of either social meaning and social change *or* scientific meaning and scientific change but as indicators of the ongoing traffic *between* social and scientific meaning and, accordingly, between social and scientific change.

Part of our motivation for this "glossary," then, is our belief that Williams' mode of analysis can provide social historians of science with a powerful technique to add to their repertoire for the ongoing project of mapping the complex interactions between science and culture. The individual contributions to this volume are not themselves directed at such a project, but they provide the raw materials that, collectively, constitute a necessary first step. It is our hope and expectation that others will be able to employ these materials in developing the linguistic tools that have already begun to emerge as critical components of a fully social analysis of science.

But evolutionary biology is also a field in which theoretical and conceptual development is ongoing—as, indeed, the relative instability of meaning of so many of its key terms attests. For those actually engaged in the current development of this field, attention to semantic multiplicity will appeal to a diametrically opposite value. Just as "the naive positivistic equivalence between object and event, or utterance, presupposes a single necessary theoretical outcome" (Beer, 1987, p. 43), a relative looseness in terminology may be correspondingly essential to the maintenance and fostering of speculative multiplicity. To the extent that the abundance of multiple meanings of the key terms of this field reflects the ways in which evolutionary theory is a developing rather than a finished theory, attention to that multiplicity might itself be argued to provide a valuable theoretical and conceptual stimulus.

Consider, for example, current debates revolving around the term "group selection." As David Sloan Wilson shows in his essay, group selection is commonly used in a number of very different meanings in the scientific literature, and much of the controversy that currently rages around the existence or importance of group selection actually depends on these differences. Clarification of these differences can serve not only to facilitate communication but also to help focus debate on specific empirical issues, and thus to pave the way for more fruitful scientific discussion. Similar claims can be made for the terms "mutualism" and "cooperation," "competition," and "altruism" as well as for "gene" and "species."

In a number of cases we have commissioned more than one essay on a topic, in some instances in an effort to distinguish historic variation in meaning from contemporary ambiguities. Richard Burian, a philosopher, presents a historical review of the concept of adaptation, for example, while the biologist Mary Jane West-Eberhard offers a survey of contemporary understandings. Similarly, the essay on competition by Robert McIntosh is intended to provide a historical complement to Keller's focus on contemporary usage, much as Diane Paul's essay on fitness provides a historical complement to John Beatty's philosophical exploration of conceptual and theoretical tensions among currently available definitions of the term.

Other essays were commissioned to explore differences in meaning that derive from divergences in disciplinary perspectives. Michael Wade's essay on epistasis, for example, focuses on the difference between the way biochemists, on the one hand, and statistical evolutionary biologists, on the other, use the term; he draws on this distinction to clarify a long-standing argument between R. A. Fisher and Sewall Wright over the importance of interaction effects in evolution. In like manner, both Richard Lewontin's discussion of genotype and phenotype and Diane Paul's discussion of heterosis contribute to the clarification of debates that have arisen between evolutionary geneticists and developmental biologists.

For those readers primarily interested in the scientific content of the essays, we must emphasize that this collection of essays is not intended to provide definitive or correct definitions of key terms in evolutionary biology, nor is it intended to adjudicate between competing claims about what might constitute a "correct," or even preferable, definition. Our aim, rather, is to provide a rough map of some of the territory of dispute and change. Similarly, this book will not provide either a profile of, or an introduction to, the field as a whole. Not all terms currently of theoretical importance are included, because not all such terms are subject to multiple or significantly changing meanings. Rather than an exhaustive survey, this book should be regarded as an introduction to a new kind of analysis. For scientists and philosophers, the major value of such analyses lies in their capacity to probe semantic origins of debates and even, perhaps, to identify particular technical problems for future research.

We offer these essays, then, as a rough guide to past and present meanings of many of the key terms in the discourse of evolutionary biology. The book is designed as a resource both for those working in this field as scientists, philosophers, historians, and sociologists and for those with a larger interest in the interrelations of culture, language, and science. Awareness of the multiplicity of meaning present in the structure and growth of central concepts in evolutionary biology should serve to facilitate communication, clarify technical debates, and move us toward an understanding of the different commitments involved in different meanings of these terms. Our understanding of the operation of language in the structure of scientific concepts and arguments has barely begun. It is our final hope that that this volume will serve as a stimulus for the further exploration of the dynamics of polysemy, ambiguity, and uncertainty in the structure and development of scientific representations of nature.

ADAPTATION: HISTORICAL PERSPECTIVES

Richard M. Burian

ORIGINALLY, the term "adapted" and its cognates meant apt, or suited to, some particular purpose or other. This keyword entry concerns uses of the term "adaptation" in evolutionary contexts. The term has other biological uses, covering, for example, physiological and biochemical adaptations of organisms to stress, altitude, nutrition, and so on. In evolutionary contexts, however, the term deals with *trans*generational changes in the properties, propensities, or capacities of organisms.

The term "adaptation" has two primary meanings in evolutionary contexts. One concerns *evolutionary processes:* here, "adaptation" means those transgenerational alterations of the features and capacities of organisms in a lineage that enable them to solve (or improve on previous solutions of) problems posed by the environment, problems of internal integration, and the problem of reproducing. According to the theory of evolution by means of natural selection, such changes are produced, at least largely, by the differential survival of organisms with (small) net advantages over conspecific organisms in respect to these "problems." Other views about the processes of adaptation are possible and have played important roles in the history of evolutionary biology. Around 1900, for example, mutationists argued that natural selection was neither cumulative nor creative; it was no more than a sieve, choosing between sharply distinctive variants produced by mutation. On most such accounts, the direction of evolutionary change was explained by the orderliness of mutational processes; according to many mutationists, therefore, internal sources of order accounted for the evident adaptive structuring of organisms. This was one reason for the affinity of mutationists for orthogenetic theories.

The second main use of "adaptation" concerns *features of organisms:* a trait or capacity counts as an adaptation if it is the product of a process of adaptation. In one recent usage reflecting a commitment to natural selection (Sober, 1984a, cf. Waddington, 1957a), a trait is an adaptation if and only if it is the product of selection *for* the trait in question. A related use of "adaptation," sometimes confused with the one just described, treats a trait (or complex of traits) as an adaptation if and only if it contributes to an organism's adaptedness—its relative fitness or likelihood of reproductive survival. This last definition does not connect adaptations with their evolutionary history. As will be argued below, it is desirable in evolutionary contexts to retain the connection of the term "adaptation" to evolutionary processes (cf. Brandon, 1978 and 1990, chap. 1).

These competing definitions reflect the history of the term in ways that call for comment. Charles Darwin followed the natural theologian Paley in requiring an explanation for the remarkable structural and behavioral "contrivances" (Paley's term) by means of which plants and animals satisfy their needs (see Nuovo, 1992). In *On the Origin of Species* (1859b), Darwin held that one could recognize the functions of most organs and behaviors and that, with important exceptions (which provided crucial evidence for evolution), their design—that is, the fit between structure and function—was remarkably good. Indeed, he considered it to be one of his most important accomplishments to have provided a wholly naturalistic explanation of the ways in which adaptations (contrivances) are brought about; in doing so he removed the explanation of adaptations from the purview of natural theology and placed it in the domain of biology. He argued that even the origin of "organs of extreme perfection and complication," such as the eye, could be explained by a process of descent with modification (pp. 186ff.) and that, in general, the modification of an imperfect into a better organ is achieved by natural selection. One important feature of that process on which Darwin insisted is the "probability of conversion from one function to another." He cited a large number of examples of this sort, such as the conversion of the swimbladders of fish to the lungs of amphibia, of the "ovigerous frena" (a secretory organ in pedunculated cirripedes) to respiratory branchiae in Balanidae, and so on.

Because of conversion of function, neither the history of an organ nor the processes by which it was shaped to perform its present function can be determined by an analysis of its present features or functions. Accordingly, one must separate the contribution that an organ, behavior, or character makes to the survival or well-being of its bearer (its contribution to adaptedness in current language) from the historical origin of that trait. History is far more difficult to determine than current function; as G. C. Williams (1966, p. vii) wrote a century later, "Evolutionary adaptation is a special and onerous concept." Its use requires comparative evidence and difficult reconstructive inference regarding the process of design.

One of the glories of Darwin's *Origin* is the rich apparatus it brings to bear on such historical inference. Darwin made it plausible, on the one hand, that even organs such as the eye arose by stepwise preservation and alteration of ancestral organs with quite different original functions and, on the other, that one could find traces of the history of an organ or behavior in its *mal*adaptedness to its present uses.

Darwin's separation of the origin from the present adaptive value of an organ or behavior freed him from supposing that organs are "perfectly adapted" (Ospovat, 1981; Burian, 1983) or that they must now clearly benefit an organism to count as adaptations. It also enabled him to use vestigial and odd features of organs as clues to their history; an organ that has features not connected with its present function may, *ipso facto,* provide traces of past form that provides crucial evidence regarding prior functions. He also separated contributions to the well-being or "quality" (present adaptedness) of an organism from contributions to its ability to obtain mates. (This was the basis of his distinction between natural and sexual selection, central to a dispute with Alfred Russel Wallace that has been revived, in substance, in the recent literature.)

Because traits with high adaptive value need not have been "selected for," it is important to distinguish terminologically between the contribution of an organ (or trait, complex of traits, behavior, etc.) to the current *adaptedness* or the *reproductive survival* of an organism and the claim that that organ is an *adaptation*—that is, that it derives from an evolutionary process that has perfected it in virtue of its contribution to adaptedness.

During the "eclipse of Darwinism" near the turn of the century (Bowler, 1983), doubt prevailed about the adaptive value of many traits now clearly recognized as contributing to well-being or reproductive success. Poulton, Punnett, and others, for example, debated the adaptive value of mimicry in butterflies (see the brief surveys in Kimler, 1983, 1986, and Turner, 1983). Such doubts complicated attempts to connect the present adaptive value of traits with the evolutionary process(es) that produced them and to explain present features of organisms by selective processes. Thus the conceptual problems surrounding the relation of process to product and the choice of proper criteria for applying the term "adaptation" to particular traits were particularly severe, as is clear from the literature of the period (cf., e.g., Kellogg, 1907).

The separate directions taken by evolutionary theory and ecology during the first half of this century turned, in part, on conceptual differences over adaptation. A dominant ecological tradition treated adaptive processes as occurring at higher levels (community, species, group). The sharp articulation of a group adaptationist position by V. C. Wynne-Edwards (1962), according to which population numbers were regulated so as to maximize the likelihood of group survival, occasioned Williams' (1966) influential insistence on an intimate conceptual linkage between selection "of alternative alleles in Mendelian populations" (p. vii) and the adaptations (selected design features) of organisms. Williams' analysis sought to show not only that this linkage was crucial to the analysis of traits as adaptations but also that no other mode of selection (e.g., group selection) could plausibly play a major role in evolution.

One other theme is important for full understanding of Williams' argument. The integration of population genetics into the synthetic theory of evolution brought the concept of (expected relative) fitness as (expected) reproductive success into prominence. This new concept introduces a further conceptual problem. It is difficult to assess the specific effect of a trait on the ecological adaptedness of its bearers; it is harder to assess its net effect on reproductive success and the interplay between such adaptedness and reproductive success (Burian, 1983). Furthermore, it is difficult for population genetics to separate the effects of a trait from the effects of traits with which it is linked. (See, e.g., Clutton-Brock and Harvey, 1979, on confounding variables and Brakefield, 1988, for a discussion of difficulties in the familiar case of industrial melanism.) As a result, correlations between a trait (or trait state) and expected reproductive fitness are often substituted for an analysis based on the design or function (contribution to organismic well-being) of the trait. By allowing traits or trait states to count as adaptations when they are merely correlated with increased reproductive success and by allowing increased (expected) fitness to stand in for adaptive value, one reduces the empirical content of explanations of adaptation and greatly weakens the theory of natural selection.

Terminological and conceptual problems are deeply intertwined here. In an effort to untangle these, Williams (1966) argued that one must show how design features of a putative adaptation have contributed to its establishment in transgenerational time. He also required that one distinguish adaptive features of populations (a fleet herd of deer) from those of individuals (a herd of fleet deer). This stance set the frame for subsequent debates over adaptation, adaptive value, fitness, and units of selection.

The remainder of this entry deals primarily with the conceptual difficulties raised by supposing that high adaptive value of a trait guarantees that it is an adaptation. T. H. Clutton-Brock and P. H. Harvey, for example, though their use of comparative methods is extremely sophisticated, exemplify the assumption that traits contributing to present adaptedness are adaptations. They define an adaptation "as a difference between two phenotypic traits (or complexes of traits) which increases the inclusive fitness . . . of its carrier" (1979, p. 548). Such usage is subject to Williams' strictures: it is quite possible that such traits are "free riders" and that they did not arise directly by a process of adaptation (or even, as Clutton-Brock and Harvey emphasize, by any genetic process at all).

The identification and the evolutionary roles of the target of selection—that is, what was "selected for"—are both at stake here (Waddington,

1957a, p. 65; Mayr, 1965b, 1988; cf. Brandon, 1982 and 1990, for one technique of determining the target of selection). This requires (1) distinguishing adaptations from "exaptations" (phenotypes coopted for their present use via conversion of function; Gould and Vrba, 1982); (2) understanding the interrelations between the benefit a trait confers on the organism as an interactor (e.g., Hull, 1980; Brandon, 1982, 1990; Dawkins, 1982b) and the way it is transmitted from one generation to the next (replication plus development); and (3) connecting the present state of the trait with its history. An example bearing on all three of these issues concerns the early evolution of the organs that became wings in insects; Kingsolver and Koehl (1985) provide suggestive evidence that thermoregulatory protuberances, undergoing allometric change with change in body size, were coopted to become wings. Should this be correct, early gliding wings were exaptations, subsequently improved for their new functions by a process of adaptation.

Definitions or procedures that conflate contributions to inclusive fitness with contributions to the resolution of specific environmental problems or with derivation by an adaptive process would force one to treat some complexes of traits (e.g., those produced by the T-allele in the mouse) as adaptations in spite of the fact that they produce organismic liabilities. The complex of traits produced in T-allele heterozygotes (the homozygote is lethal) increases the inclusive fitness of its bearers; therefore it must count as an adaptation insofar as inclusive fitness is the criterion. But the increase in inclusive fitness is due to meiotic drive which, in specifiable circumstances, increases the representation of the heterozygotes in subsequent generations *in spite of deleterious effects of those traits at the organismic and group levels.* At best, on this account, the T-allele confers genic, but not organismic, advantage. Given the Clutton-Brock and Harvey definition, the evidence concerning adaptive value mistakenly classifies the traits associated with the allele as (organismic) adaptations. Knowledge of causal history is required for a correct classification.

On the account here advocated, one cannot identify evolutionary adaptations without an analysis of historical pathways (or a satisfactory surrogate, perhaps based in part on the comparative techniques employed by Clutton-Brock and Harvey, 1979). The fact that possession of a trait complex correlates with increased adaptedness or inclusive fitness of its bearers is not sufficient to establish that it is an (evolutionary) adaptation, for that fact leaves unresolved questions regarding the historical pathway by which the trait arose. Whatever one's terminology, it is necessary to distinguish among the present adaptive value of a trait, its linkage with other traits, its effect on inclusive fitness, the historical sequence that produced it, and the causal processes determining its history. The concern with terminology in this domain arises from the manifest confusion in the literature arising from failure to apply these distinctions consistently; it is for this reason that the conceptual issues underlying these distinctions are of great importance and deserve close and careful attention.

Recent controversies over the "adaptationist program," largely initiated by Stephen Jay Gould and Richard Lewontin (1979), have turned on the question whether advocates of the synthetic theory of evolution are, or should be, committed to the following claims: (1) arbitrary traits are usually adaptations or, rarely, side effects of adaptive processes; (2) when traits are not optimal, it is nearly always because of trade-offs among competing demands; (3) adaptive selection works on relatively separable features of the organism, virtually always at the organismic level (cf. Mayr, 1983).

None of these commitments need be part of a sound adaptationist program in the sense discussed by West-Eberhard (on ADAPTATION). But the deep and complex biological issues that occasioned the debate remain: does selection operate at many levels, producing many levels of adaptations? If so, how does this affect the nature of adaptive processes and of adaptations? To what extent, and how, are the features and behaviors of organisms compromises between conflicting adaptive constraints? Can we clarify the relationships among adaptations and adaptive values, the levels of selection, the determination of (expected) fitness, and the analysis of evolutionary processes? Clarification of fundamental concepts is by no means sufficient to accomplish these ends, but it is a necessary part of the effort to advance evolutionary theory beyond its current state.

ADAPTATION: CURRENT USAGES

Mary Jane West-Eberhard

IN CONTEMPORARY evolutionary biology an "adaptation" is a characteristic of an organism whose form is the result of selection in a particular functional context (see Williams, 1966; Futuyma, 1986). Accordingly, the process of "adaptation" is the evolutionary modification of a character under selection for efficient or advantageous (fitness-enhancing) functioning in a particular context or set of contexts. The word is sometimes also applied to individual organisms to denote the "propensity to survive and reproduce" in a particular environment (general adaptation) (see Mayr, 1988). Ernst Mayr (1986) suggests substituting the term "adaptedness" for this usage.

The use of "adaptation" by evolutionary biologists thus differs from that in some other areas of biology, where the term can refer to short-term physiological adjustments by phenotypically plastic individuals (adaptability) or to a change in the responsiveness of muscle/nerve tissue upon repeated stimulation.

According to strict usage in evolutionary biology, it is correct to consider a character an "adaptation" for a particular task only if there is some evidence that it has evolved (been modified during its evolutionary history) in specific ways to make it more effective in the performance of that task, and that the change has occurred due to the increased fitness that results. Incidental ability to perform a task effectively is not sufficient; nor is mere existence of a good fit between organism and environment. To be considered an adaptation a trait must be shown to be a consequence of selection for that trait, whether natural selection or sexual and social selection—whether the selective context involves what Darwin called "the struggle for existence," or competitive interactions with conspecifics.

Several kinds of evidence can contribute to determining whether or not a characteristic of an organism is an adaptation (after Curio, 1973, elaborating on suggestions of Tinbergen, 1967). The first is correlation between character and environment or use. A character shows evidence of being an adaptation if: (a) the same form or similar forms occur in similar environments in a number of different species, especially in unrelated species (due to convergence); (b) variant forms of a character in a number of related species (e.g., of a single genus) accord with differences in the environments of the respective species, or with the mode of usage of the character in different species; (c) variant forms appearing in different life stages during ontogeny accord with differences in the environment or behavior of the respective life stages; or (d) for complex characters in a particular context, the more their component aspects can be related point by point to function in that context (the goodness of "design" of Williams, 1966, pp. 12ff.).

The second kind of evidence used in determining whether a characteristic is an adaptation is that which results from altering a character. An organ or behavior is experimentally altered or eliminated, in order to see how this affects its efficiency in a particular function or environmental condition.

A third kind of evidence is obtained through comparison of naturally occurring variants (individual differences). The efficiency or reproductive success of different forms or morphs within a species are compared in the situation(s) where they are hypothesized to function as adaptations.

All of these approaches provide evidence for or against the hypothesis that the structural peculiarities of a trait owe their existence (spread and persistence) in a population to their contribution to fitness via performance of a particular task.

An example can serve to illustrate some of the difficulties in applying the adaptation hypothesis to particular cases. The elaborately sculptured and species-specific forms of the head and thoracic horns of male beetles have been imagined to be adaptations for fighting, for digging, and for influencing female choice of mates. Observations of behavior, however, demonstrate that the structural details of beetle horns and the differences between related species correspond to inter-specific differences in the particular ways they are wielded during battles between males; their special features are not used in special ways during courtship or digging, although they are occasionally used to hold females or to enlarge holes occupied by beetles (Eberhard, 1979, 1980). Thus the available evidence supports the hypothesis that beetle horns are adaptations for fighting, and that they are only incidentally or secondarily used during mating and digging. It could be argued, however, that the structural peculiarities observed are developmental or pleiotropic results of traits evolved in other contexts (the "exaptations" of Gould and Vrba, 1982), and that the high degree of correlation with behavior (which is difficult to consider merely coincidental) has been produced by selection to use these incidentally common structures to the individual's advantage in fights; by this interpretation horn morphology would be a nonadaptation and the form of behavior an adaptation.

It is not always easy to apply the distinction between adaptation and incidental use, even given information on present employment and evolutionary history. Suppose an incidental use or secondary function were to persist while the original, evolved function disappeared (e.g., horns came to be used exclusively for digging even though they had not been modified in that context). Strict adherence to the above definition would not permit horns to be considered an adaptation for digging even though digging had become the exclusive context for their use and even though they might be maintained (rather than lost) under selection in that context. The concept of "pre-adaptation" has been applied to such cases, in which a trait has evolved in one context and has come to be used (function) in another.

Suppose a horn used secondarily but exclusively in digging undergoes some small modification enhancing the digging function. Can it then be considered an "adaptation" for digging? Evidently it can, although this points up another difficulty in the distinction: how much modification is necessary to consider a character an adaptation in a particular context? What, indeed, is a "character," as opposed to a feature or modification of a character? (See CHARACTER.) The designation of an aspect of the phenotype as a character (whether an adaptation or not) is always somewhat arbitrary: is digging behavior, along with horn morphology, part of a single co-selected trait? This would classify the pre-adapted horn as part of a new "adaptation."

Curio (1973) argues that when exactly the same character is employed in more than one context and contributes to fitness in all contexts it should be regarded as an adaptation only for that context where it makes the greatest contribution to fitness. Such an argument can lead to contradictions in applying the above criteria, for example, if the form of a character has been shaped in the past primarily by a function presently of less importance (in terms of fitness) than another use (which by Curio's criterion would be the primary adaptive context even if not effecting evolutionary modification of the character). In most discussions, the historical criterion (rather than fitness difference) would predominate: the character would be considered an "adaptation for" the function in which it was originally or primarily shaped by selection. Even when multiple uses are completely contemporaneous in their fitness effects Curio's criterion seems difficult to apply, given that, insofar as the same form can serve multiple functions, the sum of all (even minor) contributions to fitness could influence form in the face of counterselection (in other contexts) favoring alternative forms. These considerations regarding multiple functions apply as well to questions of selection at different levels of organization, whereby the same trait may simultaneously affect, for example, the survival or replication rate of individuals and groups, and hence the population frequencies of their constituent genotypes.

Given current usage of the word "adaptation," it is clear that not all observable evolved characteristics of organisms are properly regarded as adaptations. In their efforts to explain peculiarities of form, biologists often attempt to apply a hypothesis of adaptation with insufficient empirical support. Several authors have argued in favor of parsimony in the use of this term (e.g., Williams, 1966; Curio, 1973; Gould and Lewontin, 1979). They stress the importance of considering alternative explanations for particular and even complex characters, especially the hypotheses that form can be vestigial (the product of selective forces no longer operating) or the incidental result of developmental processes evolved under selection for other aspects of the phenotype.

Stephen Jay Gould (1984b) has proposed that covariance of characters could be accepted as "positive evidence" of nonadaptation, and has erected a dichotomy of "automatic sequelae" (nonadaptations) versus selected traits (adaptations). This criterion of nonadaptation tacitly requires some analysis of adaptation, however, because it is impossible to tell from covariance alone which of several developmentally associated traits has been most important in the spread and/or maintenance of the set. Furthermore, one cannot assume that covariant aspects have not been modified independent of each other. For example, Gould (1981) interpreted the male-like female display morphology and behavior of the genitalic displays of female hyenas as a nonadaptation, evolved by selection in males and only incidentally or secondarily expressed in females. However, female genital displays are known to function as appeasement gestures (Wickler, 1966; Eibl-Eibesfeldt, 1970), and if modified or somewhat specialized due to selection on females, they would qualify as adaptations. This would be true even if a set of characters used in this way originated via a regulatory mutation that allowed them to be expressed in females as well as in males (where the original set had been formed by selection). Indeed new adaptations may sometimes originate as coadapted character sets, whose expression has been shifted between sexes or life stages (via heterochrony) and then modified in the new context (see West-Eberhard, 1989).

Gould (1984b) also argued that "ecophenotypic responses" to environmental conditions cannot be regarded as adaptations, because they are not "genetically mediated," but this criterion for nonadaptation (environmental influence in phenotype determination) cannot hold unequivocably: plasticity itself can be seen as an adaptation. Furthermore, ecophenotypic responses are always products of gene-environment interaction and thus are genetically mediated (see West-Eberhard, 1989). By Gould's criterion, all environmentally cued, facultatively expressed phenotypes would presumably be classified as "nonadaptations," including the winter pelage of hibernating mammals, the restive walking behavior of the swarming phase of migratory locusts, and the ability of chameleons to match the background coloration of their resting places.

Developmental mechanism per se does not provide enough information to determine whether or not a trait is an adaptation, though it might provide information on how nonadaptive traits are maintained (e.g., via covariance with adaptive traits), and even on how adaptive traits originate. An aspect of the phenotype that is a secondary "by-product" of selection for another aspect (in the sense of being either completely covariant with it or a less commonly expressed product of the same genotype) may have the following relationships to adaptation and selection.

a. The observed frequency and form of the secondary aspect of the phenotype may be completely owing to characteristics evolved under selection for a covariant aspect, in which case the character would not be regarded as an adaptation.

b. More than one covariant aspect of the phenotype may contribute simultaneously to fitness in different functional contexts (e.g., pleiotropic effects of a single gene) from the time of their (simultaneous) origin and be concurrently favored by selection. I would call both positively selected traits adaptations, even if one of them made a greater contribution to the fitness and spread of the covariant set and its underlying genes, because both aspects contribute to the rate of spread of the set in competition with alternatives; Curio (1973) would term only the greater contributor to fitness an adaptation.

c. The initial spread or frequency of the secondary aspect of the phenotype in the population may have been entirely due to selection for a covariant aspect, but its form and/or frequency of expression may have been modified in the context in which it is expressed. In this case a phenotype not originally an adaptation has become an adaptation by evolution in its own context.

To classify a pleiotropic or secondary effect as a non-adaptation requires showing not only that it is (a) only expressed together with a developmentally related trait that is a proven adaptation, but also evidence that (b) concurrent positive selection, and (c) independent modification do not apply.

Overly facile application of the term adaptation encourages the assumption that all characters are adaptive; for this reason, some authors have urged restraint on use of the term. It remains the case, however, that persistent attempts to discern the adaptive significance of phenotypic traits—to apply an adaptation hypothesis—have been a primary and fruitful occupation of evolutionary biologists since before Darwin. There is still controversy over the importance of selection and adaptation versus nonadaptation in the evolution of phenotypes. Although adaptation cannot be assumed, some authors argue that it should be regarded as the most important (commonly supported) hypothesis for the spread and persistence of organismic traits: "The experimental study of adaptation has unravelled adaptive values in such unobtrusive and inconspicuous details of organismic organization that one should think of a character as having survival value until the contrary has been demonstrated" (Curio, 1973, p. 1046). Richard Lewontin (1978, p. 125) gave the following compelling reason for continuing to pursue the "adaptationist" program that seeks to explain characters in terms of their evolved functions, in spite of its difficulties: "Even if the assertion of universal adaptation is difficult to test because simplifying assumptions and ingenious explanations can almost always result in an ad hoc adaptive explanation, at least in principle some of the assumptions can be tested in some cases. A weaker form of evolutionary explanation that explained some proportion of the cases by adaptation and left the rest to allometry, pleiotropy, random gene fixations, linkage and indirect selection would be utterly impervious to test. It would leave the biologist free to pursue the adaptationist program in the easy cases and leave the difficult ones on the scrap heap of chance. In a sense, then, biologists are forced to the extreme adaptationist program because the alternatives, although they are undoubtedly operative in many cases, are untestable in particular cases."

ALTRUISM: THEORETICAL CONTEXTS

Alexander Rosenberg

IN *Sociobiology: The New Synthesis* (1975, p. 578) E. O. Wilson defines altruism as "self-destructive behavior performed for the benefit of others." More specifically, sociobiology treats behavior as altruistic whenever the behavior increases the reproductive fitness of another at the expense of one's own reproductive fitness. At the outset of his touchstone of contemporary behavioral biology, Wilson identifies altruism as the "central theoretical problem of sociobiology," and asks, "how can altruism, which by definition reduces personal fitness, possibly evolve by natural selection?" (1975, p. 3). The problem is apparently one of explaining how the actual is possible. Altruism, like cooperation in general, is an obvious feature of human and infrahuman behavior. Indeed, sociality requires it. And yet if altruism reduces fitness, in the evolutionary long run it should have been expunged, not enhanced. So we face a choice between exempting human and some infrahuman behavior from the constraint of natural selection or finding a way of rendering it consistent with Darwin's theory. This is Wilson's problem.

The biological problem of altruism is vexed by a prior terminological controversy. Altruism as commonly understood is by definition action that advantages another by design. It is an "etiological" concept, which carries a definitional commitment to a motive—an intentional cause. But this motive is missing in Wilson's definition. And few sociobiologists suppose that the occurrences of actions and their intentional causes are explainable by natural selection. Nevertheless, sociobiological altruism might be relevant to motivated altruism. It may explain a genus—altruism motivated or unmotivated—of which motivated altruism is a species.

But even if what the sociobiologist means by altruism has nothing to do with motivated altruism, Wilson's stipulative definition still describes an important phenomenon with which evolutionary theory must come to terms. For other-regarding behavior—no matter its cause—seems endemic to mammalian life and essential for human society. We need an explanation of how it can be consistent with the theory of natural selection.

In addition to differences about motivation, we should bear in mind another significant difference between biological altruism and motivated altruism. What ordinary language means by altruism is sacrifice of self-interest—that is, the interests of the individual. Biological altruism involves not what we would ordinarily recognize as the interests of the individual but, rather, the interests of its offspring, immediate and distant. For motivated altruism's disservice to the interests of the self, biological altruism substitutes disservice to the future survival of the line of descent, the biological lineage of which the self is a member; the coin in which biological altruism is measured is evolutionary fitness—differential rates of reproduction. Sometimes the interests of the self and of its descendants coincide. But not always. For a sociobiological account of altruism to shed much light on motivated altruism among humans it must link the interests of a lineage with that of an individual.

The problem of altruism, and more generally of cooperation among humans, is one Charles Darwin broached in *The Descent of Man* (1871). Broadly, his answer to the question of how other-regarding behavior is possible over the long haul of natural selection appealed to the group as the unit of selection. Darwin linked reproductive fitness to motivated altruism via the emotion of sympathy. Identifying the immediate cause of other-regarding behavior as the feeling of sympathy, Darwin wrote:

> In however complex a manner this feeling may have originated, as it is one of high importance to all those animals which aid and defend each other, it will have been increased, through natural selection; for those communities, which include the greatest number of the most sympathetic members, will flourish best and rear the greatest number of off-spring.
>
> . . . an instinctive impulse, if it be in any way more beneficial to a species than some other or opposed instinct, would be rendered the more potent of the two through natural selection; for the individuals which had it most strongly developed would survive in large numbers. (pp. 82, 84)

Thus altruism persists among individuals, despite its cost in individual fitness, because it enhances fitness of the group. This account of individual cooperation by appeal to group selection hinged on an important explanatory success of *On the Origin of Species* (1859c). There Darwin noted that "many instincts of very difficult explanation could be opposed to the theory of natural selection,—cases in which we cannot see how an instinct could possibly have originated . . . I will not here enter on these several cases, but will confine myself to one special difficulty, which at first appears to me insuperable, and actually fatal to my whole theory. I allude to the neuters or sterile females in insect communities . . . from being sterile, they cannot propagate their kind" (p. 257). Neuter castes are relevant to the problem of altruism, because they represent the extreme case of sacrificing reproductive fitness for the advantage of others. Darwin argued that "some insects and other articulate animals in a state of nature occasionally become sterile; and if such insects had been social, and it had been profitable to the community that a number should have been annually born capable of work, but incapable of procreation, I can see no very great difficulty in this being effected by natural selection" (p. 258). Sterility is the sacrifice of all of the organism's reproductive interests to the benefit of the community and its fertile members. It is the height of biological altruism, although in ordinary terms it hardly counts as self-destructive behavior. Darwin's explanation of the sterile castes reveals that in the case of the social insects the individual is not the organism but the colony or hive that it belongs to. This supra-organismic individual is not itself biologically altruistic. The colony's fitness-maximizing strategy, however, involves the production of sterile castes that aid nonsterile ones.

Among organisms much below man in cognitive powers, there is of course no role for sympathy as the cause of self-sacrifice. And Darwin did not know anything about how genetic information can control such behavior. But at the level of organization at which sympathy can emerge Darwin also hinted at another factor that would encourage the natural selection of sympathetic feelings and the action that stems from them: kinship. In *The Descent of Man* he wrote,

> It is evident in the first place, that with mankind the instinctive impulses have different degrees of strength; a young and timid mother urged by the maternal instinct will, without a moment's hesitation, run the greatest danger for her infant, but not for a mere fellow-creature . . . [The social virtues] are practiced almost exclusively in relation to the men of the same tribe; and their opposite are not regarded as crimes in relation to the men of other tribes. No tribe could hold together if murder, robbery, treachery, etc. were common; consequently such crimes within the limits of the same tribe are . . . branded with everlasting infamy . . . ; but excite no such sentiment beyond these limits. (1871, p. 87, 93)

Subsequent discussion of the problem of altruism has focused on these two tendencies in Darwin's thought: the role of the group as the unit of selection and the influence of kinship on behavior. Group selectionist accounts of altruism and other individual fitness-reducing traits fell out of favor among biological theorists after the rediscovery of Gregor Mendel at the turn of the century. R. A. Fisher, J. B. S. Haldane, and Sewall Wright all recognized the biological possibility of group selection for traits that are individually maladaptive, but held that the process requires conditions unlikely to be found in nature. Accordingly, they held group selection to be unsatisfactory as an explanation for altruism or other cooperative instincts. Even the emergence of sterile castes in the social insects need not require group selection: as Darwin himself suggested, it could be the result of an optimal individual reproductive strategy, one that involves producing a mix of fertile offspring and sterile offspring that aid the survival and reproduction of the fertile ones.

For a long time the role of altruism as a cause of group selection was more widely accepted outside of the circle of mathematical population geneticists. Among advocates of the efficacy of group selection in the evolution of cooperation were Peter Kropotkin (1902), A. M. Carr-Sanders (1922), W. C. Allee (1955), Konrad Lorenz (1966), and V. C. Wynne-Edwards, whose book *Animal Dispersion in Relation to Social Behavior* (1962) reopened widespread discussion of group selection as a Darwinian explanation of altruism in particular and of cooperation in general. Wynne-Edwards was struck by the fact that populations appear to regulate their sizes to levels well below the environment's carrying capacity, and that when populations approach carrying capacity, members lower their offspring numbers. Wynne-Edward's explanation involved the concept of altruistic self-restraint on the part of individual organisms who reproduce at rates below those optimal for their individual fitness, but optimal for the survival of the group of which they are members. Individual fitness maximization involves consumption of resources at the expense of other members of the group. Altruistic self-restraint ensures the survival of the whole group. It should be noted that this explanation for population growth limits had already been undermined in favor of an individual fitness-maximizing strategy in David Lack's study of clutch size among finches (1954). Lack noted that having smaller clutch sizes could be a more adaptive strategy than having larger ones, because this enables the parent to divide investment in offspring into small numbers of larger bundles, thus increasing the likelihood that individual offspring will survive to adulthood. The availability of explanations for apparent altruism that appeal to its advantages for individual selection has motivated much of the discussion since the work of Lack and Wynne-Edwards.

After G. C. Williams' attack on the notion of group selection, even the benefits of altruism for selection at the level of the individual became controversial. Williams (1966) held that adaptational significance should not be accorded to a trait at any level above that necessary to explain the persistence of the trait in question. Following this principle, he argued that the locus of selection is always the individual gene—so that if altruism is to be given an evolutionary explanation it must be shown how it can be adaptive for the individual allele. And as J. Maynard Smith showed at about the same time (1964), selfish individuals or genes could subvert altruistic groups, no matter how these altruistic groups emerged, by individual or group selection. Start with a group of altruists, regardless of how they may have emerged, and introduce an egoist to the group. Over time the egoist will traduce the cooperators, and because the egoist's strategy advantages it at the expense of the others, over the long haul the egoist and its egoistic offspring will reproduce at greater rates and eventually swamp the altruists in the group. Altruism is an "invadable" strategy. Given variation and selection, strategies that can be invaded will be.

By the late 1960s Darwin's other idea, that altruism might be selected for because of kinship between the altruist and the recipient of altruism, came to dominate discussions of how the emergence of cooperation might be explained. The idea is due principally to W. D. Hamilton (1964), though it can be found in G. C. Williams and D. C. Williams (1957). Hamilton argued that nature will select for the strategy that leaves the largest number of copies of the gene that codes for it. In the case of sexually reproducing organisms this will usually be direct offspring—sons and daughters. But it need not be. A strategy that sacrifices a son or daughter to save two siblings, three nephews or nieces, or a parent will ensure the survival of just as many or more copies of the individual's genes as will a strategy of saving a son or daughter. Accordingly, nature will select for "inclusive fitness," which combines the organism's fitness with the fitness contribution (adjusted by some coefficient of relatedness) of each of its kin—the other organisms with whom it shares copies of the same genes. If nature engages in "kin selection" by selecting for "inclusive fitness," then altruism may emerge as an adaptive strategy for an individual that is part of a group of kin. Not only will feeding, protecting, teaching, and otherwise devoting resources to direct offspring be an adaptationally optimal strategy for individuals, but so will devoting resources to kin in proportion to their consanguinity. Thus we may explain Darwin's observation in *The Descent of Man* that the social virtues are always practiced by men of the same tribe.

There are two problems with this neat picture. First, how do individual kin-altruists identify other animals as close enough kin, and what do they do in cases of uncertainty? This problem is especially difficult for males, because the uncertainty of paternity is always greater than the uncertainty of maternity, and may sometimes be substantial. (For a discussion of this problem as it touches sociobiology, see Alexander, 1979.) Second, why does an altruist provide resources at its own expense to *unrelated* organisms? This latter is the important question for sociobiology. As human societies have evolved they have been characterized more and more by cooperation, and less and less by kinship. Either human society slips the leash of selection for the fittest strategies, or else we need to find an explanation for the emergence of altruism other than its inclusive fitness.

It is easy to hypothesize that altruism emerges as a reciprocal strategy, in which individuals are cooperative in order to secure cooperation from others on a later occasion. The term "reciprocal altruism" was introduced in an account of the emergence of non-kin altruism by Robert Trivers (1971). Altruistic actions offered in the expectation of reciprocation are, however, vulnerable to disappointment; indeed, it appears that the optimal strategy for a fitness maximizer is to accept altruistic benefits from others, but to decline to make them. Accordingly, altruism in the expectation of reciprocation will be displaced in the long run by egoism.

Trivers and others had noticed that among animals, including humans, the problem of reciprocal altruism reflects what game theorists and others have called a "prisoner's dilemma." Suppose two partners in crime are apprehended and are then separately interrogated by the police. Each is offered the following "deal": if neither confesses, both will be imprisoned for two years; if one confesses and the other stands mute, the confessing prisoner will spend one year in jail and the nonconfessing prisoner will be sentenced to ten years; if both confess, each will be imprisoned for five years. What should each partner do? Confession is the "dominant" strategy—the strategy that gives the best payoff no matter what the other partner does. Not to confess means that one can be played for a sucker, and by confessing one at least stands a chance to free-ride on the other partner's refusal to confess. The situation is a dilemma, however, because the dominant strategy for each partner leads to an outcome less desirable than another attainable one. Both prefer the two-year sentences each will receive if both stand mute to the five-year sentences each will receive if both confess. But any reason to think that standing mute is a good strategy for oneself is a reason to think that the other partner will also decide to stand mute, and this is an inducement to confess and get off with only one year in prison. The point can be illustrated graphically (see the accompanying figure).

		Partner 2	
		Don't confess	Confess
Partner 1	Don't confess	2 (b) 2	1 (d) 10
	Confess	10 (a) 1	5 (c) 5

The preference rankings for the two partners are as follows:

Partner 1: (a) > (b) > (c) > (d)
Partner 2: (d) > (b) > (c) > (a)

Notice that although both rank outcome (b) second, the dominant strategy for each results in the attainment of outcome (c), which both rank third. Whence the dilemma: for either partner the only reason to act so as to attain outcome (b) is just another reason to confess so as to stand a chance to attain the most preferred alternative, and therefore increases the likelihood of outcome (c). The payoff to free-riding and the cost of being taken for a sucker both exclude altruism as an available strategy to the rational agent maximizing utility. To the extent that nature selects for maximizing fitness, it will discourage altruism and encourage free-riding whenever the payoffs in nature are like those in the prisoner's dilemma. When are the payoffs in nature like this? Not in jail time or money or utilities, of course, but in numbers of offspring the individual leaves as a result of its choice and its opponent's choice. These payoffs arise whenever opportunities to cooperate have a structure that results in the two preference rankings given above. The actual payoffs can be any amount of any commodity, as long as the two players have the stipulated order of preferences. This will happen whenever there is a chance of being taken for a sucker combined with an opportunity to free-ride: possible cases include signaling the presence of predators or silently fleeing instead, sharing food or hogging it, respecting unprotected territory or encroaching on it, and raising offspring or procreating without staying around to raise the offspring. Opportunities for altruism by one agent are opportunities for free-riding by another, and it appears that the costs and benefits make free-riding an adaptational strategy for individuals and altruism a maladaptive one. If occasions for social interaction are like prisoner's dilemmas, then there is a great impediment to the emergence of altruism or cooperation. The problem is worse than Darwin imagined.

It remained for Hamilton, together with Robert Axelrod, to suggest a way in which reciprocal altruism might evolve even when the structure of altruistic opportunities reflects the payoffs of the prisoner's dilemma. The first thing to note is that opportunities to cooperate or not come repeatedly in nature. The prisoner's dilemma is a situation animals face over and over again with a relatively small number of other animals. Only when we reach complex industrialized human societies does the frequency with which any two people can expect to play a prisoner's dilemma game a second time fall to anything close to zero (and, of course, if altruism is ever an adaptive strategy it surely must have become one before the onset of contemporary industrial society). It is in the search for an optimal strategy for repeated or "iterated" prisoner's dilemma situations that Hamilton and Axelrod

(1981) found a potential solution to Wilson's problem of how altruism might have emerged.

Employing computer simulations of Axelrod's (1984), they showed that, under certain circumstances, in an iterated prisoner's dilemma, the optimal strategy for the individual is one called "tit-for-tat": cooperate in game one, and then in each subsequent round do what the other player did in the previous round. Thus, if the other player tries to free-ride in round one, the best response is to refuse to cooperate in round two. If in round two the other party changes strategies and cooperates, in round three one should return to cooperating as well.

Axelrod's computer simulation pitted players using a variety of strategies for playing the prisoner's dilemma against each other. If, among a number of players using different strategies, the ones with the lowest payoffs are eliminated, say, after every five turns, then in the end after enough turns, all remaining players will be using tit-for-tat. In the long run no strategy generates a higher payoff than tit-for-tat. Tit-for-tat is an effective strategy in part because it is clear—opponents can easily tell what strategy a player is using; it is nice—it begins by cooperating; and it is forgiving—it retaliates only once for each attempt to free-ride on it. If playing tit-for-tat in an iterated prisoner's dilemma is a significant individually adaptive strategy and can be transmitted from generation to generation (genetically or otherwise), then in the long run, reciprocal altruism can be established even among animals that have neither kinship ties nor even common membership in the same biological species. Tit-for-tat is altruism in the expectation of reciprocation, with the threat of retaliation just in case cooperation is not forthcoming. Note that when a group of players play tit-for-tat among themselves, reciprocating regularly, they and their strategy are not invadable by players using an always free-ride–never cooperate strategy. Players who do not cooperate will do better on the first round with each of the tit-for-tat-ers, but will do worse on each subsequent round, and in the long run will be eliminated. In Maynard Smith's (1982) terms, tit-for-tat is an evolutionarily stable strategy: if it gets enough of a foothold in a group it will expand until it is the dominant strategy, and once it is established it cannot be overwhelmed by another strategy.

It is important to bear in mind that tit-for-tat is an optimal strategy for maximizing the individual's payoff (evolutionary or otherwise) only under certain conditions. Among these conditions are some that are easily satisfied in evolutionary contexts and others that are much harder to be sure of. Most important is the requirement that the number of games any two players play with each other cannot be known to either. If two players play an iterated prisoner's dilemma a certain number of times, and the number is common knowledge to both, then though they may agree to cooperate on each round, on the last round each will certainly find it irrational to cooperate, for the last round is like a single-case prisoner's dilemma game. But this means that the next-to-last round now becomes the last chance to free-ride on the other party's cooperation, and the last chance to be taken for a sucker. Therefore, in this penultimate round, neither party will cooperate—and so on, for each game working back to the very first one. Common knowledge about the exact number of iterations unravels the optimization of the tit-for-tat strategy and destroys any chance of cooperation. Even knowledge of the probability that some future game will be the last game can lead to a breakdown, if the payoff to free-riding on that game is worth the risk that further games will be played after all. Another important condition reflects the trade-off between immediate gains and long-term gains. For tit-for-tat to be the best strategy, the long-term payoff to cooperation must be greater than the short-term gain from free-riding. Tit-for-tat is a strategy that sacrifices current-round opportunities to free-ride for future benefits from reciprocation by the other player. As the economist would put it, the discounted value of future payoffs to cooperation must be greater than the value of present payoffs to free-riding. Still a third condition for tit-for-tat to get started among large numbers of players is that players be able to recognize each other and recall previous games in the iteration. This requirement is of course unnecessary when the game is played repeatedly with only one other partner. And if tit-for-tat becomes fixed among a small number of players, it will not be invadable as the number of players grows, even though the ability to identify players and remember previous rounds does not increase. Moreover, there is no requirement that the tit-for-tat strategy be the conscious result of calculation. Unlike motivated altruism, tit-for-tat governs behavior, not its causes.

How might tit-for-tat strategies actually emerge and spread in nature, especially among organisms of limited cognitive power? The best the sociobiologist can do by way of answering this question is to point to the power of nature to provide variations in behavior and to point to environmental exigencies as being sharp enough to select the most optimal behavioral routines no matter how fortuitously they may emerge. If there are iterated prisoner's dilemmas in nature, and if the adaptational advantages of tit-for-tat strategies are powerful enough, nature will find them, for it has world enough and time.

It is of course unclear whether apparently cooperative behavior evidenced by animals within and across species actually constitutes reciprocal altruism, and still less whether it is strategic behavior in accordance with a tit-for-tat rule. But the tit-for-tat solution to iterated prisoner's dilemmas does solve Wilson's problem: it shows how evolutionary altruism is possible in at least some possible environments ruled by natural selection for individual fitness. The work of converting this abstract possibility into an explanation of actual cooperative behavior, let alone motivated altruism, has yet to be completed. The story will have to link the adaptational advantages of reciprocal altruism for fitness maximization to the actual institutions of cooperation and the widespread occurrence of motivated altruism. It may establish these links via Darwin's mechanism: sympathy and other social motives persist because the behavior they cause contributes to the fitness of the individual and its lineage. Or sociobiological theory may find less obvious connections between the self that common sense tells us altruism sacrifices and the line of descent that biological altruism may benefit.

ALTRUISM: CONTEMPORARY DEBATES

David Sloan Wilson and Lee A. Dugatkin

EVOLUTIONARY BIOLOGISTS have a penchant for borrowing familiar words from everyday life to construct their formal terminology. Although often dismissed as "unscientific," this practice is more aptly described as "risky"—capable of both revealing and obscuring great insights. On the one hand, we wish to understand how behaviors that we describe with familiar words evolved, and this effort is facilitated by using familiar words in evolutionary theory. On the other hand, familiar words usually have a variety of meanings that are surrounded by a dense atmosphere of values and associations. They lack the precision, uniformity, and neutrality that scientific terms are supposed to have. Evolutionary terms derived from familiar words therefore often have only a veneer of these properties, which can be peeled back to reveal all the complexity that exists for the words in common language.

The word "altruism" and the associated words "cooperation," "selfishness," and "spite" are used by evolutionary biologists to describe behaviors that affect not only the fitness of the actor but the fitness of other individuals as well. One standard way that these terms are defined is shown in the accompanying figure. Cooperation benefits both self and other, altruism benefits other(s) at a cost to the self, and so on. Benefits and costs are, of course, always defined in the currency of fitness. Thus it seems that all behaviors can be neatly classified by the four terms. The evolutionary definitions also seem to match tolerably the everyday meanings of the words.

Despite numerous references to the "strict terminology of evolutionary biology" (e.g., E. O. Wilson, 1975, p. 572), however, these definitions are so ambiguous that single behaviors can and often do shift categories depending on a number of underlying and usually unstated assumptions. There seems to be an irresistible temptation, moreover, to coin new terms or add modifying adjectives, some of which verge on oxymorons: "pseudo-altruism" (Pianka, 1983), "self-interested refusal to be spiteful" (Grafen,

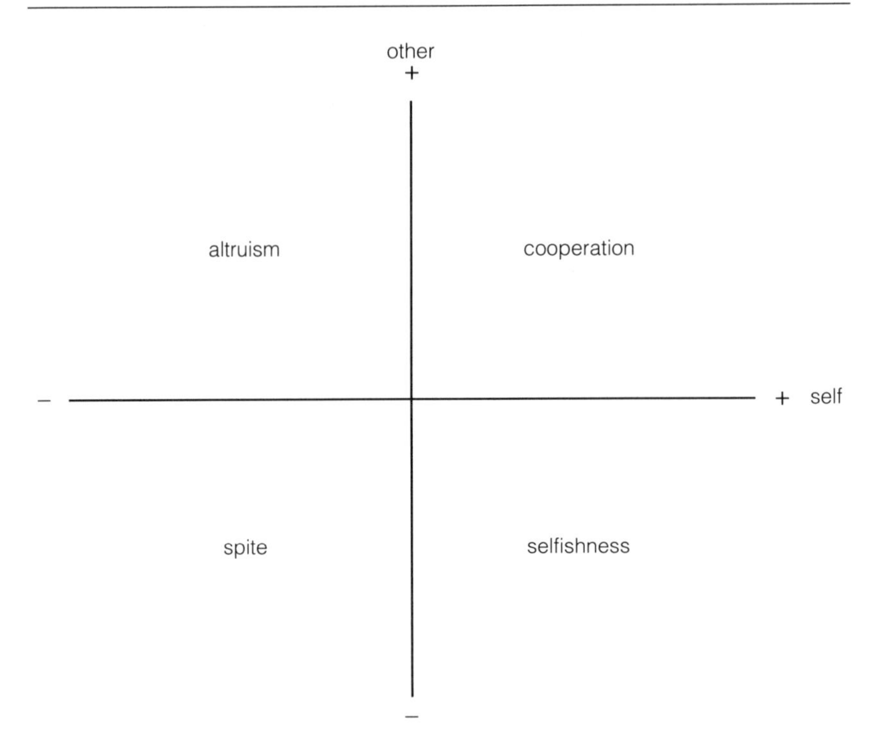

One standard set of definitions of altruism, cooperation, selfishness, and spite. The positive and negative signs refer to increasing and decreasing effects of a behavior on the fitness of self and others.

1984), "by-product mutualism" (Brown, 1987), "return benefit altruism" (Trivers, 1985), "pseudoreciprocity" (Conner, 1986), "quasi-altruistic selfishness," "imposed altruism" (West-Eberhard, 1975), and "joint stock individualism" (Kropotkin, 1908). What are the sources of this ambiguity?

Absolute fitness versus relative fitness. "Positive" and "negative" in the figure usually refer to the absolute fitness of individuals, yet natural selection maximizes the fitness of genotypes *relative to other genotypes in the population.* To predict what evolves we must know how absolute fitness translates into relative fitness, and this translation is not always straightforward. Consider, as example 1, a genotype that has an effect d on itself and an effect r on a small group of recipients, chosen at random from a very large population. Because the recipients are a small and random sample, effects on them on average do not alter gene frequencies. The term r can be ignored and (assuming simple genetics) the genotype evolves whenever $d > 0$. Here, maximizing absolute fitness of self translates directly into maximizing relative fitness. Traits labeled by "selfishness" and "cooperation" in the figure evolve; traits labeled by "altruism" and "spite" do not. Notice that evolution seems to be an individualistic process in this example, because effects on others appear irrelevant. This blurs the distinction between "selfishness" and "cooperation," and some authors prefer to invent terms such as "mutual selfishness" to replace "cooperation," which as an everyday word implies concern for the welfare of others.

Alternatively, consider example 2, a genotype that has an effect d on itself and an effect r, not on a small group of recipients, but on all other individuals in the population (e.g., Hardin, 1968). Now the genotype evolves whenever $d > r$ and the effect on others cannot be ignored. Maximizing absolute fitness of self does not translate directly into maximizing relative fitness; behaviors labeled "cooperative" and "spiteful" may or may not evolve, depending on the relative magnitude of effects on self and others. For this reason the four-word taxonomy of behaviors based on absolute fitness in the figure does not make the crucial distinction with respect to what evolves and is often replaced by a two-word taxonomy based on relative fitness; a behavior is "selfish" when $d > r$ and "altruistic" when $r > d$ (e.g., D. S. Wilson, 1980). The two taxonomies are seldom distinguished, however, creating a lamentable diversity of meanings among authors. Considering the two examples together, a behavior for which $r > d > 0$ can be labeled selfish, cooperative, or altruistic!

Local versus global frames of comparison. A behavior for which $r > 0 > d$ is labeled "altruistic"—and is selected against—in both of the examples described above. Such behaviors can nevertheless evolve if a global population is subdivided into many local groups, and if altruistic and selfish genotypes are sufficiently segregated from each other. In this case, example 3, the selfish genotypes have the highest relative fitness in all groups containing both types, but individuals in predominantly altruistic groups enjoy a higher fitness relative to individuals in predominantly selfish groups (see GROUP SELECTION).

Unfortunately, even clear-cut examples of altruism, with a negative absolute effect on the actor and a positive absolute effect on others, can be redefined as selfish by changing the frame of comparison. Many authors do not calculate the relative fitness of genotypes within single groups, but rather average the fitness of genotypes across all groups. "Fitness" now includes the costs and benefits of expressing the behavior *and* the probability of being in groups with many altruists. Because the altruistic genotype evolves, its averaged fitness must be greater than the averaged fitness of selfish types, despite the fact that it is selected against within each group. Thus a behavior can appear altruistic when relative fitness is calculated at the level of local groups, and selfish when relative fitness is calculated at the level of the global population. In this fashion, *any* behavior that evolves can be described as selfish, simply by averaging the fitness of genotypes across all circumstances (Wilson, 1983a; Wilson and Sober, 1989).

Short-term versus long-term effects. Similarly, a behavior that appears "altruistic" when judged by its short-term effects can be relabeled selfish when judged by consequences over the longer term. Robert Triver's (1971) concept of reciprocal altruism is founded on the idea that benefits to others at the expense of the self can evolve if the actor is "repaid" by the recipient's altruism in the future. Viewed this way, the initial act seems like part of a self-interested economic transaction.

Fitness currencies. In all of the examples described above, costs and benefits are defined in units of fitness. But "fitness" is used by evolutionary biologists in at least three different ways. *Classical fitness* includes the survival and reproduction of individual organisms. In this usage, benefits to others tend to be classified as cooperative or altruistic, even when the others are genetic relatives. *Inclusive fitness* includes the effects of behaviors not only on the actor but on others whose genes are (in part) identical by descent. Thus an individual that benefits a sibling at the expense of itself can be maximizing "its" inclusive fitness. In this sense, a behavior that is labeled "altruistic" in the currency of classical fitness can be relabeled "selfish" in the currency of inclusive fitness. Finally, *gene fitness* includes the effects of behaviors on any gene coding for the behavior in the actor and in all other individual organisms in the population. As phrases such as the "selfish gene" (Dawkins, 1976) and "genetic selfishness" (Alexander, 1974) suggest, this currency implicitly classifies everything that evolves as "selfish."

Fitness versus motivation. In common language, "altruism" and its associated words are defined not only on the basis of effects but also on the basis of the actor's motives. A person who uses others as tools to achieve personal ends is considered selfish even when his effects on others are positive. Evolutionary biologists use the words for their intuitive appeal, yet define them purely on the basis of fitness effects. But how can the words retain any intuitive appeal when stripped of their most salient characteristic?

This question has not been seriously examined by evolutionary biologists. The answer appears to be that, in a cryptic way, the evolutionary definitions do rely on something similar to motivation—not the cognition of the animal doing the behaving, but the cognition of the evolutionist as he or she calculates what evolves in the model. To see this, return to our first example, in which a genotype has an effect d on itself and an effect r on a small set of recipients chosen at random from a very large population. On average the effect on others does not alter gene frequencies and can be ignored in calculating what evolves in the model. It is the seeming irrelevance of effects on others that makes the word "selfish" appear apt, just as a person is selfish when he doesn't "care" about his effects on others, even when the effects are positive. Nunney (1985) explicitly makes this comparison to justify his own terminological framework.

The problem with this kind of intuitive appeal is that it depends entirely on how the model is parameterized. Continuing with example 1, if the actor and all individuals affected by the behavior are defined as a group, and if gene frequency change within this group is monitored, the genes responsible for the behavior decline in frequency whenever $d < r$—as outlined in example 2. Since we know that the trait evolves globally in example 1 whenever $d > 0$, some other evolutionary force must exist to counter negative selection within each group when $r > d > 0$. This force is between-group selection; by chance, some groups contain more genes responsible for the behavior than others, and these groups produce more progeny—as outlined in example 3. The term r must be considered to calculate both within-group and between-group selection, and only cancels out when both forces of evolution are considered together. This is merely a different parameterization of example 1, but in a way that makes effects on others a critical variable, and which makes the word "altruistic" an appropriate description of behaviors for which $r > d > 0$.

The intuitive appeal of words such as "altruism" and "selfishness" as evolutionary terms depends on the degree to which effects on others seem to "matter" in the evolution of behavior, yet this in turn depends on the particular way that gene frequency change is monitored. As a result, each conceptual framework has its own classification of behaviors that appears "natural" within that framework, but which conflicts with the classifications of other frameworks. The intuitive appeal of each framework provides useful insights about the evolutionary process, yet going from one framework to another requires a *translation*. Translating from one framework to another is made difficult by the fact that they use the same words to speak their different languages.

It might seem that the use of these four familiar words as scientific terms is hopelessly confusing and should be abandoned. On the contrary, the multiple definitions illustrate an interesting form of pluralism in science that is valuable to the extent that multiple conceptual frameworks are desirable in the study of a complex process such as evolution (Wimsatt, 1980). The confusion enters when scientists fail to appreciate the heterogeneity of their discipline and assume a uniformity of meaning that does not and probably cannot exist.

ALTRUISM: SOME THEORETICAL AMBIGUITIES

Marcy K. Uyenoyama and Marcus W. Feldman

EVOLUTIONARY BIOLOGISTS generally restrict the term "altuistic" to behavior that involves the sacrifice of a certain amount of fitness on the part of one organism (the donor) in exchange for augmented fitness on the part of a conspecific (the recipient). J. B. S. Haldane (1932), in the first quantitative exploration of the evolution of such "socially valuable but individually disadvantageous" characters, assumed that such traits reduce the fitness of their carriers relative to noncarriers, but increase offspring production by the group. Subsequently, the groundbreaking study by W. D. Hamilton (1964) addressed the effect of genetic relationship between donor and recipient on the evolution of altruism.

Since these initial explorations, an extensive theoretical literature has developed to address the evolutionary origin of genetically determined altruism (reviewed by Michod, 1982). Although these studies operationally define altruism, we regard them more as examples of particular kinds of interactions than as definitions. Only relatively recently have formal definitions of altruism been proposed that attempt a wider purview than the particular modeling approach used. All currently used definitions of altruism, both formal and operational, involve some benefit accruing to the recipient at a cost to the donor (reviewed by Uyenoyama and Feldman, 1980). Costs and benefits influence the course of evolution only to the extent that they modify fitness. We discuss two definitions of altruism, proposed by S. Karlin and C. Matessi (1983; see also L. Nunney, 1985) and by M. K. Uyenoyama and M. W. Feldman (1980). The major difference between these definitions concerns the context, meaning the composition of the group, within which interactions that affect fitness occur. No disagreement exists over the findings obtained from specific quantitative models; rather, it is only the recognition of certain behaviors as "altruistic" that has become somewhat controversial.

Karlin and Matessi (1983) define costs and benefits associated with altruism using the functions $f_A(x)$ and $f_S(x)$, which denote the probabilities of survival to reproductive age of altruistic (A) and selfish (S) members of a group of N individuals of which x are altruists and $N - x$ nonaltruists. This definition applies a "mutation test" (Nunney, 1985, p. 217), which examines the effect of switching to the opposite social type. The conversion of a particular altruist into a nonaltruist reduces the number of altruists in the group from x to $x - 1$, and changes the fitness of that individual from $f_A(x)$ to $f_S(x - 1)$. Altruism bears a cost to the donor if

$$c(x) = f_S(x - 1) - f_A(x) > 0, \tag{1}$$

for all possible values of x ($x = 1, 2, \ldots, N$). The mean fitness among all other members of the group is a weighted average of $f_A(x)$ and $f_S(x)$ before the conversion of the altruist, and $f_A(x - 1)$ and $f_S(x - 1)$ after the conversion. Altruism confers a benefit on the recipient if

$$\begin{aligned} b(x) = {} & (x - 1)[f_A(x) - f_A(x - 1)] \\ & + (N - x)[f_S(x) - f_S(x - 1)] > 0, \end{aligned} \tag{2}$$

for all values of x. A trait is recognized as altruistic under this definition only if both $c(x)$ and $b(x)$ are positive.

By contrast, our (1980, p. 395) definition of altruism requires knowledge of group structure within the population: "A group is the smallest collection of individuals . . . such that genotypic fitness . . . is not a . . . function of the composition of any other group . . . A genotype or allele is said to be altruistic if (a) relative to members of the same group the fitness of the genotype or allele is less than that of other genotypes or alleles, and (b) relative to members of other groups the fitness of members of each group is a monotone increasing function of the frequency of the genotype or allele." A group is a collection of interactants. Although altruism improves the fitness of all members of the group relative to members of *other* groups with fewer altruists, individuals that perform altruism reduce their fitness relative to members of the *same* group that do not perform altruism.

Stated in terms of Karlin and Matessi's (1983) fitness functions, our definition requires that both

$$f_S(x) - f_A(x) > 0 \tag{3}$$

$$\text{and} \quad \frac{df_A(x)}{dx}, \frac{df_S(x)}{dx} \geq 0, \tag{4}$$

for all x. Condition (3) implies that individuals that perform altruism have reduced fitness relative to nonaltruistic members of the same group. Condition (4) indicates that both social types benefit from the presence of altruists in the group.

Although Karlin and Matessi (1983) do not restrict $f_A(x)$ and $f_S(x)$ to monotone functions, all examples known to us in the literature have this property (see also Matessi and Karlin, 1984, 1986). Under this assumption, (2) is always satisfied; in this sense, our condition (4) is a more restrictive form of (2). Comparison of (1) and (3) reveals the major difference between the two definitions. The conversion of an altruist into a nonaltruist changes both the composition of the group and the fitness function of that individual [$f_A(\cdot)$ to $f_S(\cdot)$]. Our definition compares the fitnesses of altruists and nonaltruists in the *same* group. Under the assumption of monotonicity (4), $f_S(x)$ exceeds $f_S(x-1)$, implying that (1) is slightly more restrictive than (3). Cases that satisfy

$$f_S(x) > f_A(x) > f_S(x - 1) \tag{5}$$

would qualify as altruism under our definition but not under Karlin and Matessi's.

The preceding comparison indicates that the two definitions are very closely related, differing only if the appearance of one additional altruist substantially increases fitness. We illustrate the similarities and differences between the two definitions by applying them to test cases. We consider first an example in which the definitions disagree, and second an example which is recognized as altruistic under both definitions or nonaltruistic under both definitions, depending on the length of the period over which the social interactions occur.

Consider an individual who prints a quantity of money, keeping for himself a smaller fraction than he donates to each of his neighbors (modified from Nunney, 1985). Nunney (1985), advocating the "mutation test" of altruism, holds that the act is not altruistic because the donor increases his own wealth, even though others derive greater benefits. In contrast, we recognize this example as altruism because the donor ends with less than his neighbors. In our view, the infusion of money merely increases the inflation rate, reducing the real value of every individual's holdings. If success is measured in terms of *relative* wealth, then the gain by the donor does not by itself imply improved success.

With respect to evolutionary dynamics, the abundance of organisms may be determined by externally imposed factors, including the availability of resources for which individuals within the population compete. Under such conditions, evolution entails changes in the frequencies of the different types rather than extinction of certain forms. It is *relative* competitive ability rather than any absolute measure that determines reproductive success.

Is reciprocal altruism altruism? Robert Trivers (1971, p. 35) states that all theories that permit the evolution of altruism "are designed to take the altruism out of altruism," meaning that the evolution by natural selection of any trait requires that it increase the contribution of its carriers to future generations. Reciprocal altruism refers to acts that are directed toward unrelated individuals, which excludes kin selection as a possible mechanism for their evolution. We consider a form of cooperation among nonrelatives in order to illustrate the significance of context for definitions of altruism.

Recent discussions of the potential for the evolution of altruism among unrelated individuals have focused on reciprocity in the context of the iterated prisoner's dilemma (Axelrod, 1981, 1984; Axelrod and Hamilton, 1981; Peck and Feldman, 1986; Feldman and Thomas, 1987; Axelrod and Dion, 1988). The classical prisoner's dilemma (see Axelrod, 1984, chap. 1) involves pairs of individuals playing a game in which each individual may cooperate or defect. The payoffs to each player (shown in the accompanying table) depend on the strategies adopted by both members of the pair. Mutual cooperation benefits both players more than mutual defection ($R > P$); further, the payoff under mutual cooperation is assumed to be greater than the payoff under equal probabilities of exploiting and being exploited [$R > (S + T)/2$]. Nevertheless, defection is always rewarded, whether the opponent adopts the strategy of cooperation ($T > R$) or defection ($P > S$).

Using the payoffs in the table, we test whether cooperation satisfies either of the two definitions of altruism. In a single game, the pair constitutes the group, and x in (1) through (4) is the number of cooperators in the pair. Let $f_A(\cdot)$ and $f_S(\cdot)$ refer to payoffs to the cooperator and defector, respectively, with $f_A(0)$ and $f_S(2)$ undefined. Under the stated restrictions, cooperators reduce their own fitness in accordance with (3):

$$f_S(1) = T > S = f_A(1), \tag{6}$$

Payoff matrix for the prisoner's dilemma (payoff to player X listed first)

		Player Y	
		Cooperate	Defect
Player X	Cooperate	R,R	S,T
	Defect	T,S	P,P

Note: $T > R > P > S$; $R > (S + T)/2$; and $R > 0$.

and cooperation improves the fitness of members of the pair in accordance with (4):

$$f_A(2) = R > S = f_A(1) \quad (7a)$$

$$f_S(1) = T > P = f_S(0). \quad (7b)$$

Further, (7a) and (7b) ensure that cooperation confers benefits under Karlin and Matessi's definition [see (2)]; in addition, cooperation bears a cost [see (1)]:

$$f_S(0) = P > S = f_A(1) \quad (8a)$$

$$f_S(1) = T > R = f_A(2). \quad (8b)$$

Cooperation in a game consisting of a single trial is indeed altruistic under both definitions.

Now consider iterated trials of prisoner's dilemma in which players may adopt either tit-for-tat (TFT) or all-defect (all-D), which are considered in the literature to be the most cooperative and most selfish strategies, respectively. TFT cooperates on the first trial and subsequently does what the opponent did on the previous trial; all-D defects on every trial. Defining x as the frequency of TFT in the population, $f_A(x)$ and $f_S(x)$ correspond to the cumulative payoffs to players adopting the TFT and all-D strategies, respectively; expressions for these fitnesses are given in Feldman and Thomas (1987). Peck and Feldman (1986) found that if the number of consecutive trials is sufficiently large, TFT will exclude all-D from the population, provided that the initial frequency of TFT is sufficiently high; in all other cases, TFT disappears from the population (see Axelrod, 1981, for an analogous game-theoretic result). The persistence of TFT in the population requires that $f_A(x)$ exceed $f_S(x)$ over some range of x; in this range, TFT does not satisfy our definition of an altruistic trait [see (3)]. Nor does TFT satisfy Karlin and Matessi's (1983) definition of altruism: for continuous $f_A(x)$ and $f_S(x)$, violation of (3) implies that $f_S(x - dx)$ cannot exceed $f_A(x)$ for arbitrarily small dx.

This example serves to illustrate the central role of social context in determining the meaning of actions. Both the size and composition of the group in which interactions occur change with the number of iterated trials. Under both definitions, cooperation is altruistic in a single trial, but not over the course of sufficiently many trials.

Natural selection refers to the differential contribution to future generations. If the relative chance of survival of a specific type depends only on that type, and not on the makeup of the population, we call it *frequency-*

independent viability selection. If the different mating types produce different numbers of offspring we call it *fertility* selection. If specific members of one sex are preferred as mating partners by members of the other sex, we call it *sexual* selection. The description of the evolutionary process is simplest under frequency-independent viability selection because each individual's fitness depends only on its own type. With fertility or sexual selection, however, fitness has a meaning only in the context of *groups* of individuals corresponding to mating pairs or competitors of the same sex. Thus the relative fitness of a specific type depends on the frequencies of the other types, and changes as they change. These are examples of frequency-dependent selection.

The context in which the contribution of an individual to future generations is assessed is particularly important for altruism. For example, under kin selection, in which groups correspond to families, altruists have reduced fitness compared with that of their nonaltruistic siblings. If families with altruists contribute more to the next generation, the average fitness of altruists in the population *as a whole* may in fact be greater than that of nonaltruists (reviewed by Michod, 1982).

The fact that an assessment of adaptive significance demands knowledge of context is best illustrated by the development of the theory of the evolution of the sex ratio. We (1980) recognize as group selection any form of selection characterized by internal groups. Sex ratio selection represents a form of group selection under this view. Individuals are either male or female, but the sex ratio is an emergent property of a *group* of individuals. Limiting the evolutionary perspective to two generations (parents and offspring), Charles Darwin (1871, chap. 8) perceived little individual advantage of the preferential production of one sex, although the population sex ratio may have implications for group defense. R. A. Fisher's (1930a, chap. 6) extension of the evolutionary perspective to three generations (parents, offspring, and grandoffspring) revealed a selective advantage accruing to parents that produce the rarer sex. The development of the theory of sex ratio evolution provides another example of the importance of context.

Under the assumption that female fertility limits offspring production, male reproductive success is limited by the number of broods sired. Reproductive value as applied to sex ratio evolution (Fisher, 1930a, chap. 6), rather than to age-dependent selection (chap. 2), refers to the relative per capita fertility of the sexes. Relative to females, the reproductive value of a male is proportional to the ratio of the number of reproductive females to that of reproductive males. Under our definition, the set of reproductive males forms a group. Viewed from the perspective of the parental generation, the fertility of offspring depends on their sex: parents that produce more of the rarer sex have more descendants in the grandoffspring generation.

Ignoring the group structure in this case renders the sex of the offspring apparently selectively neutral. It is only through the extension of the evolutionary perspective to include three generations and the recognition of the group structure, together with the differences in reproductive value that it entails, that the selective significance of the brood sex ratio becomes apparent.

We began by describing two definitions of altruism that have been used in evolutionary theory, and ended with a general discussion of the significance of context in determining both the course of evolutionary change and one's perception of the selective mechanism. The major difference between the two definitions of altruism discussed concerns the context in which the effects of altruism on donors and recipients are assessed. It is not our intention to use semantics rather than science to advocate one viewpoint over another: both definitions are unambiguous and internally consistent. Rather, it is the significance of social context in determining the meaning of actions that emerges as the most important lesson of this exercise.

CHARACTER: HISTORICAL PERSPECTIVES

Lindley Darden

A CHARACTER is a property of an organism—for example, the color of a flower. Characters occur in different states—for example, red and white flowers—and thus provide the variation potentially important in evolutionary change.

The "unit-character" concept, the view of an organism as composed of independently variable characters, was an alternative to the earlier view of the organism as exhibiting the whole essence of the species. Lewontin (1974b) and Mayr (1982a) argue that Charles Darwin and Gregor Mendel shared an important difference from their predecessors: each concentrated his study on individual variations in characters, rather than holding the older view of the organism as a whole, a representative of a species type. The latter "essentialist" view of organisms was held, for example, by C. V. Naudin, a hybridist in the 1850s.

Naudin believed that each species had a "specific essence," and that after a hybrid cross the maternal and paternal essences separated. Although he observed results of hybrid crosses similar to Mendel's, Naudin "believed that the species segregates as a whole" (Olby, 1985, p. 51). The contrast between Naudin's view and that shared by Darwin and Mendel is important because of the role the unit-character concept plays in determining the kind of data that can be collected about variation, the empirical generalizations that can be formed based on that data, and the nature of the theoretical explanations. In particular, neither the theory of natural selection nor the Mendelian theory of the gene could have been formulated had organisms been viewed as having "essences" rather than as being composed of independently variable characters. The move to the view that organisms have individual characters, whose *variations* within a population are important objects of study, was indeed an important shift in nineteenth-century biology.

In his discussions in *On the Origin of Species* (1859b), Darwin focused on the individual variations of characters. A population exhibiting different characters provided the raw material upon which natural selection worked. In his 1868 hypothesis of pangenesis, in contrast to his emphasis in the *Origin,* Darwin did not focus specifically on individual *characters.* Each hereditary unit (gemmule), he claimed, was produced by one cell, was passed on at fertilization, and grew into that same type of cell in the developing embryo. The hereditary units Darwin postulated, therefore, did not cause types of characters, such as red color in flowers. Instead, a gemmule grew into each cell in the flower petal. As many types of gemmules were needed as there were uniquely located cells in the organism. Data about characters provided no evidence about *types* of gemmules (Darwin, 1868, vol. 2, chap. 27). Darwin's hereditary theory thus was based on a *one unit–one cell* concept, rather than the *one unit–one character* concept so important for early Mendelism.

Hugo de Vries (1889), who regarded his own hypothesis of intracellular pangenesis as a modification of Darwin's, made an important conceptual shift that brought him closer to Mendel's viewpoint than to that held by Darwin in 1868. De Vries' pangens produced cell parts, such as pigment granules. Thus copies of the pangen for red pigment, for instance, were found in the nuclei of all the cells of a red flower and so produced the character of red flower color. As a result visible characters, such as red color in poppies, could be counted and used to make inferences about types of pangens. Characters were to be differentiated, according to de Vries, if they were independently variable. Each independently variable character was caused by a different pangen. The unit-character concept, implicit in Mendel's work (1865) and stated explicitly by de Vries (1889), represented an important conceptual shift: it affected the kind of data collected, as well as the kinds of inferences about unobservable hereditary units that could be made based on that data.

De Vries' 1900 discussion began with a restatement of the unit-character concept, which he had introduced in his 1889 *Intracellular Pangenesis,* published the year before: "According to pangenesis the total character of a plant is built up of distinct units. These so-called elements of the species, or its elementary characters, are conceived of as tied to bearers of matter, a special form of material bearer corresponding to each individual character" (1900, p. 107). His statement made explicit not only the view that the observable organism was to be considered as composed of separable characters but also his own assumption that underlying material particles caused those characters. Other early Mendelians, however, were not committed to an underlying material cause of the unit characters. William Bateson, in particular, stressed that the nature of the cause of the characters, whether material or not, was completely unknown (1902, p. 5).

Although one might question the early geneticists' ability to discriminate the different unit characters clearly, they developed a workable criterion: a unit character is individuated by noting whether it can vary independently. They thus depended on the occurrence of natural variation for their supply of experimental material. Bateson, in discussing the "new conceptions" of Mendelism, stated explicitly: "Each such character, which is capable of being dissociated or replaced by its contrary, must henceforth be conceived of as a distinct unit-character; and as we know that the several unit-characters are of such a nature that any one of them is capable of independently displacing or being displaced by one or more alternative characters taken singly, we may recognize this fact by naming such unit-characters *allelomorphs*" (1902, p. 22). Bateson's term "allelomorph," which later was shortened to "allele" and used to refer to alternative states of a gene, was thus originally introduced to refer to one of a pair of alternative, observable unit characters (e.g., red or white flowers). Seeing the organism as composed of unit characters and seeing varieties as having alternative characters were both important conceptual developments in the early stages of Mendelism.

As Mendelism developed, various distinctions were sorted out, including differences between hereditary and nonhereditary character variations; removal of a sharp dichotomy between "continuous" and "discontinuous" variations (Darden, 1977); and clear differentiation of the genotype from the phenotype (Churchill, 1974). Further developments in Mendelism enabled geneticists to formulate the multiple factor hypothesis, which "complicated" the oversimplified view that one gene caused one character. A given character might be affected by several genes, whose locations could be mapped along chromosomes (Morgan, 1926).

Multiple gene interactions in the production of characters figured more prominently in, for instance, Sewall Wright's theoretical work in population genetics than it did in the work of other mathematical population geneticists, such as R. A. Fisher (Provine, 1986b). Just as much of the work in evolutionary biology in the 1920s and 1930s moved to theoretical claims about genes rather than characters, work in molecular biology also focused attention at the genetic level rather than at the level of gross phenotypic characters. The *one gene–one enzyme* hypothesis of George Beadle and Edward Tatum (1941) enabled characters to be viewed at the lower level of proteins. Hence the study of variations in populations could be conducted at that lower level of organization. Work by Richard Lewontin and J. L. Hubby (1966) showed a surprising amount of variation in protein molecules in natural populations.

Characters that can clearly be attributed to the presence or absence of a protein, such as enzymes that function in the production of a pigment for flower color, show an unproblematic correlation between phenotypic character and molecular variation. But for many complex "characters," such as the shape of the chin, sorting out the sources or limits of independent variation may require extensive research into genetics and embryology. It is not yet known how genes and the proteins they produce control the development of many gross phenotypic characters. "Character" is thus not an unproblematic, easily operationalized concept, even in contemporary biology.

CHARACTER: CURRENT USAGES

Kurt Fristrup

AN OBSERVABLE feature, a correlated set of features caused by a single developmental or ecological process, a historic event in the evolution of a feature: all of these are "characters." These usages, ranging from simple observables to technical inferences, share common aspects. Characters are always defined with reference to a sample of biological entities (organisms, populations, species). Characters are named to express a perceived or hypothetical likeness among the entities in the sample. Characters provide the basis for relating or comparing these entities. By recognizing likeness, we define a character, but its importance lies in emphasizing variations: differences in the character values assigned to entities. In fact, characters are never defined unless an observed or probable variation exists.

Likeness can represent analogous responses to common environmental factors. The wings of bats, birds, flying fish, insects, and some seeds share some characters: structural and functional attributes. Here, characters refer to properties that are presumed to be independent of historical context. They play explanatory roles in models and theories, as in the theory of *natural selection*. Likeness can be also inherited: this information can be used to infer ancestral relationships among individuals (genealogy) and species (systematics). These characters are historical (see HOMOLOGY). They may identify constraints on the form of hypothetical or undiscovered members of a lineage, but they do not otherwise play explanatory roles.

Observed likeness is often due to both processes, so observations are usually employed to identify units of functional or historical likeness required to construct and test models of natural selection or phylogeny. Furthermore, raw observations often contain redundancies that need to be removed. "Character" is variously applied to units at different stages in these analyses. Character can mean a prescription for observation, an independent unit of information (Sneath and Sokal's [1973] unit character), or the hypothetical units/events that play a causal role in the process being studied. For example:

> In a study of relationships of proteid salamanders, Hecht and Edwards (1977) were faced with the following problem. The genera *Necturus* and *Proteus* have been related by several morphological shared and derived characters by other workers, but Hecht and Edwards (1977) found nine of these to be the result of paedogenesis. Actually, these were not nine independent characters as used by previous authors but at least at this level of discrimination, they were really only one character, the state of paedogenesis. The nine characters were the expression of a single growth process. (Hecht, 1976)

> A character is thus a theory, a theory that two attributes which appear different in some way are nonetheless the same (or homologous). As such, a character is not empirically observable; hence any (misguided) hope to reduce taxonomy to mere empirical observation seems futile. (Platnick, 1979, p. 543)

Platnick's characters are (unobservable) historic events in the evolution of observable features.

Additional differences in usage exist (Ghiselin, 1984; Colless, 1985; Rodrigues, 1986); Colless terms these semantic categories:

- *character-part*—a part of an individual: *Joe's blue eyes.*
- *character-variable—fundamentum divisionis,* basis of comparison: Joe's *eye color* is blue.
- *character-attribute*—an attribute of an individual: Joe is *blue-eyed.*

Character-attribute and character-variable are closely related usages of character. They denote a description or a measurement that applies to members of a population. Character-part is quite distinct: it refers to a set of related objects. Confusion of parts with attributes and variables is viewed by all three authors to be the principal difficulty with usage of character.

Characters as parts. "Part" is an important concept in the construction and operation of machines. Clocks are assembled from parts; parts exist prior to (and independent of) the existence of the clock. Parts are simple to delineate, and their functions (design criteria, actual performance) are known or easily determined.

Strict application of this concept to biology is problematic, although recent advocates of genes as the *units of selection* describe segments of DNA in similar fashion (Williams, 1966; Dawkins, 1982b). These "genes" aside, the delineation of "parts" of organisms is difficult. The distinct (and problematic) usage of character-part implies that the parts are related by descent (reproductions of a common ancestral part). They are natural units, not artifacts of our observations.

Character-part usage often is imprecise, but some of the literature in cladistic systematics appears to adopt character-part. In this view, character essentially designates descent from a particular common ancestor, and a character is labeled by the description of the corresponding feature in that ancestral taxon (Platnick, 1979; Nelson and Platnick, 1981). Because the inferred ancestors of modern snakes were quadrupedal (tetrapods), snakes really "have" legs (Platnick, 1979, p. 343), appearances notwithstanding. Leggedness is hypothetical, and a snake's legs can be found on their supposed ancestors. If new evidence—perhaps unrelated to legs in any way—caused observers to place snakes in another group, snakes would no longer have legs. Those parts would no longer define a portion of their ancestry.

Characters as variables. Many biologists associate or equate the terms character and variable. Increasing usage of *character-variable* has paralleled the widespread introduction of numerical techniques in evolutionary biology.

- Character—any quantity or code assigned to each of a group of organisms. See Variable. (Bookstein et al., 1985, p. 252)
- The actual property measured by the individual observations is the character or variable. (Sokal and Rohlf, 1981, p. 9)
- Character is a feature of organisms that can be evaluated as a variable with two or more mutually exclusive and ordered states. (Pimentel and Riggins, 1987, p. 201)

"Feature" denotes a description or measurement of a part, quality, or action observed in an entity. This description specifies the conditions under which the observation was made, including the developmental and behavioral state of the entity and reference to other features in the entity's form or behavior. Even on one entity, consistently identifying the same feature is not trivial. The feature may appear for only a brief interval during development. Further confusion is possible for features that are repeated in an organism's form (certain hairs on a mammal, or particular nucleotides in a genome). Behavioral features present additional challenges: obtaining sufficient observations to ensure representative sampling, controlling for social context and reproductive condition.

The criteria used to define characters—relating features—are established by convention, as Crovello (1970) emphasizes. Examples of these criteria are similarity of form, correspondence of internal structure, congruence of external associations, developmental congruence, and functional relation to other features. Alternative or additional criteria are frequently specified to select measurements or descriptions that reflect similarity of function or similarity due to common ancestral relationship.

The coding of relationships among individual features may embody information about distance or ordering among features or may simply label distinct classes of features. Thus character codes reflect the kind of information available or selected for emphasis. Quantitative characters usually specify ordering and distances. Ranked characters and transformation series specify order, and sometimes direction. Qualitative characters indicate some affinity among features, with no information regarding ordering or distance. Qualitative characters often arise because our sensory skills allow us to perceive classes of features that would be difficult to specify or distinguish quantitatively.

Characters are sometimes defined with the aid of statistical procedures; individual character values are determined by manipulating measurements (characters themselves) according to a prescribed formula. This practice, used to reduce complexity and produce independent variables, is common in studies of morphology and development. Bookstein et al. (1985) present detailed procedures for analyzing size and shape. They measure distances between homologous landmarks, and use multivariate statistical techniques to extract size and shape characters. Raup (1966, 1967) develops a geometric model of coiled shells, and subsequent research has used the parameters in his model as characters; individual character values are obtained by fitting the model to measurements of the shell. Baum (1988) discusses statistical methods of establishing discrete characters from quantitative observations.

Species characters in systematics. Several issues arise regarding usage of "character" that are particular to applications in classification and systematics or ecology and natural selection. Systematists seek characters that indicate common origins. Features that have independently evolved similar appearance (convergent similarity) provide false indications of common ancestry. Characters that might be prone to this kind of misinterpretation are typically excluded or given far less weight in the analysis.

Systematists must also deal with individual variation when they describe characters for populations or species. Mayr (1963, p. 59; 1969, pp. 121–122) excludes characters that exhibit variation within species; his *taxonomic* or *species characters* refer to features that do not vary within species, but do vary among species. Others have attempted to code or represent individual variability. Jardine and Sibson (1971) proposed methods for representing individual variation and regard this information as integral to the process of inferring phylogenies. Baum (1988), among others, discusses coding variable characters in discrete form.

The special demands of systematists often result in particularly restrictive definitions: "A character is a feature of organisms that can be evaluated as a variable with two or more mutually exclusive and ordered states. There are two underlying assumptions of the definition: the organisms are the ingroup and outgroup of any given study, and the characters are intrinsic features that have homologous states among those organisms. This definition imposes two other requirements on the data: order and independence" (Pimentel and Riggins, 1987, p. 201).

The absence of features can make specifying characters and character states problematic. A feature may be absent in two or more populations because it never existed in their ancestral lineages. This is not evidence of common ancestry for these populations; it conveys no genealogical information. Absence can also arise because the feature once existed in a common ancestor, and was subsequently lost (secondary loss). This represents useful genealogical information if we can establish the uniqueness of the loss. Mayr (1969) and Hecht (1976), among others, caution that secondary loss is more frequent and more subject to misinterpretation than the evolution of a new feature.

A common, but controversial, practice in classification and systematics involves adjusting the influence of characters in proportion to estimates of their reliability or significance. There are three methodological bases for this practice. One basis is the impact of "conditional" characters; for example, eye color and related variables have meaningful values only if eyes are present. These conditional characters are sometimes given less weight so that they do not artificially inflate the differences between taxa with and without eyes. Another basis is variability in evolutionary rates. Characters that evolve slowly have greater weight in a computation of distance or dissimilarity between pairs of taxa (Mayr, 1969; Jardine and Sibson, 1971). Examples of heavily weighted characters include complex characters, correlated complexes of characters, and characters that are unaffected by ecological shifts or are unrelated to specific habits. Examples of lightly weighted characters include monogenic or oligogenic characters, "loss" or regressive characters, characters subject to strong selection, and characters that are highly variable within taxa. The third basis for weighting is unequal support for assessments and codings of character values (Neff, 1986). Greater weight is attached to those characters that are clearly identifiable in all taxa, and for which patterns of homology seem well established.

An unusual extension of the term character applies to ecological and distributional associations that are presumed to reflect intrinsic propensities (Mayr, 1969; Sneath and Sokal, 1973; Vrba and Gould, 1986). Examples of these "extra-individual" characters are diet, habitat, parasites, and hosts. At the level of populations, examples are population density, population dynamics, and geographical distribution.

Characters and natural selection. Characters play an explanatory role in ecology and natural selection by focusing attention on features that affect population dynamics and individual or species *fitness*. The historical origins of features are deemphasized, and the related features do not have to be homologous.

Convergent similarity—a bane of systematics—represents one of the most significant indications of the action of natural selection. For example, Rensch (1959, p. 43) reviews "certain rules of climatic character gradients." Within groups of warm-blooded vertebrates (largely birds and mammals), several trends correlate with cooler climates: body size increases, external appendages are smaller relative to overall size, epidermal melanins decrease, the relative size of several internal organs tends to increase, the number of young produced per clutch or litter increases. These "rules" demand attention only to the extent that they apply across a wide variety of vertebrate groups. In environmental physiology (Schmidt-Nielsen, 1983), several studies have established the functional efficiency of various structures and behaviors, including aspect ratios (body shape), surface roughness, muscle performance, surface-to-volume ratios, and the energetics of different gaits. Clearly, convergent similarity of function is important in defining these characters.

Natural selection changes the distribution of character values if offspring character values correlate with parental character values (heritability). However, heritability does not need to be demonstrated before characters can be defined. In many areas of research (as in paleontology), it is either impractical or impossible to measure heritability, yet characters play key roles in describing and explaining evolution.

One impetus for a general definition of characters has been the attempt to promote a hierarchical view of selection and evolution. This view treats a range of biological units (genetic elements, "groups," species) as individuals, just as organisms are treated. One part of this literature introduces potentially important qualifiers: a distinction between emergent and aggregate characters (Vrba and Eldredge, 1984; Vrba and Gould, 1986). A debate is in progress regarding the importance of this distinction in distinguishing cause and effect in the evolution of characters (Lloyd, 1988; Lloyd and Gould, 1992).

We can identify some themes common to all usages of "character." Characters are recognized by comparing entities to discover similarities. However, a character is important because it allow us to express differences—expressed as character values or codes—within the scope of the recognized similarity. Given these considerations, it makes no sense to discuss characters without specifying the population of entities to which they apply. Finally, recognition of characters is always influenced by perceptual acuity (instruments, biochemical analyses), experience, and prior beliefs.

The varied uses of "character" reflect the different goals of historical reconstruction and research in ecology and natural selection. Additionally, a source of confusion has been failure to distinguish between characters as parts, attributes, or variables; are characters natural units or artifacts of observation and description? In both systematics and ecology, there is often a considerable gulf between observables and the units that play causal roles in our models. Some use character to refer to unprocessed observations; others introduce additional restrictions or analyses to produce characters that more closely resemble the information they would most like to have.

Community

Peter Taylor

All evolution occurs in an ecological context, and, as Charles Darwin noted, the workings of that context can be almost "infinitely complex" (1859b, p. 61). At present, however, the structure and dynamics of this ecological context have not been well integrated into evolutionary theory. In the theory of population genetics organism-organism and organism-environment relationships are compressed into the fitness conferred on an organism by its characters. Center stage can then be occupied by the genetic basis and differential representation of characters within single species. Speciation becomes a process of genetic divergence of populations, and the environment's role is simply as a barrier to gene flow. The different uses of the term "community" in theories of ecological organization help identify the difficulties to be tackled if evolutionary theory is to become more ecological.

An ecological community consists of the populations of different species co-inhabiting a site—a lake, the leaf litter layer in a forest, a dung pat, and so on. This broad definition immediately raises several issues. To begin, let us note the connotations that "community" carries over from its colloquial use:

1. *Boundaries.* Who or what is included and excluded? (Is the community defined to be only the fish in the lake or also their planktonic food and avian predators? Does it encompass the feeder streams or stop at the edge and surface of the lake?)

2. *Means of integration.* What are the roles for members of the community or the internal relations that enable the community to persist, if, in fact, it does?

3. *Effects of membership.* What are the benefits to members or the effects on them from being in the community?

4. *External relations.* What is the relationship with other communities and outside influences?
5. *Historical development.* A community must have an origin. What is the history leading to the community's present state?

These interrelated issues of boundaries, components and levels, integration and stability, consequences of context for components, external relations, and change are woven through ecological theory. There is also an ongoing tension between seeking theoretical generalizations and being satisfied with many particular descriptions. Ecologists have addressed these issues in different ways, as evidenced in the specific slants and explanatory weight given to the concept of community.

Plant communities, associations, and populations. Early in the twentieth century, Frederic E. Clements proposed that the community (or "formation") of plants colonizing an uninhabited site is like a developing organism; it passes through a predictable succession of stages, each providing the conditions for the following stage, and results finally in a stable, self-sustaining "climax" (Clements, 1916). The interactions among the species (especially competitive interactions) and the changes the species effect on the habitat provide the cause for this development. At the same time the climax is determined by the habitat and climate, which suggests the possibility of a "physiological" analysis of the complex organism responding to its external environment. In practice, Clementsian quantitative studies of communities (which I take to include measurement and comparison of diversity of species at different sites; see Pielou, 1975) have not achieved the analytic power of physiology proper; they provide description and classification more than insight about mechanisms.

In the responses of Henry Gleason to Clements' community concept we find almost all the opposing terms that have arisen in debates in American and British ecology since Clements. For the integrated complex organism, Gleason substituted a shifting association of individual species. The properties of this association are very contingent on the physical environment and the patterns of immigration from the surrounds, that is, the association depends on particular conditions that are not controlled by it. In Gleason's view, the ecologist should expose regularities by analyzing variable individual responses to variable environmental conditions, rather than by delineating and classifying communities and successional sequences. This follows because "succession is an extraordinarily mobile phenomenon, whose processes are not to be stated as fixed laws, but only as general principles of exceedingly broad nature, and whose results need not, and frequently do not, ensue in any definitely predictable way" (Gleason, 1927, p. 299).

Gleason's view gained adherents in the 1950s when plant ecologists, using multivariate statistical methods, found that vegetation data can be better described as continua; delineation into communities is more or less arbitrary (McIntosh, 1967). Plant community ecologists have continued to develop methods of exposing patterns in vegetation data, and using those patterns aim to generate hypotheses about the underlying causal gradients. But to design multivariate techniques that efficiently and without distortion expose the causal factors, ecologists need prior knowledge or clear hypotheses about those factors; in their absence, multivariate analyses retain a primarily descriptive flavor (Austin, 1987).

Theory, as contrasted with description, has been more actively developed in plant population ecology, where the individualistic thrust of Gleason's view has been pushed further. The result has been a discounting of the issue of ecological organization. Plant population ecologists focus on demographic strategies of individual plant species—how they colonize, grow, survive, disperse, and so on (Harper, 1977; Dirzo and Sarukhan, 1984). Or, pushing the reduction still further, some of them analyze the plant "behavior" (form and physiology) that is optimal, in the given conditions, for energy capture and growth (Givnish, 1986). For most plant evolutionists, therefore, the individualistic approach has eliminated the need for theory about ecological organization above the individual level. Plant associations may display regularities, but they are seen as contingent and perhaps temporary outcomes of underlying processes, such as colonization and growth. In demographic strategy theory the ecological context is acknowledged by studying the plasticity of those strategies necessitated by environmental variability (Sultan, 1987). In optimal strategy theory the methodological assumption is that characters are adaptive; biologists aim to identify what trade-offs shape the evolution of the strategies of individual plant species. In both demographic and optimal strategy theories the larger multispecies context simply provides the boundary conditions for each specific case.

Plant ecology has come to use the term "community" descriptively, while explanations, especially with respect to evolution, are now advanced in terms of populations or variants within populations of individual species. Nevertheless, for the evolutionary theorist pursuing the issue of ecological organization, the contrast of Clementsian versus Gleasonian formulations can serve as a checklist; many of the contrasting positions on the general issues listed earlier have been highlighted. Should ecological organization be theorized in terms of *integrated* entities (e.g., communities) or associations having hard-to-define or permeable *boundaries?* Do we focus on explaining apparently *stable* situations or situations in flux? Do we look for determining factors *internal* to the community (interactions among the species) or *externally* (in the climate, soil, etc.), in the biological conditions or in the physical? Is *change* orderly (succession) or an unpredictable outcome of contingent processes? Finally, should we search for *general* principles about ecological structure and change or concentrate on describing or analyzing *particular* situations?

Community and system. Similar "axes of controversy" have become apparent in ecology more generally (Schoener, 1987). However, for longer than was the case in plant ecology, a central aspect of the Clementsian community concept, the organismic metaphor, has been very productive of theory about ecological organization. To develop this point we need to extend our view to include a modern relative of the organismic metaphor, the concept of system in what I shall call its "strong" form. Although the term "system" may be taken simply to denote that there are many elements interacting, it usually has stronger connotations: systems are entities with natural boundaries and having coherent internal dynamics that govern their development and responses to external influences.

G. Evelyn Hutchinson criticized the ecology of his contemporaries in the 1940s for being overly concerned with classification (of types of communities, biological interactions, etc.). If it meant anything to think of a community as an organism, Hutchinson wanted to be able to analyze the "metabolism" of the community-organism, and he outlined two approaches to that analysis, which correspond to the two subsequent trajectories of the system concept (Hutchinson, 1948). The first approach is biogeochemical, constructing detailed quantitative budgets of elements cycling through systems—for example, *P* in a lake or *C* in the biosphere. Biological and chemical processes are tightly linked; organisms have a balancing effect on the chemical cycles; and biological productivity can be related to changing concentrations of available nutrients. The second approach, the biodemographic stream, follows the more abstract and speculative lines pioneered by G. F. Gause, V. Volterra, and A. J. Lotka (Kingsland, 1985), who analyzed simple mathematical models of population regulation through reproduction, competition, and predation. The biogeochemical and the biodemographic approaches, notwithstanding their differences, are united by a theoretical proposition: Groups of organisms are systems having feedback loops that ensure self-regulation and persistence. It may be observed that the systems view is not sharply distinguished from the earlier organismic views by this formulation. Both require nature to be divisible into integrated wholes, and both assume that the balancing or self-regulation of these wholes can arise, either from natural selection operating at a variety of levels (from the individual up to the biogeochemical cycle) or equivalently as a result of stable systems at any level outpersisting ones with destabilizing components.

The biogeochemical stream has developed into systems or ecosystem ecology, which focuses on measurement of nutrient and energy flows between compartments in an ecosystem, in contrast to experiments on well-controlled parts of the system. (An ecosystem consists of the ecological community in combination with the chemical and physical processes in its environment.) When translating measurements into diagrams and computer models, systems ecologists are pragmatic in their placement of boundaries and internal aggregations—for example, species are lumped into trophic compartments. More generally, theoretical principles have been of less importance than organizing measurements on a huge variety of systems, an emphasis beginning in the late 1960s with the International Biological Program. Hierarchy theorists (O'Neill et al., 1986) are a recent exception to this generalization. They claim, first, that at different spatial and temporal scales a system will be stable in its composition or in its processes (an organismic or system view), and second, that data analytic methods, including multivariate statistical techniques, can expose the natural boundaries and aggregations (measurement again is central here). In their view communities, especially when decomposing organisms have been omitted, are probably not a natural or coherent subsystem of an ecosystem and so are an insufficient basis for ecological theory.

In contrast, community ecology, as Hutchinson's biodemographic stream developed into, has focused on parts of ecosystems, namely, on communities of interacting populations. The boundaries of communities can be drawn so that they include simply a host-parasite pair or include several species, in a "guild" of animals requiring a similar type of food or other resource, or in a web of predators and prey. From the late 1950s Robert MacArthur championed the search for general theoretical propositions about regulation of population sizes and distributions (in space and in characters) through interspecific interaction (chiefly competition for limiting resources), and he popularized the use of models, often mathematical, to formulate and investigate these propositions (MacArthur and Wilson, 1967; Kingsland, 1985). The MacArthurian strategy of modeling is that a model is useful for generating qualitative and general insights about communities. For example, in island biogeography theory the number of species on an island corresponds to a balance between immigration (declining with distance from the mainland) and local extinction (increasing with species number and decreasing with the size of the island). Discrepancies between the model and observed patterns in nature imply that some additional biological postulates are needed. Qualitative insights and discrepancies together enable ecologists to generate interesting questions for investigation, such as how habitats differ among islands that have different numbers of species even when the islands are of similar size and distance from the mainland.

Although rarely discussed in these terms, a strong system view is also essential to MacArthurian community ecology. Using simple models to study a pair or a guild of competing species as a community requires the dynamics of the community to be more or less independent of its context—other species, resources, and nutrient cycles. Thus, either the community has a separate time scale or location in space, or the external interactions are weak or change little compared with the interactions within the community. Similarly, in mathematical models variables are distinguished from parameters; the system is represented in the variables, while the external context enters only through the parameters. Without this separation mathematicians cannot apply their tools, especially the analysis of the stability of equilibria, to communities. Furthermore, this system view promotes a conception of complexity as decomposable into loosely linked systems. Although context-independence might be achievable in well-controlled laboratory experiments, it is likely to represent a special situation when it occurs in the field. Nevertheless, to MacArthur, the future of ecology lay in using a naturalist's intuition to identify interesting patterns and a theorist's imagination to explain them. To minimize the effects of particular location and history, he deliberately focused on situations that persisted over time, in which influences on the species seemed to be in balance or equilibrium (Kingsland, 1985).

Natural selection fits readily into the MacArthurian framework. Provided that competition does not lead to extinction of a population, the frequency of characters evolves in response to the same intra- and interspecies interactions as govern the ecology. Possible outcomes of such interactions include character displacement or niche divergence among similar populations. This is the domain of evolutionary ecology, including coevolutionary studies (Futuyma and Slatkin, 1983). An early and popular idea of evolutionary ecology is that selection pressures can be dichotomized into *r*- and *K*-selection. These correspond to selection of characters enabling rapid colonization of newly available sites (*r*-selection) versus characters promoting persistence in well-established communities (*K*-selection) (MacArthur and Wilson, 1967). Studies of demographic strategies, however, yield a more complex picture. Yet, as with individualistic plant ecology, theory in this field depends on the interacting species being effectively the only players and the ecological conditions not changing at a rate faster than evolutionary change can track (Levins, 1968; Herrera, 1986).

Against and beyond communities as units of ecological organization. A well-developed, nonsystem alternative to MacArthurian community ecology has existed since the 1950s in the work of the Australian ecologists H. G. Andrewartha and L. C. Birch (1984). Only in the 1980s, however, have both the generality and the usefulness of the MacArthurian ideal of expressing ecological theory in simple models been widely and vigorously questioned (Strong et al., 1984). The critics employ comparisons of actual patterns of species coexistence or morphological differences with the patterns generated by a "null" model, that is, a model lacking the species interactions of the MacArthurian model. When the null model fits the observations just as well as the model based on species interactions, the MacArthurian approach is called into question. Such results also fuel an antitheoretical or particularistic view (Simberloff, 1982): Many factors operate in nature and in any particular case at least some of these will be significant. A model cannot capture these and still have general application; instead, ecologists should investigate particular situations and experimentally test specific hypotheses about these situations, guided by and adding to knowledge about similar cases. According to such a particularistic view—as with the individualistic approach in plant population ecology—evolution requires no theory of ecological organization.

Admonishment to test hypotheses tells us little, however, about how theory is generated in the first place; whereas models used in an exploratory fashion can stimulate new formulations and questions (Caswell, 1988). For example, mathematical exploration of how complexity of communities relates to their persistence or stability has helped shift theory onto new grounds. Ecological theory originally emphasized the possibility of achieving equilibrium in a community; accordingly, ecological complexity was held to result from the underlying stability of the ecological system. But mathematical analysis shows that complexity is destabilizing—unless the complexity is nearly decomposable (e.g., consists of loosely linked subsystems). Subsequently, a "landscape" view emerged. In this view a community can persist in a landscape of interconnected patches even though the community is transient in each of the patches (DeAngelis and Waterhouse, 1987). A related view emerges from the construction of model systems by addition and elimination of populations: Complexity can persist at far greater levels than found in decomposable systems, even when any particular system is transient. In this "constructionist" perspective, investigations of ecological complexity should include the historical development of that complexity and continuing species turnover, not just analysis of the system's current configuration. Furthermore, because the pool of populations entering the modeled system is drawn from surrounding patches, the relations with these surroundings need to be included (Taylor, 1989).

Another important example of the stimulating effects of models for ecological theory is the recent development of models that distinguish among the characteristics and spatial location of individual organisms. These models generate certain observed ecological patterns where large-scale, aggregated models have failed, for example, to accommodate patterns of change in size distribution of individuals in a population over time. In addition, some individual-based models connect the community with its ecosystem context by keeping track of nutrients and other resources; these show that changes in just one physical condition, such as soil type or chemical composition of leaf litter, can generate different successional sequences in the communities (Huston et al., 1988).

By reintroducing historical contingency, local context, and individual detail, such exploratory modeling is undermining aspirations for general theory about ecological mechanisms—an ironic development if one considers its roots in MacArthurian ecology. Indeed, not only in modeling but in community ecology more generally there has been a shift to the view that communities are contingent constructions, permeable to invasions and reconstruction, and that analysis of the responses of individual species is required—a Gleasonian view. (The field has by no means, however, surrendered to particularism; the legacy of MacArthur and the responses to critics of community as a unit of organization are well represented in Diamond and Case, 1986, and Gee and Giller, 1987.) The shift is evident, for example, in patch dynamic studies in which the scale and frequency of disturbances (that create open "patches") is emphasized as much as the species interactions during periods between disturbances (Pickett and White, 1985). Studies of succession and of the immigration and extinction dynamics for islands also now pay attention to the particulars of species dispersal and of the habitat being colonized that determine successful colonization for different species. These studies no longer use distributional data to advance some single theory of succession (Gray et al., 1987).

The challenge that theorists of ecological organization now face is to find rules or regularities of historical construction that can account for observed patterns in ecological organization. This must be done with a level of generality that does not discount the particularity of different species and physical conditions. Well-founded generalities about the construction of ecological organization would allow evolutionary theorists to shift the focus from genetics, differential representation of characters, and individual populations, and better capture the complexity of evolution in its ecological context.

Evolution of communities. To address the evolution of communities, we must put aside for a moment the view of communities as contingent and permeable, because all proponents of natural selection of communities (or ecosystems) have taken the existence and integrity of communities for granted—that is, they have a strong system view.

Prior to the 1960s ecologists held natural selection to be capable of shaping the properties of populations, communities, and ecosystems, but they did not critically analyze what such a claim entailed. In the 1960s G. C. Williams and others convinced biologists of the difficulties of linking functional organization or integration at any level above the individual, such as at the community level, to natural selection operating at that level.

Following Williams, community evolution becomes an artifact of the evolution of individuals (or even smaller units of selection) pursuing their self-interest. In the last decade the idea of higher levels of selection has regained some credibility. David Sloan Wilson and others have shown that individual self-interest is compatible with natural selection of community (or population) characteristics, provided that there exists recurrent periods of dispersal or spread of the communities, for example, by transport on a carrier species. In these periods certain communities are spread differentially by virtue of characteristics of the community, for example, because of the benefits to the carrier species (see GROUP SELECTION). Moreover, in such situations the evolution of characters cannot be understood without reference to the character's indirect effects transmitted through all the linkages in the community (Wilson, 1983a).

The debate on community evolution has not addressed Hutchinson's idea (mentioned earlier) that stable systems could outpersist ones with destabilizing components. Yet, suppose that we define natural selection broadly, as the differential representation of entities over a period of time by virtue of some property or character that differs among those entities. Hutchinson's system evolution is then natural selection of systems according to differences in their stability. Unlike Wilson's scheme, no dispersal stages are assumed in this form of higher-level natural selection.

However, just as characters are not transmitted but must be developed through ontogeny (Oyama, 1985), ecological organization cannot be spread or dispersed without processes of reconstruction of the organization as it spreads or after dispersal. Although an organism has integrity through ontogeny, the organismic or system view of ecological organization is, as we have seen, under question from many sides. The conditions necessary for community evolution to occur along the lines just discussed—in particular, the separation of ecological and evolutionary time scales—are not likely, therefore, to be widespread. The existence and evolution of systems become special outcomes to be explained rather than a starting point or shortcut for theory. Such a perspective does not, however, necessarily validate individualistic positions. Instead, ecologists could work to identify and theorize about the structure of the ecological context in which organisms are mobilizing resources to make their living. They could build theory that allows the structure to have history, that is, to be changing in structure, at the same time as it constrains and facilitates living activity—and its evolution.

COMPETITION: HISTORICAL PERSPECTIVES

Robert McIntosh

ALTHOUGH IDEAS OF competition among organisms antedated Charles Darwin, students of evolution, genetics, and ecology derived their ideas of, and interest in, competition largely from *On the Origin of Species*. In it, Darwin described competition as universal (1859a, p. 60) and as the chief component (pp. 205, 220) of the struggle for existence and of natural selection. "Compete," "competition," and "competitor" are used eighty-one times in the *Origin*, sometimes in conjunction with the familiar phrases "struggle for life," "struggle for existence," and the equally loaded words "battle" and "war." This juxtaposition is blamed by some for subsequent confusion about the meaning of the term "competition." Darwin, however, specifically stated that he used "struggle for life" in a "large and metaphorical sense," including effects on individuals and on their success in leaving progeny. He wrote (p. 116) that the struggle for life occurs: (1) between individuals of the same species, (2) between individuals of different species, and (3) with physical conditions. Although Darwin did not formally define competition, he consistently described it (pp. 69, 78, 140, 175, 205, 320, 400) as a component of the struggle for life involving a relationship between organisms (1 and 2 above), not a relationship between an organism and the physical environment (3) as described by some subsequent writers on evolution and ecology. Darwin also distinguished competition from other interactions between organisms, especially parasitism and predation. He contrasted (p. 69), for example, competition between plants with their destruction by animals. He recognized (p. 175) three types of interaction between organisms: (1) with a species on which an organism depends, (2) with a species by which an organism is destroyed, and (3) with a species with which an organism comes into competition. He clearly distinguished "enemies" from "competitors," and he wrote (p. 77) of the structure of an organism as related to those with which it comes into "competition for food or residence" or, in contrast, as related to organisms from which it has to escape or on which it preys.

Early students of evolution and genetics referred to a struggle for life and competition but did not formally define competition, elaborate on Darwin's usage, or pursue detailed studies. Mayr (1988) writes that competition "disappeared from the consciousness of evolutionists when geneticists began to dominate evolutionary thinking" (p. 143). Whatever the cause, extended discussion of competition or citation in the indices of early books on evolution are infrequent.

Early ecologists, however, did recognize competition as an important process in the organization of communities, even though they left its meaning unspecified. The American plant ecologist Frederic E. Clements (1905) published what may be the earliest formal definition of competition for biologists—"the relation between plants occupying the same area and dependent upon the same supply of physical factors" (p. 316). Clements observed that green plants, unlike animals, all use the same components of their environment and that competition between plants is indirect, one plant affecting the physical environment by using light, nutrients, or water which, in turn, affects adjacent plants. Clements' text added elements to the formal definition that were elaborated in the first extended survey of competition (Clements, Weaver, and Hanson, 1929, p. 317): competition occurs "when the immediate supply of a single necessary factor falls below the combined demands" of the competitors. Most subsequent definitions turned on this idea of common demand of two or more organisms for a resource in limiting supply, although consensus on concept and terminology was long delayed (see RESOURCE).

Competition was independently considered in the 1920s in the context of theoretical mathematical models and experimental studies of population growth. Raymond Pearl and L. J. Reed (1920) rediscovered the "logistic" equation of population growth, which is commonly represented by the differential equation:

$$\frac{dN_1}{dt} = rN_1\left(\frac{K_1 - N_1}{K_1}\right)$$

where N_1 is number of individuals of species 1, t is time, r is the maximum rate of population increase in an unlimited environment, and K_1 is a limiting population. The logistic equation represents the population growth rate of a single species, which approaches 0 as N_1 approaches the limiting value K_1. A. J. Lotka, working in Pearl's laboratory, and V. Volterra, working independently, developed equations for competition between two species and subsequently extended it to many species (Kingsland, 1985). The Lotka-Volterra competition equations are:

$$\frac{dN_1}{dt} = r_1N_1\left(\frac{K_1 - N_1 - N_2}{K_1}\right)$$

$$\frac{dN_2}{dt} = r_2N_2\left(\frac{K_2 - N_2 - N_1}{K_2}\right)$$

where t, N, r, and K are defined as in the logistic equation. The burden of the logistic and competition equations is that populations of competing organisms are controlled by a common factor such as food or other essential of life.

G. F. Gause (1934), a Russian entomologist influenced by the work in Pearl's laboratory, set out to test the mathematical models of competition in laboratory experiments. He enunciated what became variously known as Gause's law, or the Lotka-Volterra principle, that "one of the species in a mixed culture drives out the other entirely" (Gause, 1934, p. 113). Hardin (1960) reviewed these eponyms and urged substitution of the more descriptive phrase "competitive exclusion principle." Paradoxically, Hardin described the ambiguities of the principle as its chief merit.

With or without formal mathematical exposition, and without complete consensus on their definition or mechanism, competition and competitive exclusion, under a variety of names, have occupied the attention of ecologists from the 1930s to the present. The link of competition with regulation of populations was emphasized by the Australian ecologist A. J. Nicholson (1933, p. 140), who asserted that any factor that produced population balance was "almost necessarily some form of competition" and was correlated with the number of individuals—that is, was "density-dependent." Unfortunately, Nicholson's definition of competition included "the ease with which they are found by enemies," confounding competition with predation.

Ecologists clarified the meaning of competition by focusing attention on the role of limiting resources. Clements and Shelford (1939, p. 139), reiterating earlier definitions by Clements, wrote, "competition may be defined inclusively as a more or less active demand in excess of the immediate supply of material or condition on the part of two or more organisms." A. C. Crombie (1947) defined competition as "the demand, typically at the same time, of more than one organism for the same resources of the environment in excess of immediate supply" (cf. Milne, 1961, p. 44). Crombie used the term "interference" as a type of competition in which one organism inhibits another "(through direct attacks, conditioning the environment, consuming food, etc.)" (p. 49).

E. P. Odum (1953) continued the connection of competition and population growth in the first edition of his influential textbook. Odum adapted from sociology and introduced to ecologists a symbolic classification of interactions between species based on three possible effects on population growth: (0) if neutral, (+) if positive, (−) if negative. One of the possible combinations of these is symmetrically negative (−/−) and is designated as competition. Predator/prey and host/parasite interactions are asymmetrical (+/−). The plus and minus symbols accord with the subjective implications of some earlier definitions and the signs of the theoretical mathematical equations of Lotka and Volterra (Williamson, 1972). Odum (1953) added to the negative effects of the shortage of necessary resources other negative effects such as the secretion of harmful chemicals (antibiosis, allelopathy). In a second edition (1959; cf. Milne, 1961) he extended, but did not clarify, his definition by adding "mutual predation," "susceptibility to carnivores, disease, etc." (p. 231).

Nicholson (1954) returned to the subject and described two categories of competition—"scramble" and "contest." "Scramble" refers to competition for a resource in limiting supply, without direct interaction among the organisms, in which each secures some portion of the resource. "Contest" occurs when an organism interacts directly with another and restricts its access to the resource. Thus some organisms get as much of the resource as they can secure and others get little or none.

Thomas Park (1954a, pp. 178–181) similarly recognized two categories of competition under different names. "Exploitation," like Nicholson's "scramble," is "a more or less active demand" for needed resources that operates only if resources are limited and limiting. "Interference" competition, like Crombie's (1947) similar usage and Nicholson's "contest," occurs when "populations compete for limited resources through mutual interference, which differentially affects multiplication and survival."

In spite of some clarification, ecologists continued to express dissatisfaction concerning use of the term "competition." L. C. Birch (1957) reviewed the meanings of competition in ecology, genetics, and evolution. He lamented that the word had largely lost its usefulness as a scientific term and had resulted in much misunderstanding and confusion. He advanced a meaning in the "strict sense": "competition between animals occurs when a number of animals (of the same or different species) utilize common resources the supply of which is short; or if the resources are not in short supply, competition occurs when the animals seeking that resource nevertheless harm one another in the process" (p. 16).

M. H. Williamson (1957, p. 423) turned to the Oxford English Dictionary and evolved a definition: "two species are in competition when they have a controlling factor in common and conversely if two species are in competition they have a controlling factor in common." This definition is reminiscent of A. J. Nicholson (1933) in its emphasis on the controlling factor and is described by A. Milne (1961, p. 48) as "the most baffling meaning." The continuing problem of defining competition was evident in a 1960 symposium, "Mechanisms in Biological Competition," at which the participants were unable to agree on a definition, although all agreed on the desirability of *experimental* inquiry into the ill-defined process (Milthorpe, 1961).

Milne (1961, p. 60) offered his own "single strict definition." "Competition is the endeavour of two (or more) animals to gain the same particular thing, or to gain the measure each wants from the supply of a thing when that supply is not sufficient for both (or all)." Even Milne's comprehensive review of prior definitions did not resolve the problem. None of the various definitions offered satisfied all ecologists. H. G. Andrewartha (1961, p. 174) decried competition as a "panchreston" like humors and elan vital and wrote, "I hope that we ecologists may soon be able to add competition to this list of abandoned carcasses." J. L. Harper (1961, p. 1) also advocated abandoning the term competition and replacing it with "interference" to describe "the hardships caused by the proximity of neighbors (usually feeding at the same trophic level)."

Robert MacArthur, the principal figure in the revival of theoretical mathematical ecology in the 1960s, wrote (1972a, p. 256), "I also believe that no precise definition can be related to competition and that the current precise definitions are premature." Nevertheless, he offered (1972b, p. 21) a wide, if tentative, definition: "two species are competing if an increase in either one harms the other." He gave three examples: (1) "species A and B can fight"; (2) "A can reduce B's food supply"; (3) "A can by its own losses increase B's predators."

Although he preferred mathematical definitions, Williamson (1972, pp. 109, 112) offered two verbal descriptions of competition: (1) "competition will be in respect of the factors that control the population because of their variation with density, that is, they are density-dependent" (p. 109); (2) "the criterion for competition will be that of affecting each other's numbers downwards. This indicates that they share controlling factors" (p. 112).

One difficulty in establishing the meaning or terminology of competition is that the term is applied to diverse categories. In some uses it clearly applies to interactions between individuals and the effect on their growth and size or shape. In others it applies to the effects on reproduction and the consequences for population growth. The effects are felt by individuals but are manifest in population growth rate and are related to fitness. Competition may also be attributed to levels beyond populations within a species. Species are commonly said to compete, or to have competed, with resultant morphological, habitat, or geographic displacement, or even extinction. Higher taxa such as families or phyla are sometimes described as competing, and even floras or faunas are said to compete to replace other like categories. Not uncommonly entire complex aggregations such as communities or ecosystems are described as competing with each other, resulting in successional or geographic displacement. The definition or mechanism of competition at levels beyond the species is not clear. Most studies of competition involve individuals or populations of a species (intraspecific) or pairs or groups of related species (interspecific). In instances where widely divergent taxa use the same resource (e.g., rodents, birds, and insects on seeds) the term "diffuse" or "generalized" competition may be used. However, demonstration of such competition is unusual.

Evolutionary studies, genetics, and ecology remained substantially independent of each other in the early decades of this century. J. B. S. Haldane (1932) recognized competition between adult animals and between intrauterine embryos but referred (p. 126) to "weeding out of individuals in competition with their environment." J. S. Huxley (1942, p. 34), in *Evolution and the Modern Synthesis,* commented that competition in rare species is more likely to be between the individual and its environment. He concurred (p. 484) with Haldane in contrasting intraspecific competition with competition with the environment. Ernst Mayr (1942, p. 271) asserted that knowledge of competition and predation was slight and wrote, "In fact it is surprising how badly ecologists have neglected the questions." It is not clear, however, whether Mayr's remarks reflected the failings of ecologists or the segregation of the disciplines. Ruse (1982, p. 176) wrote that evolutionists acknowledged ecology but did not "incorporate ecological thought in any systematic formal way into evolutionary theorizing." He commented that the geneticists Theodosius Dobzhansky and F. A. Ayala ignored the ecological background of evolutionary theory. I. I. Schmalhausen (1949, p. 61) was unusual among evolutionists in providing an explicit definition of competition: "when members of one species are subjected to the same abiotic or biotic danger there is possible natural selection of some who escape limitation." His elements of competition included struggle—for food, with enemies and parasites, and with severe climatic conditions—equating it to Darwin's struggle for life in its broadest scope and not in accord with concepts of competition then current among ecologists.

The gap between ecologists on the one hand and geneticists and evolutionists on the other narrowed in the 1950s as the importance of competition came to be recognized by other biologists. Mayr (1963, p. 664) provided a definition of competition, "the simultaneous seeking of an essential resource of the environment" without at first specifying "in limited supply"—a qualification added only in the 1982 edition (Mayr, 1982a).

H. H. Ross (1962) adapted the definitions of ecologists, defining "direct" competition as occurring when organisms use a commodity in short supply simultaneously, side by side, and in the same way. Animals in "indirect" competition use the same commodity but at different times or use different parts of it. Ayala (1970a) disputed the generality of the competitive exclusion principle, reviewed the meanings of competition, and explicitly differentiated competition from the more inclusive natural selection. He adopted the ideas on competition advanced by Nicholson (1954) and Park (1954a) and approved of the definition given by Birch (1957). T. V. Grant (1971), under a heading "Ecological Interaction," described competition as use of a resource needed by two or more organisms which is present in limiting amounts. Not all writers on evolution, however, were entirely clear about, or accepting of, definitions of competition widely adopted among ecologists (e.g., Simpson, 1949; Stebbins, 1966). Some continued to equate competition with predation or the more inclusive struggle for life (Darlington, 1980; Conrad, 1983).

R. D. Holt (1977) identified "apparent" competition, which occurs when two prey species are limited by a common predator. If one prey increases in density the density of the predator increases, leading in turn to an increase in the mortality rate of the other prey. The two species are in competition, according to J. M. Emlen (1984), via an intermediary organism, namely, the predator.

It can be said that definitions of competition in evolutionary and ecological literature are converging. Most definitions are adaptations of the early uses of Clements as summarized by Birch (1957). Competition is predicated upon collective demand for a common resource when the available supply is inadequate for all of the organisms. The ideas and terminology suggested by Nicholson (1954) are widely used: "scramble" competition, also called exploitation, passive, or consumptive competition, is the use of a common resource when the supply is inadequate for all and is limiting to the competitors. "Contest" competition, also called by some interference or active competition, is the use of a common resource where fighting or other direct harmful behavior, or a chemical inhibition, limits access to the resource. Although considerable agreement on the definition of competition has been achieved, dispute about its mechanism and significance for ecological and evolutionary dynamics of populations continues.

COMPETITION: CURRENT USAGES

Evelyn Fox Keller

A PARTICULAR problem arises for anyone inquiring into the systematic neglect of cooperative (or mutualist) interactions and the corresponding privileging of competitive interactions, evident throughout almost the entire history of mathematical ecology. When we ask practitioners in the field for an explanation of this historical disinterest in mutualist interactions, their response is usually one of puzzlement—not so much over the phenomenon as over the question. How else could it, realistically, be? Yes, of course, mutualist interactions do occur in nature, but not only are they rare, they are necessarily secondary—indeed, it is often assumed that they are in the service of competition: such phenomena have at times actually been called "cooperative competition." The expectation of most workers in the field that competition is both phenomenologically primary *and* logically prior is so deeply embedded that the very question has difficulty getting airspace: there is no place, as it were, to put it. My question thus becomes: what are the factors responsible for the closing off of that space?

Part of the difficulty in answering this question undoubtedly stems from the massive linguistic confusion in conventional use of the term "competition." One central factor can be readily identified, however, and this is the recognition that, in the real world, resources *are* finite and hence ultimately scarce. Scarcity, in the minds of most of us, automatically implies competition— both in the sense of "causing" competitive behavior and in the sense of constituting, in itself, a kind of de facto competition, independent of any actual interactions between organisms. So automatic is the association between scarcity and competition that, in modern ecological usage, competition has come to be defined as the simultaneous reliance of two individuals, or two species, on an essential resource that is in limited supply (see, e.g., Mayr, 1963, p. 43). Because the scarcity of resources can itself hardly be questioned, such a definition lends to competition the same a priori status.

This technical definition of competition was probably first employed by V. Volterra (1926), A. J. Lotka (1932), and G. F. Gause (1932) in their early attempts to provide a mathematical representation of the effects of scarcity on the population growth of "interacting" species, but it soon came to be embraced by a wider community of evolutionary biologists and ecologists—partly, at least, in an attempt to bypass the charge of ideologically laden expectations about (usually animal) behavior, and in fact freeing the discourse of any dependence on how organisms actually do behave in the face of scarcity. The term "competition" now covered apparently pacific behavior just as well as aggressive behavior—an absurdity in ordinary usage, but protected by the stipulation of a technical meaning. As Ernst Mayr explains,

> To certain authors ever since [Darwin], competition has meant physical combat, and, conversely, the absence of physical combat has been taken as an indication of the absence of competition. Such a view is erroneous . . . [T]he relative rarity of overt manifestations of competition is proof not of the insignificance of competition, as asserted by some authors, but, on the contrary, of the high premium natural selection pays for the development of habits or preferences that reduce the severity of competition. (1963, pp. 42–43)

Paul Colinvaux goes one step further, suggesting that "peaceful coexistence" provides a better description than any "talk of struggles for survival": "Natural selection designs different kinds of animals and plants so that they *avoid* competition. A fit animal is not one that fights well, but one that avoids fighting altogether" (1978, p. 144).

But how neutral in practice is the ostensibly technical use of competition that is employed both by Mayr and by Colinvaux? I want to suggest two ways in which, rather than bypassing ideological expectations, it actually preserves them, albeit in a less visible form—a form in which they enjoy effective immunity from criticism. In order not to be caught in the very trap I want to expose, let me henceforth denote competition in the technical sense as "Competition" and in the colloquial sense (of actual contest) as "competition."

The first way is relatively straightforward. The use of a term with established colloquial meaning in a technical context permits the simultaneous transfer and denial of its colloquial connotations. Let me offer just one example: Colinvaux' own description of Gause's original experiments that were designed to study the effect of scarcity on interspecific dynamics—historically, the experimental underpinning of the "competitive exclusion principle." He writes: "No matter how many times Gause tested [the paramecia] against each other, the outcome was always the same, complete extermination of one species . . . Gause could see this deadly struggle going on before his eyes day after day and always with the same outcome . . .

What we [might have] expected to be a permanent struggling balance in fact became a pogrom" (p. 142). Just to set the record straight, these are not "killer" paramecia, but perfectly ordinary paramecia—minding their own business, eating and dividing, or not—perhaps even starving. The terms "extermination," "deadly struggle," and "pogrom" refer merely to the simultaneous dependence of two species on a common resource. If, by chance, you were to misinterpret and take these terms literally to refer to overt combat, you would be told that you had missed the point: the Lotka-Volterra equations make no such claims; strictly speaking, they are incompatible with an assumption of overt combat; the competitive exclusion principle merely implies an avoidance of conflict. And yet the description of such a situation, only competitive in the technical sense, slips smoothly from "Competition" to genocide.

The point of this example is not to single out Colinvaux, which would surely be unfair, but to provide an illustration of what is a rather widespread investment of an ostensibly neutral technical term with a quite different set of connotations associated with its colloquial meaning. The colloquial connotations lead plausibly to one set of inferences and close off others—while the technical meaning stands ready to disclaim responsibility if challenged. (See Keller [1987] for a discussion of Hardin's [1960] use of the same slippage in arguing for the universality of the "competitive exclusion principle.")

The second and more serious route by which the apparently a priori status of competition is secured can be explored through an inquiry into the implicit assumptions about resource consumption that are here presupposed and the aspects of resource consumption that are excluded. The first presupposition is that a resource can be defined and quantitatively assessed independent of the organism itself; and the second, that each organism's utilization of this resource is independent of the presence or activity of other organisms. In short, resource consumption is here represented as a zero-sum game. Such a representation might be said to correspond to the absolutely minimal constraint possible on the autonomy of each individual, but it is a constraint that has precisely the effect we are focusing on—namely, establishing a necessary link between self-interest and competition. With these assumptions, apparently autonomous individuals are in fact bound by a zero-sum dynamic that guarantees not quite an absence of interaction but the inevitability of purely competitive interaction. In a world in which one organism's dinner necessarily means another's starvation, the mere consumption of resources has a kind of de facto equivalence to murder. Individual organisms are locked into a life and death struggle not by virtue of their direct interactions but merely by virtue of their existence in the same place and time.

It is worth noting that the very same (Lotka-Volterra) equations readily accommodate the replacement of competitive interactions by cooperative ones, and even yield a stable solution. This fact was actually noted by Gause himself as early as 1935 (Gause and Witt, 1935), and has been occasionally rediscovered since then, only to be, each time, reforgotten by the community of mathematical ecologists. The full reasons for such amnesia are unclear, but it does suggest a strong prior commitment to the representation of resource consumption as a zero-sum dynamic—a representation that would be fatally undermined by the substitution (or even addition) of cooperative interactions.

Left out of this representation are not only cooperative interactions but *any* interactions between organisms that affect the individual's need for and utilization of resources. Also omitted are all those interactions between organism and environment that interfere with the identification and measurement of a resource independent of the properties of the organism. Richard Lewontin (1982) for example, has argued that organisms "determine what is relevant" in their environment—that is, what *is* a resource—and actually "construct" their environment. But such interactions—either between organisms or between organism and environment—lead to payoff matrices that are necessarily more complex than those prescribed by a zero-sum dynamic—payoff matrices that, in turn, considerably complicate the presumed relation between self-interest and competition, if they do not altogether undermine the very meaning of self-interest.

Perhaps the simplest example is provided by the "prisoner's dilemma." But even here, where the original meaning of self-interest is most closely preserved, Robert Axelrod (1984) has shown that under conditions of indefinite reiterations, a "tit-for-tat" strategy is generally better suited to self-interest than more primitive competitive strategies.

Interactions that effectively generate new resources—or either increase the efficiency of resource utilization or reduce absolute requirements—are more directly damaging to the principle of self-interest itself. These, of course, are the kinds of interactions that are generally categorized as special cases: "mutualist," "cooperative," or "symbiotic" interactions. The view of these as special cases tends to persist even in the most recent literature, where a new wave of interest in mutualism can be detected among not only dissident but even a few mainstream biologists. Indeed, numerous authors are hard at work redressing the neglect of previous years (see, e.g., Boucher, 1985c, for discussion and references).

Finally, interactions that affect the birth rate in ways that are not mediated by scarcity of resources are also excluded by this representation. Perhaps the most important of these omissions for interspecific dynamics are those of mutualist interactions, and for intraspecific dynamics, I would point to sexual reproduction—a fact of life, as I have argued elsewhere (Keller, 1987) that potentially undermines the core assumptions of radical individualism.

A second problem with the language of competition arises in evolutionary theory quite generally. I refer to the widespread tendency to extend the sense of "competition" to include not only the two situations distinguished earlier (conflict and reliance on a common resource) but also a third situation in which there is no interaction at all. Here "competition" denotes an operation of *comparison* between organisms (or species) that requires juxtaposition not in nature, but only in the biologist's own mind. This extension, where "competition" can cover all possible circumstances of relative viability and reproductivity, brings with it, then, the tendency to equate competition with natural selection itself.

Charles Darwin's own rhetorical equation between natural selection and the Malthusian struggle for existence surely bears some responsibility for this tendency. But today's readers of Darwin like to point out that he did try to correct the misreading his rhetoric invited by explaining that he meant the term "struggle" in a "large and metaphoric sense"—including, for example, that of the plant on the edge of the desert: competition was only one of the many meanings of struggle for Darwin. Some authors have been even more explicit on this issue, repeatedly noting the importance of distinguishing natural selection from a "Malthusian dynamic." Lewontin has written: "Thus, although Darwin came to the idea of natural selection from consideration of Malthus' essay on overpopulation, the element of competition between organisms for a resource in short supply is not integral to the argument. Natural selection occurs even when two bacterial strains are growing logarithmically in an excess of nutrient broth if they have different division times" (1970, p. 1).

Such attempts—by Lewontin and, earlier and more comprehensively, by L. C. Birch—to clarify the distinction between natural selection and competition (what Engels called "Darwin's mistake") have done little to stem the underlying conviction that the two are somehow the same, however. Thus, in an attempt to define the logical essence of "the Darwinian dynamic," Bernstein et al. (1983) freely translate Darwin's "struggle for survival" to "competition through resource limitation" (p. 192), thereby claiming for competition the status of a "basic component" of natural selection. G. C. Williams (1986) describes a classic example of natural selection in the laboratory as a "competition experiment," a "contest" between a mutant and a normal allele, in which he cites differential fecundity as an example of the "competitive interactions among individual organisms" that cause the relative increase in one population (pp. 114–115).

The question at hand is not whether overtly competitive behavior or more basic ecological scarcity is the rule in the natural world; rather, it is whether or not such a question can even be asked. To the extent that distinctions between competition and scarcity, on the one hand, and between scarcity and natural selection, on the other, are obliterated from our language and thought, the question itself becomes foreclosed. As long as the theory of natural selection is understood as a theory of competition, confirmation of one is taken to be confirmation of the other, despite their logical (and biological) difference.

DARWINISM

Michael Ruse

DARWINISM IS A term much like Christianity and Marxism, in that everybody "knows" what it means, and yet on not very close inspection it turns out that everybody's meaning is slightly different. Recognizing that to simplify is to falsify, and yet not to simplify is to remain incoherent, I suggest that there are essentially two meanings that go under the name of Darwinism.

The first meaning, the more general one, is of a kind of world picture or *Weltanschauung*. It is of a sort of philosophy in the vernacular sense—perhaps even something akin to a faith or religion. Without intending to be pejorative, it is properly characterized as a metaphysical notion, in the sense of a framework for understanding the world rather than of something read directly from the surface of nature. The key idea behind this picture is of a kind of change or development—everything, certainly in the material world but usually also in the world of thought and human culture is seen to be in a state of Heraclitean flux or becoming. The cosmos evolves. Moreover, despite the fact that expectations are usually if not invariably allowed, not only is change seen to be undirectional, but almost always such change is seen to be value-impregnated, that is to say progressive. In the organic world, humankind generally is regarded as the endpoint of already-accomplished change, although many have seen humans as a mere stepping stone to even better organisms—members of a crowning race, as Tennyson (1851), echoing the evolutionist Robert Chambers (1844), called them—and just as many have worried that we humans today have fallen back from earlier peaks.

Leaving aside for the moment the exact relationship between Darwinism, as thus characterized, and the life and work of Charles Robert Darwin, what we can certainly say is that this metaphysical Darwinism as a concept or idea existed before Darwin wrote and continued after him, up to the present. In the nineteenth century, two (probably *the* two) leading exponents of the notion were the Englishman Herbert Spencer and the German Ernst Haeckel. For both men, Darwinism was an overall philosophy that manifested itself particularly in the moral and social realm (Richards, 1987).

Spencer's ideas in this direction, which became popularly known as *Social* Darwinism, struck responsive chords in many parts of the world, particularly the United States (Russett, 1976; Jones, 1980). The reason for this lay primarily in the fact that Spencer (1852, 1893), starting with Malthusian views on human struggles for existence (which he thought would only be compounded by misguided state welfare systems), argued for fairly extreme laissez-faire socioeconomic policies. This conclusion found receptive ears in late nineteenth-century capitalist quarters. However, proving the point that Darwinism in this sense could legitimate beliefs of very different styles, we find that the founder of U.S. Steel, the Scottish-American industrialist Andrew Carnegie, was spurred by his Spencerian beliefs into founding public libraries, so that the poor but gifted child might nevertheless rise in society through self-help, thus ensuring the onward progressive rise of the human species.

For all of its surface materialism, this kind of Darwinism harbors strong idealistic elements given its teleological flavor of upward progress (Bowler, 1984b). These elements are most obvious in German writings on the subject. Whereas for the British, the key to change was personal effort, with no absolute guarantees of successful progressive change, for Haeckel (1866) there was more of a Hegelian inevitability to change, with the successful evolution of humankind, especially Germanic humankind, having almost providential undertones. Scholars still divide, bitterly, over the extent to which German Darwinism contributed (if at all) to the rise of National Socialism (Gasman, 1971; Kelly, 1981). There was certainly no simplistic cause and effect relationship, although influenced by pan-Germanic yearnings for one Volk under one Kaiser, there is in Haeckel's writings an empathy for state controls quite alien to British Social Darwinism. (Haeckel, like Hegel before him a professor at Jena, was such an enthusiast for Bismarck that he proposed him, successfully, for the degree of Doctor of Phylogeny.)

In this century, this kind of Darwinism has been much less prominent, although one should not ignore the effects it had on developing cultures throughout the world, for example, in China (Pusey, 1983). Cultural unease with the suggestion that one's ancestors are monkeys was balanced with the undoubted veneration that Charles Darwin showed to his like-evolutionary-minded grandfather, Erasmus Darwin. In the West, many members of the eugenics movements in the 1920s and 1930s saw their ideals as being in part a reflection of the need to protect nature's stern creative laws against the debilitating effects of modern society, whether these manifest themselves through the unhappy side effects of scientific medicine (which protect the biologically inadequate until after they have reproduced their kind) or through cheap and effective transportation (which enable the biologically inadequate to flood newly forming or expanding societies, to the point of moral and economic collapse) (Kevles, 1985).

Today, eugenics is as little in favor as the pan-Germanic yearnings of yore. It is important to emphasize, however, that although we are now quick to label such movements "fascist" or "right-wing," in their day the supporters often thought of themselves as forward-looking advocates of reform. J. B. S. Haldane, for instance, as one of the leading evolutionists of this century, found that he could be both a eugenicist and a Marxist (Clark, 1968)! It is true that, at present, the most visible forms of metaphysical Darwinism are to be found embodied in the right-wing socioeconomic policies of various powerful governments of the West. But there have always been others who have tried to extract a more gentle, less harsh message from Darwinism. This tradition goes back at least to the writings of the Russian geographer and anarchist, Peter Kropotkin, who in *Mutual Aid* (1902) argued that biology promotes harmony and good feeling between fellow (human) species members. The most prominent representative of this school in recent times was Julian Huxley, who as founding director of UNESCO devoted his efforts to the cause of global harmony.

In part metaphysical Darwinism (of whatever ilk) has flourished because it seems to many to provide a more adequate world view for modern-day industrialized society than do traditional ways of thinking, specifically those of orthodox Christianity. Metaphysical Darwinism has enjoyed a particular resurgence for those who feel that not only has Christianity failed but so also have other secular alternatives to Christianity (such as Marxism). They argue that one must therefore start all inquiry and understanding with the developmental nature of things, especially including humans and their intellectual achievements.

The most noteworthy philosopher drawn to this position is Sir Karl Popper (1972) and the most eminent biologist is the Harvard sociobiologist Edward O. Wilson (1978). Both of these men express admiration for Spencer ("a thinker of great courage and originality," Popper, 1972, p. 256), both are progressionists, and both think their metaphysical commitments have moral and social implications. Wilson (1984), for instance, as part of his beliefs about our obligations to future human species members, finds himself drawn increasingly into questions of global ecology.

Finally, what of Charles Darwin's own relationship to this general form of Darwinism? Some (generally scientists) have argued that he had little or nothing to do with metaphysics, Darwinian or otherwise (Gould, 1977b).

In their opinion, such an ideology is a travesty of all he stood for. Others (generally humanists) see much in Darwin that is Darwinian in all senses of the word (Ospovat, 1981). Such difference of opinion leads one to suspect that there may have been confusion in Darwin's own mind, and this, I believe, is close to the truth. There were major reasons why Darwin would have been uncomfortable with a simplistic metaphysical Darwinism, based on a fairly direct unilinear view of life's progress. Although he responded sympathetically to the best embryology of his day (especially that of Karl Ernst von Baer), he was never very comfortable with the position that saw profound analogies between the development of the individual and the evolution of the group (symbolized by Haeckel's so-called biogenetic law—"ontogeny recapitulates phylogeny"). Even more significant, branching and divergence was always a fundamental component of Darwin's evolutionary picture. For this reason, especially, he developed his own idiosyncratic notion of relativistic progress, where organisms evolve in ever more specialized ways down particular branches. Darwin spoke in terms of "comparative highness," a notion that today finds a remarkable echo in biological thinking about evolutionary "arms races" (Dawkins, 1986).

Yet, rightly or wrongly, branching was never taken as incompatible with some kind of overall direction—Haeckel himself was the master painter of the "tree of life." Darwin was very much a child of his time, and he like other Victorians thought that, ultimately, life's forces point to the supremacy of our own species. Somehow, comparative progress translates into absolute progress. Hence, although he rather downplayed all of this in the first edition of *On the Origin of Species* (1859), by the time of the third edition of the *Origin,* in 1861, his commitment to progress was clearly in view. Indeed, when he came to write the *Descent of Man* (1871), Darwin argued freely to the virtues of capitalism, the eugenical dangers of modern medicine, the superiority of whites (especially males), and the like (Greene, 1977).

The second meaning of the term Darwinism (and of the related term, Darwinian) is a more strictly scientific meaning, and is unambiguously tied to Darwin himself. To grasp the notion fully it is best to start with a trichotomy noted by several recent commentators (e.g., Ayala, 1985). There are three things an evolutionist must do if a complete picture is to be presented. He or she must speak to the *fact* of evolution, to the *paths* of evolution (phylogenies), and to the *mechanism(s)* of evolution. In the *Origin,* as all have agreed, Darwin did much to establish the fact of evolution. Nevertheless, he was certainly not the first evolutionist, and hence it is not on this that the scientific notion of Darwinism rests primarily. Nor is it on Darwin's success in tracing phylogenies, for neither in the *Origin* nor elsewhere did Darwin do much to discover paths. Rather, we must turn to the question of mechanisms, that part of the *Origin* most central to Darwin's heart and on which rests his prime claim to fame.

Darwinism in our second sense, the scientific sense, accepts with Darwin that the key mechanism of evolutionary change is natural selection or (as it was also called) the survival of the fittest (see HETEROSIS)—a mechanism that depends crucially on the struggle between organisms for food and space. Because together with this kind of struggle goes the intraspecific struggle for mates, most Darwinians (in the second sense) accept also the subsidiary mechanism of sexual selection—although some have been inclined to roll the two together or downplay the latter. Darwin also believed ardently that selection works solely for the benefit of the individual, and Darwinians tend to agree, although again with exceptions (Brandon and Burian, 1984).

Above all, Darwin's mechanism of natural selection was intended to explain that which British natural theology found so significant: adaptation. Although the raw stuff of evolution, the new variations always appearing in populations of organisms, are random (in the sense of not appearing in response to organic need), the winnowing effect of selection ensures that evolved features are designlike or adapted (Limoges, 1970).

It is remarkable how little enthusiasm so many of Darwin's contemporaries felt for natural selection. The doubters or the indifferent included some who were extremely close to Darwin personally, including Thomas Henry Huxley (1893). Those who did take to natural selection tended to be those for whom a mechanism was urgently needed. These were (especially) biologists working on fast-breeding organisms, such as Henry Walter Bates (1862) and the man who discovered natural selection independently of Darwin, Alfred Russel Wallace (1866). Both of these men studied lepidoptera and turned natural selection to good advantage. A like pattern continues to hold today. Enthusiastic Darwinians (that is, in the sense of those who cherish natural selection) tend to work on fruit flies rather than on dinosaurs (Endler, 1986).

There were two reasons for the lukewarm reception of natural selection. On the one hand, there were doubters who, while agreeing with Darwin on the importance of adaptation, disagreed that natural selection is adequate to the task of explanation. These critics ranged from outright theists, who wanted no truck with any natural processes, to those who proposed alternative mechanisms such as Lamarckism (the inheritance of acquired characters) (Bowler, 1983). With the failure of the alternatives, and with a general move toward naturalism, these critics today are less vocal and less numerous. (I except the so-called creation scientists, religious fundamentalists, who want evolution replaced by literal interpretations of Genesis. They think adaptation very important and that natural selection is a quite inadequate mechanism; see Morris, 1974, and Ruse, 1988b.)

On the other hand, perhaps more interesting were those critics of Darwin who devalued selection because they were not convinced that adaptation is as significant as Darwin and the English natural theologians believed. Some, like Huxley, were simply insensitive to adaptation. Others, generally from the Continent (or, like the British anatomist Richard Owen [1849], influenced by continental ideas), were themselves the holders of a natural theological legacy, a rival to the British variety. This is one associated with *Naturphilosophie,* a movement that finds the significant mark of the organic in repeated patterns or themes (what Owen called "homologies"), as God or Nature plays out variations on one basic blueprint or *Bauplan.*

It was this clash between two natural theologies, adaptationism (or, as it is also known, utilitarianism) and *Naturphilosophie* (or transcendentalism), rather than between nonevolutionism and evolutionism, that separated the Frenchmen Georges Cuvier and Etienne Geoffroy Saint-Hilaire in their celebrated debate before the French Academy in 1830 (Appel, 1987). In this respect, Walter Faye Cannon (1961) and Michel Foucault (1966) were right in the provocative claim that evolution owes as much, if not more, to Cuvier than it does to his opponents. The great critic of evolutionism, the father of comparative anatomy, stood much closer to Darwin on the essential adaptive nature of organisms than did pre-Darwinian evolutionists such as Jean Baptiste de Lamarck.

By the time of Darwin, no biologist could be entirely indifferent to either adaptation or homology. Indeed, Darwin congratulated himself on his explanation of the isomorphisms between organisms. This phenomenon, which he called the "Unity of Type," was something that he thought fell right out of organisms' descent from a shared ancestor. (In other words, for Darwin the ancestor and the *Bauplan* were the same.) Not all were satisfied that matters could be settled this quickly, however. There are still questions about why the *Bauplan* should be as it is, and why its traces persist down through evolutionary lineages, despite selection. Are there certain "constraints" that limit the effectiveness of any mechanism directed toward adaptation?

Interestingly, this debate between strict Darwinism (as it is now being conceived) and its critics continues today, particularly in the controversy over the so-called theory of punctuated equilibria (Eldredge and Gould, 1972). Ostensibly, this is a debate about the fossil record—whether (with phyletic gradualism) it shows smooth continuous evolution or whether (with the critics) it shows uneven, jumpy evolution. Truly, however, the debate is in major part a replay of the nineteenth-century debate (Ruse,

1989). The gradualists are gradualists primarily because they are committed adaptationists, and they cannot see how discontinuous or very rapid change would stay in adaptive focus. Significantly, the arch-gradualist Richard Dawkins (1986) argues that adaptive complexity is the mark of the living and that only selection can explain it. He calls himself a "transformed Paleyist." Nongradualists like Stephen Jay Gould argue strenuously against ubiquitous adaptationism, and they argue that the rapid rearrangement of constraints, followed by periods of stability, could lead to some of the uneven aspects of evolutionary history. Significantly, Gould has written favorably of transcendentalists and even allied himself with them (Gould, 1971, 1983).

Some thinkers, perhaps Darwin himself and certainly some twentieth-century evolutionists such as R. A. Fisher (1930b), have subscribed to both forms of Darwinism (see EVOLUTION). Fisher was an ardent adaptationist, thinking selection (combined with Mendelian genetics) an adequate explanation of all organic change. Yet he also saw the organic world as revealing an overall upward drive, with humans at the top. His self-styled fundamental theorem of natural selection was supposed to function as a kind of second law of thermodynamics in reverse, bringing increasing order and complexity. Recently, working from entomology and sociobiology, E. O. Wilson (1975) has also shown the ability to endorse both Darwinisms.

Yet despite clear areas of overlap, in major respects and intent the two Darwinisms are very different. Most metaphysical Darwinians were not scientific Darwinians—both Spencer and Haeckel downplayed selection—and certainly not all scientific Darwinians were metaphysical Darwinians. Indeed, some of the former (such as G. C. Williams, 1966) argue that the two Darwinisms are incompatible—for instance, the scientific version precludes the progressionism that the metaphysical version demands. Furthermore, the two Darwinisms are different in type as well as in content. For all of its origins in natural theology, scientific Darwinism is grounded in the empirical world in a way that the other form of Darwinism is not. Huxley used to quip that Spencer's idea of a tragedy was a beautiful theory destroyed by an ugly little fact.

The death of Darwinism—any version—is pronounced with monotonous regularity. Yet both metaphysical and scientific Darwinism show remarkable vitality. I am confident that, one hundred years from now, this will still be the case.

Environment

Robert N. Brandon

The two basic concepts in the theory of natural selection are that of fitness or adaptedness and that of environment. Selection is the differential reproduction that is due to differential fitness in or adaptedness to a common environment. The concept of fitness or adaptedness has received extensive attention both from biologists and philosophers of biology, but the concept of environment has been largely ignored. Perhaps that is because the concept seems to be unproblematic. The term "environment" serves to demarcate the external from the internal. The environment of an organism seems to be simply the totality of external factors, both biotic and physical, in which the organism develops and lives.

But things are not so simple. In population biology there are at least three quite disparate ways in which we can measure environmental factors, and corresponding to these different measures, three importantly different conceptions of environment. One approach is to measure some factor, or set of factors, entirely external to the organisms of interest. If we were interested in a population of some species of grass in a field, for instance, we could measure the concentrations of molybdenum in various parts of the field. Such measurements need not involve the organisms themselves. They measure the *external environment.* By means of such methods we could discover a pattern of external environmental heterogeneity like that depicted in Figure 1. More generally, the external environment is the sum total of factors, both biotic and abiotic, external to the organisms of interest. This is the operative notion of environment in ecology.

Unfortunately, the external environment is only indirectly related to evolutionary studies. The reason for this is simple; the factors we pick out to measure in the external environment may or may not influence the organism's survival and reproduction. In our example, the grasses we are studying may or may not respond to different concentrations of molybdenum. If our concern is the environment as experienced by the target organisms,

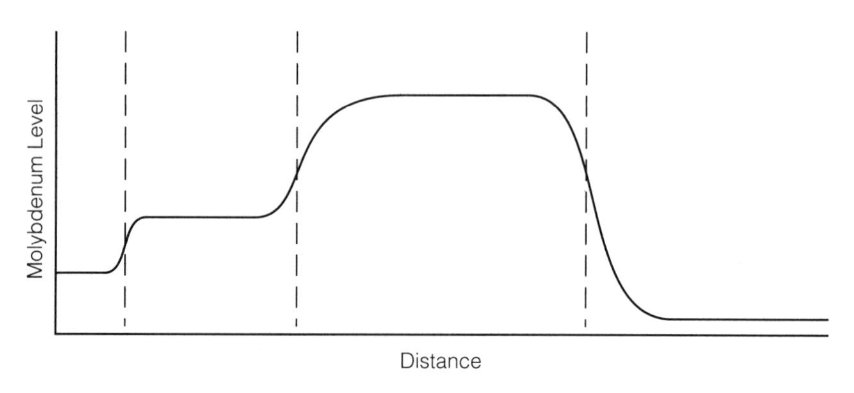

Figure 1 External Environment

then a more direct way of measuring the environment is to use the organisms themselves as measuring instruments. This has been called the "phytometer" method, pioneered by Frederic E. Clements and G. W. Goldsmith (1924). Recently it has been used extensively in ecological genetic studies by Janis Antonovics and his associates. (See Antonovics, Clay, and Schmitt, 1987; Antonovics, Ellstrand, and Brandon, 1988; see also Turkington, Cahn, Vardy, and Harper, 1979; and Turkington and Harper, 1979. These studies all involve plants, as the name "phytometer" implies, but this method is by no means restricted to plants.)

Suppose we could clone multiple copies of a seed of some genotype. We could then plant these seeds out along a spatial transect in our field and over a period of years we could measure each seed's reproductive value (*sensu* Fisher 1930a). (See Figure 2.) By this method we measure the *ecological environment*. (Note that we would want to use multiple copies of a single genotype, rather than seeds of various genotypes, so as not to confound environmental differences with genotypic differences. Alternatively, we could sample the genotypes extant in the field, clone them and plant them out along the transect. Then we would average over all genotypes.) The ecological environment reflects those features of the external environment that affect the organisms' contributions to population growth. Because different organisms (or different genotypes) may respond to the external environment differently, it follows that the scale of heterogeneity of the ecological environment depends on the organisms (or genotypes) used as measuring instruments.

If our concern is with the concept of environment as it functions in the theory of natural selection, we must go one step further. Natural selection is essentially comparative. The *selective environment* is measured in terms of the relative actualized fitnesses of different genotypes across time or

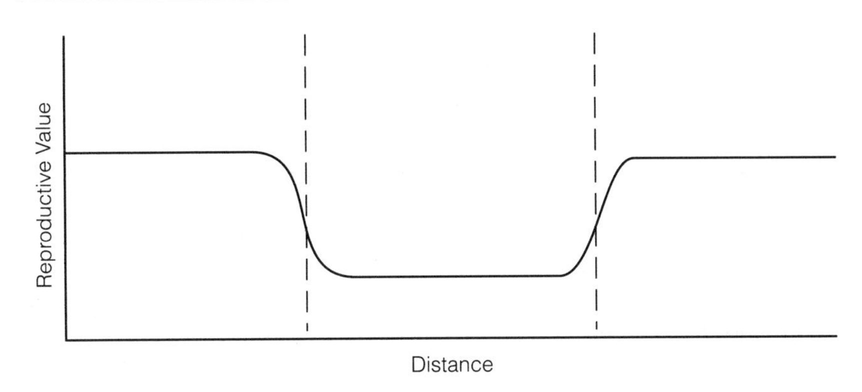

Figure 2 Ecological Environment

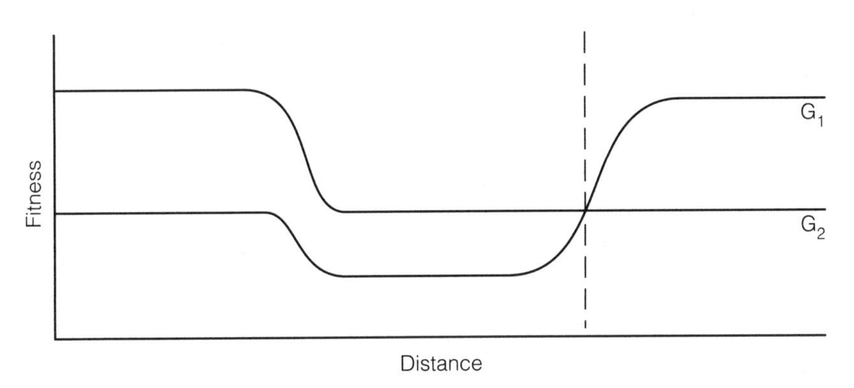

Figure 3 Selective Environment

space. For example, suppose we clone multiple copies of two genotypes, G_1 and G_2, and plant them out along our transect. The result is Figure 3. The scale of heterogeneity of the selective environment reflects the differential performance of genotypes in different regions or at different times, that is, genotype-environment interactions, and, like the ecological environment, it depends on the organisms (or genotypes) used as measuring instruments. (The distinction of external, ecological, and selective environments is introduced in Antonovics, Ellstrand, and Brandon, 1988; it is explored in greater detail in Brandon, 1990.)

It is useful to compare the relations among these three environments. The external environment consists of all factors external to the population of interest. The ecological environment reflects those aspects of the external environment that affect the target organisms' reproductive output. As Figures 1 and 2 illustrate, not all variation in the external environment is reflected in ecological environmental variation. That is, some differences in the external environment make no difference to the organisms of interest. However, one might think that external environmental heterogeneity was a necessary condition for ecological environmental heterogeneity. But whether this is so depends on whether or not we include conspecific organisms as part of the external environment. This makes a difference in cases of density-dependent population regulation. One patch in our field may be a poor spot for seedlings of, say, *Danthonia spicata,* because it is already crowded with mature *Danthonia* plants. If the population of *Danthonia* is the one of interest then this heterogeneity is important, but I do not think it should be considered heterogeneity of the external environment. From a populational point of view the external environment contains only factors external to the population of interest. Although it may be true that most ecological environmental heterogeneity reflects external heterogeneity, such variation is not a necessary condition of ecological environmental heterogeneity.

Heterogeneity of the selective environment reflects differing relative fitnesses in time or space (genotype-environment interaction). For the relative fitnesses of multiple types to vary in time or space the absolute fitness of at least one type must vary. Thus ecological environmental heterogeneity is a necessary condition of selective environmental heterogeneity. It is not, however, a sufficient condition. Variation in the ecological environment will not be reflected as variation in the selective environment if all of the competing types respond to the ecological heterogeneity in the same way, that is, if their relative fitnesses do not change.

From the point of view of the theory of natural selection the relevant environment is the selective environment. Selection occurs when differential adaptedness to a common selective environment leads to differential reproductive success. When these differences in adaptedness are heritable adaptive evolution can occur. But the process of differential reproduction across a selectively heterogeneous region is quite different. For instance, suppose we plant one seed in good soil and another in toxic soil. The first is likely to grow better and produce more seed. But this is not natural selection (I am assuming the plants are not "choosing" their soil); here, rather, the differences in reproductive success are due to environmental differences, specifically differences in the selective environment. It follows from the considerations already discussed that external environmental heterogeneity is neither necessary nor sufficient for selective environmental heterogeneity. It is not necessary because in cases of density- or frequency-dependent selection changes in the population structure alone can result in a change in selective environments. It is not sufficient because many changes in the external environment either do not affect the organisms at all or do not affect them differentially.

This decoupling of the selective environment from the external environment has important consequences for some of the major questions in contemporary evolutionary biology. I will mention only two. The first concerns the prevalence of sexual reproduction. Sex seems problematic from a genetic point of view, because there is a 50% cost to meiosis—that is, by sexually reproducing a female reduces her genetic contribution to her offspring by 50%. Theoreticians have developed many hypothetical explanations of this apparent anomaly (see Ghiselin, 1974a; Williams, 1975; Maynard Smith, 1978; Bell, 1982). We need be concerned with only one, which explains sex on the basis of the advantage of sexual reproduction in heterogeneous environments. Consider, for example, a case in which environments change from generation to generation so as to be negatively autocorrelated. Here, the argument goes, the sexual reproduction of variable progeny would be favored over the asexual reproduction of the types favored in the parental generation. (A similar argument can be made for spatial heterogeneity.)

The second major question has to do with the high level of genetic and phenotypic polymorphisms found in natural populations. This observation seems problematic if one thinks of selection acting in a uniform way across space and time. But again environmental heterogeneity has been offered as an explanation of this phenomenon. The idea is simple: if different genotypes (or phenotypes) are favored at different times or in different places, then a high level of polymorphism can be maintained (see Hedrick, 1986). Although it may be natural to think of the environmental heterogeneity posited in both of these explanations as heterogeneity of the external environment, it is easy to see that it is selective environmental heterogeneity that is relevant. Indeed an extreme form of selective environmental heterogeneity is required in both of these explanations. In the case of sex, what is required is that there be reversals in the ordinal ranking of genotypes over generational time (or through space). Otherwise an asexual reproducer of the genotype that is always (or everywhere) fittest would be favored by selection. This reversal in the ordinal fitness-rankings of the genotypes, or a "crossing type" genotype-environment interaction, is also required in the polymorphism case. Without a crossing type genotype-environment interaction, one genotype will be consistently favored. (The role of environmental heterogeneity in hypotheses concerning the evolution and maintenance of sex is discussed in Antonovics, Ellstrand, and Brandon, 1988; both this case and the case of the maintenance of polymorphisms are discussed in Brandon, 1990.)

In both of the cases just mentioned the theory behind the proffered explanation requires the concept of selective environment. Not only does making this explicit facilitate theoretical developments, it is also necessary for the empirical testing of these hypothetical explanations. One needs to know what sort of environmental heterogeneity is at issue if one is to try to measure it, and given the decoupling of the selective environment from the external environment, the intuitively appealing idea of measuring some aspect of the external environment will not suffice.

Epistasis

Michael J. Wade

THE TERM "epistasis" is often used synonymously with the phrase "gene interaction," but in fact it has two very different meanings in genetics. In molecular and biochemical genetics, genes whose products function sequentially as substrates or catalysts in a common biochemical pathway are considered to be "epistatic" to one another. In this usage, one gene is "epistatic" to another if the function of its gene product in a biochemical pathway is *conditional* upon the success or failure of the other gene operating at an earlier step in the same pathway. These kinds of effects of gene action are commonly observed in the sequential activation of genes during development. Use of the term "epistasis" to indicate this sort of biochemical contingency is very different from the other meaning given to "epistasis" by quantitative or statistical geneticists.

In statistical and quantitative genetics, epistasis is not a property of a biochemical pathway but rather a populational concept describing the relationship between the phenotypic variations among individuals and the genotypic variations among them. Epistasis is the between-locus "nonadditive" component of the genetic *variance* for a trait. This component of genetic variation measures the statistical effects of *variations* in gene combinations between individuals in relation to the total phenotypic variance between individuals in a population.

Epistasis between loci in statistical genetics requires genetic variation at each of at least two loci. Without genetic variation at two loci, the variation between individuals within a population cannot be attributed (in the statistical sense) to variations between individuals in gene combinations. Genes can be epistatic to one another in the biochemical sense but not contribute to epistasis in the statistical sense. For example, using material from isogenic stocks of the fruit fly *Drosophila melanogaster,* one can study the accumulation of reaction products of sequentially acting genes and thus characterize biochemical epistasis. However, there is no genetic variation within an isogenic stock by definition; all individuals share the same genetic constitution. One cannot partition components of genetic variance in such a stock, so there can be no statistical epistasis. Sewall Wright's principle of "universality" of epistasis at the biochemical level (1968, pp. 59–60) does not guarantee an important or even a significant role for epistasis in evolution at the populational level. It is the statistical effects of gene interactions on the phenotype that determine how gene combinations will evolve and whether or not selection can operate directly on differences between individuals in gene combinations.

Because it is the additive component of genetic variation that determines the response to selection within populations, the epistatic component of genetic variation is not usually accorded much attention in evolutionary biology. The existence of additive genetic variance, however, does not imply additive gene action (Falconer, 1981). The measurement and operational definition of epistasis, nonadditive gene effects, depends upon the statistical concepts of "main effects" and "interactions." They were introduced with the analysis of variance (ANOVA) as a means of drawing inferences from the results of controlled experiments. The statistical terms "main effects" and "interactions" are used to characterize the contribution of experimentally controlled factors to the observed variation among treatment means in randomized, factorial, replicated experiments. Just as the development and interpretation of the analysis of variance as a method for experimental inference depends upon the concepts of "main effects" and "interaction," the meaning of "additive" and "nonadditive" (i.e., "epistatic") gene effects in statistical genetics depends on them as well. It is the interpretation of the phenotypic variations among progeny from controlled breeding designs using a genetic model and the methods of analysis of variance that constitutes the formal definition of the "additive" and "nonadditive" components of genetic variance (Anderson and Kempthorne, 1954).

In experimental statistical genetics, genes are the causal factors and different levels of these causal factors are established by experimentally varying the degree of genetic relationship among the progeny in controlled breeding designs. When several different genetic factors are important in producing a particular condition, the influence of a particular genetic factor may depend upon the nature of one or more of the other factors. The methods of analysis of variance permit us to understand the individual contribution of each factor as well as the joint effects of the factor combinations. Inherent in the analysis of variance, however, there is a difference in the size of the standard errors of additive and interaction effects. The nature of this bias is such that it is more difficult to detect the epistatic as opposed to the additive effects of genes. This complicates the genetic interpretation of results in replicated, factorial experiments, although the complications diminish as the number of replications becomes very large. Nevertheless, if the additive component of genetic variation is considered the most important for predicting the response to selection within populations, then these kinds of complications can be viewed as essentially irrelevant to evolutionary discussion.

R. A. Fisher emphasized the additive effects of genes when measured in different genetic backgrounds in his several discussions of gene interactions and natural selection in large populations. Wright, however, emphasized the biological importance of the variation owing to gene interactions and discussed the "inadequacy of the simple additive concept of gene effect" (1968, p. 419) for describing gene action and evolution on many occasions, especially in regard to the interaction effects of genes on fitness. He concluded that "all genes that approach additivity in their effects on quantitatively varying characters will be favorable in some combinations and unfavorable in others in terms of natural selective value and, thus, exhibit interaction effects of the most extreme sort in the latter respect" (1968, p. 420). Thus Wright identifies fitness as a unique trait with respect to the expected predominance of genetic interaction effects at least on theoretical grounds. He stresses the point that genic effects may be additive with respect to many characters, but that fitness is a fundamentally different kind of character with strong interaction effects. This point is also made by D. S. Falconer (1981, p. 394): "Abundant evidence proves that virtually all metric characters are genetically variable in populations that are more or less in equilibrium, including characters that affect fitness. There must therefore be genetic variance of fitness. But, since selection for fitness produces no response, there can be no additive genetic variance for fitness; so all the genetic variance for fitness must be non-additive, i.e., variance due to dominance and epistatic interactions."

A second consideration raised by Wright (1968, p. 420) is that of the multiplicity of the effects of single genes on the phenotype (universal pleiotropy): "This again insures that natural selective value is a function of the system of genes as a whole rather than something that can be assigned individual genes." Here Wright raises the consideration that a multivariate analysis of gene effects also prohibits a strictly additive partitioning of phenotypic variation among genetic factors. This happens because a partitioning adequate for describing the effects of a gene on a single character becomes inadequate when characters are combined into a whole system, a single unitary phenotype.

In his review of *The Genetical Theory of Natural Selection* (Fisher, 1930a), Wright (1930) for the first time put forward this difference between himself and Fisher with respect to gene effects. "He [Fisher] assumes that each gene is assigned a constant value, measuring its contribution to the character of the individual (here fitness) in such a way that the sums of the contributions of all genes will equal as closely as possible the actual values of measures of the character in the individuals of the population" (Wright, 1930, p. 353). The phrase "additive gene effects" means that "the sums of the contributions of all genes will equal" the value of the character. In contrast, however, Wright (1930, p. 353) believes that "genes favorable in one combination, are, for example extremely likely to be unfavorable in another." That is, the additive concept of genic effect for Wright is not adequate for characterizing the relationship between a gene and fitness, because the *sign* of a gene's contribution to fitness is not constant but changes depending upon the genetic background.

The position taken by Fisher (1937, p. 108) is the opposite of that taken by Wright: "In studying the properties of a system of interaction factors it has been shown (Fisher 1918) that departures from the simple additive law of interaction will usually have effects somewhat similar to non-heritable modifications. We may therefore be confident that, even if a strictly additive interaction is not exactly realized, the mass effects of segregation in a large number of factors will closely simulate those of cumulative systems." The disparity of opinion on the importance of gene interactions is extreme. Fisher, on the one hand, dismisses gene interactions as comparable in their effect on the evolutionary process to the transient influence of "non-heritable factors" (a view common to some of the current models of the evolution of continuous characters, e.g., Lande, 1976).

Wright, on the other hand, advocated the primacy of interaction effects in evolution and developed his "shifting balance theory" as the most effective mechanism for selecting directly on gene interactions to ensure creative adaptive evolution. Wright argued that the evolutionary effects of gene interactions would be most important in *small* subdivided populations. In such small populations, random genetic drift would create variation between populations in the system of gene interactions. Differently put, Wright believed that natural selection operating in small populations admits the same kind of complications regarding the interpretation of additive and epistatic gene action as one encounters in the analysis of variance of a factorial experiment with a small number of replicates. Wright argued that it is the interactions among genetic factors that determine their effect on fitness and ultimately their evolutionary fate. That is, owing to finite population size, it is a gene's interactions with other genes that determines its relationship to fitness. A population subdivided into more or less isolated breeding groups is analogous to several small experiments being conducted on the same genetic system. The evolutionary fate of a gene in one deme within the subdivided population will be determined by the system of interacting genes characteristic of that local "experiment."

These arguments are not near resolution at the present time. Further empirical, theoretical, and perhaps epistemological research must be done before a deeper understanding of these questions emerges. An incisive evaluation of the role of epistasis in evolution will involve addressing fundamental questions in the definition and experimental methods upon which the modern synthesis in the biological sciences was based.

Eugenics

Daniel J. Kevles

The term "eugenics" was coined in 1883 by the English scientist Francis Galton, a cousin of Charles Darwin and a pioneer of the mathematical treatment of biological inheritance. Galton took the word from a Greek root meaning "good in birth" or "noble in heredity." He intended the term to denote the "science" of improving human stock by giving the "more suitable races or strains of blood a better chance of prevailing speedily over the less suitable" (Galton, 1883, pp. 24–25).

The idea of eugenics dated back at least to Plato, and discussion of actually achieving human biological melioration had been boosted by the Enlightenment. In Galton's day, the science of genetics had not yet been invented: Gregor Mendel's paper, the foundation of that discipline, was not only unappreciated but generally unnoticed by the scientific community. Nevertheless, Darwin's theory of evolution taught that species did change as a result of natural selection, and it was well known that by artificial selection farmers and flower fanciers could obtain permanent breeds of plants and animals strong in particular characters. Galton thus supposed that the race of men could be similarly improved—that through eugenics, mankind could take charge of its own evolution.

The idea of human biological improvement was slow to gather public support, but after the turn of the twentieth century, eugenics movements broke out in many countries. Eugenicists everywhere shared Galton's understanding that man might be improved in two complementary ways—by getting rid of the "undesirables" and by multiplying the "desirables." They spoke of "positive" and "negative" eugenics. Positive eugenics aimed to foster greater representation in society of people whom eugenicists considered socially valuable. Negative eugenics sought to encourage the socially unworthy to breed less or, better yet, not at all. (Entry into the vast historical literature may be obtained through Farrall, 1979; Roll-Hansen, 1988; and Harwood, 1989.)

How positive or negative ends were to be achieved depended heavily on the theory of human biology that people brought to the eugenics movement. Many eugenicists, particularly in the United States, Britain, and Germany, believed that human beings were determined almost entirely by their germ plasm, which was passed on from one generation to the next and which overwhelmed environmental influences in shaping human development. Their belief was reinforced by the rediscovery, in 1900, of Mendel's theory that the biological makeup of organisms was determined by certain "factors," which were later identified with genes and which were held to account for a wide array of human traits, both physical and behavioral ones, "good" as well as "bad." A number of eugenicists, however, notably in France, assumed that biological organisms, including human beings, were formed primarily by their environments, physical as well as cultural. Like the early nineteenth-century biologist Jean Baptiste de Lamarck, they contended that environmental influences might even reconfigure hereditary material (see Schneider, 1982).

In the interest of positive eugenics, environmentalists contended that more attention to factors such as nutrition, medical care, education, and clean play would, by improving the young, better the human race. Some urged that the improvement should begin when children were in the womb, through sound prenatal care. The pregnant mother should avoid toxic substances such as alcohol. She might even, for the sake of her fetus, expose herself to cultural enrichment, including attendance at fine plays and concerts. In the interest of negative eugenics, germ-plasm determinists insisted that socially inadequate people should be discouraged or prevented from reproducing themselves by urging or compelling them to undergo sterilization. They also argued for laws restricting marriage and immigration to keep genetically undesirable people out. Individuals with good genes were assumed to be easily recognizable from their intelligence and character. Those with bad genes had to be ferreted out, and in the early twentieth century eugenics also meant programs of research in human heredity (Kevles, 1985).

Eugenics entailed as many meanings as did terms such as "social adequacy" and "character." Indeed, eugenics mirrored a broad range of social attitudes. Many of them centered on the role in society of women, because they were indispensable to the bearing of children. On the one hand, positive eugenicists of all stripes argued against the use of birth control or entrance into the work force of middle-class women on grounds that any decline in their devotion to reproductive duties would lead to "race suicide." On the other hand, social radicals appealed to eugenics to justify the emancipation of women, contending that contraception would permit women to divorce sexual pleasure from reproduction and thus allow them to approach child bearing with a purely eugenic interest.

Social prejudices as well as dreams pervaded all of eugenics, including eugenic research. Eugenic studies claimed to reveal that criminality, prostitution, and mental deficiency (then commonly termed "feeblemindedness") were the products of bad genes. They concluded that socially desirable traits were associated with the "races" of northern Europe, especially the Nordics, and that undesirable traits were identified with those of eastern and southern Europe. Opponents pointed out that social deviancy is primarily the product of a disadvantageous social environment—notably, for example, of poverty and illiteracy—rather than of genes, and that apparent racial differences were not biological but cultural, the product of ethnicity rather than of germ plasm (see Haller, 1963; Ludmerer, 1972; Cravens, 1978; Kevles, 1985). These critics fought to bring a halt to the eugenic drive for power in many countries in the 1930s, but not before the movement had managed to pillory many lower-income people and members of minority groups in the United States and Europe. Eugenics reached its apogee of power during the Nazi regime, when hundreds of thousands of people were sterilized for negative eugenic reasons and when scientific authority joined with social hatred to send millions considered racially unfit to the gas chambers (see Weiss, 1987; Proctor, 1988).

In the 1930s various biologists also began trying to sanitize eugenics. They sought to distinguish between Galton's idea of human biological improvement, on the one hand, and the social prejudice that had invaded the conception, on the other. They realized that a sound eugenics would have to rest on a solid science of human genetics, one that scrupulously rejected social bias and weighed the respective roles of biology and environment, of nature and nurture, in the making of the human animal. They succeeded in laying the foundation for such a science of human genetics, and the field has made great strides in the half century since. Eugenics remains a dirty word, but the concept continues to tantalize biologists who see in the rapid advance of human genetics a powerful tool of human betterment, one to be embraced by individuals acting not for the race but for their own welfare.

EVOLUTION

Robert J. Richards

THE TERM "evolution," as we now use it, commonly refers to change of species over time and is usually associated with Darwin's theory of the descent of species by natural selection. The word did not always carry this set of specific meanings and associations. The term may itself be regarded as an evolved product, which reached its present state through descent with modifications induced by conceptually selective forces. Accordingly, its older meanings yet lie below the surface, still gently affecting the term's significance. Darwin's own theory of species evolution will look, to modern eyes, oddly more nineteenth-century when these genealogical relations stand clear. Three moments in the gradual alteration of the meaning of "evolution" may be distinguished: its use to describe embryological development; its bridging function in the principle of recapitulation; and its theoretically weighted deployment to refer to species change (see also Huxley, 1878; Bowler, 1975; and Richards, 1992).

Evolution versus epigenesis in embryogenesis. The Latin verb *evolvere* means to unfold or disclose. Its substantive form *evolutio* refers to the unfolding and reading of a scroll, as in Cicero's *De finibus* (I, vii, 25): "Quid poetarum evolutio voluptatis affert?" (What pleasure does the reading of the poets provide?) The *Oxford English Dictionary* suggests that "evolution" was first used in a biological context by an anonymous English reviewer in 1670, to characterize the preformationist embryology of the Dutch entomologist Jan Swammerdam (1637–1680). In his posthumously published *Historia insectorum generalis* (Dutch, 1669; Latin, 1685), Swammerdam had argued that an insect grub was generated from the semen of the female (stimulated by the semen of the male) and that the adult form lay already encased in the embryonic larva, requiring only that its outer skin be shed and its preexisting internal parts be augmented and expanded. The English reviewer, in the *Philosophical Transactions of the Royal Society* (Anonymous, 1670, p. 2078), observed that when Swammerdam referred to the "change" that insects underwent, "nothing else [is] to be understood but a gradual and natural Evolution and Growth of the parts." The great Swiss anatomist Albrecht von Haller (1707–1777) weighed the rival theories of embryological change in his commentary on the *Praelectiones academicae* (1744) of his own famous teacher at Leyden, Hermann Boerhaave (1668–1738). Aristotle, Harvey, and a few Italians proposed epigenesis, which taught that "the parts of animals are successively generated out of fluid according to certain laws." However, wrote Haller, "the theory of evolutions proposed by Swammerdam and Malpighi obtains almost everywhere" ("evolutionum theoria fere ubique obtinet a Swammerdamio . . . & Malpighio proposita") (1744, p. 489).

Haller's own extended researches on fertilized chicken eggs, in which he traced more articulated embryonic structures to their folded and translucently invisible earlier stages, provided him compelling evidence for adopting the theory of evolution. During his research and in the articulation of his new theory, Haller received constant encouragement from his countryman and friend Charles Bonnet (1720–1793), whose discovery of parthenogenesis in aphids and microscopical observations of the aphid imago just beneath the skin of the grub, coupled with his degenerating vision and accelerating religious enthusiasm, undoubtedly disposed him to champion preformationism (see Roe, 1981).

Bonnet is chiefly responsible, in the eighteenth century, for transposing individual into species evolution. In his highly influential *Considerations sur les corps organisés* (1762), he advanced "evolution [*l'Evolution*] as the principle that better conforms to the facts and to sound philosophy" (vol. 1, p. vi). Later, in his more speculative *La Palingénésie philosophique* (1769), he elaborated his complementary, though tentative, doctrine of emboîtement, or encapsulation, to produce a general theory of "natural evolution of organized beings" ("d'Evolution naturelle des Etres Organisés") (vol. 1, p. 250). Bonnet, a fervent if heterodox Calvinist, believed that God had originally created a plenitude of germs, each encapsulating a miniature adult that, in turn, carried germs containing yet more homunculi and their germs, enough to reach the Second Coming.

Shortly after Haller published his new theory of embryological evolution, it was attacked by the young German physician Caspar Friedrich Wolff (1734–1794). In his doctoral dissertation *Theoria generationis* (1759), Wolff defended an epigenetic theory against Haller's "mechanistic medicine," which explained "the body's vital functions from the shape and composition of its parts." Braced by the empirico-rationalistic traditions established at Halle in the previous generation by Christian Wolff (Richards, 1981), he undertook a study of the vascular system of the embryonic chick, which convinced him that the animal's vessels formed out of homogeneous matter under the aegis of a principle of generation, or "motor force" *(vis motrix)*, "by whose agency all things are effected" (1759, pp. 124, 106).

By the end of the eighteenth century, the epigenetic theory of Wolff prevailed. The older evolutionary theory succumbed to better microscopes, to the decline of interventionist theology, and to German enthusiasm—stimulated by Kant and Schelling—for disciplinarily independent, formative principles. Preformationism, however, did not simply go extinct; rather, it was transformed. At the turn of the century many embryologists detected in the development of the fetus, not the expansion of the already formed adult of that species, but the serial unfolding of the forms of more primitive species. The embryo seemed to recapitulate sequentially the hierarchy of species below it. And the term "evolution" itself gradually moved toward this altered meaning; it came to be used synonymously with "development" and *Entwickelung* to refer to the progressive stages of individual and then species change. By 1833 the meaning of "evolution" had shifted 180 degrees, as this passing observation of Etienne Geoffroy Saint-Hilaire (1772–1844), a colleague and supporter of Lamarck, indicates: "Of the two theories of the development of organs, one supposes the preexistence of germs and their indefinite emboîtement, the other acknowledges their successive formation and their evolution [*leur évòlution*] in the course of ages" (1833, p. 89).

The theory of embryonic recapitulation. The renowned embryologist and later opponent of Darwinian theory Karl Ernst von Baer (1792–1876) indexed the shifting meaning of "evolution" in his critical rejection of recapitulation theory. In his celebrated *Entwickelungsgeschichte der Thiere* (1828), he referred to his academic disputation of 1823, in which he had argued that "the law proclaimed by naturalists was foreign to nature, namely that 'the evolution which each animal undergoes in its earliest period corresponds to the evolution which they believe to be observed in the animal series'" ("evolutionem, quam prima aetate quodque subit animal, evolutioni, quam in animalium serie observandum putant, respondere") (pp. 202–203). By the 1820s the principle of recapitulation had gained a strong hold on the imagination of German anatomists and guided their investigations. The idea in identifiable form, though, seems to have been first expressed in the late eighteenth century by the great English autodidactical physiologist John Hunter (1728–1793). Shortly thereafter, in 1793, Karl Friedrich Kielmeyer (1765–1844), in a Kantian attempt to discover unity in nature's forces, alluded vaguely to comparable general stages in the early embryogenesis of men and birds and remarked that sense organs appeared in the individual "almost in the same order" as in the series of lower organisms (1793, p. 261); similar comparisons were made by Johann Heinrich Autenrieth four years later, in 1797. Von Baer cautiously did not mention his opponents by name, but internal evidence suggests that his targets must have been those who more fully articulated the principle, his contemporaries Lorenz Oken (1779–1851), Friedrich Tiedemann (1781–1861), Gottfried Treviranus (1776–1837), and Johann Meckel (1781–1833). Von Baer, however, later confessed in his autobiography ([1886] 1986, p. 312) that it was principally Meckel's work he had set out to destroy.

In the first volume (1821) of his *System der vergleichenden Anatomie,* Meckel argued (as Darwin later would) that the arbitrary and uncertain distinctions between varieties and species, as well as those between species and the higher classes, suggested the possibility that such "larger and smaller collections of organisms are only alterations, probably originating gradually, of one and the same Urorganism" (p. 62). The structural plan that united the animal kingdom resulted from descent from more primitive stages that commenced with an Urtype. Variations on subsequent types, he thought (pp. 344–345), might be introduced into species through individuals, by the sort of heritable modifications Lamarck had mentioned. This evolutionary history would be revealed, Meckel argued, in embryogenesis. These parallels between species evolution and individual evolution, moreover, indicated that the same developmental laws were at work in both (p. 396).

Meckel provided the most sophisticated form of the recapitulation principle, and his work displayed the demonstrative power of comparative embryology for the theory of species evolution. The spread of this approach to species change, however, was due in no small measure to the embryologist Etienne Reynaud Serres (1786–1868). Serres, a disciple of Geoffroy, regarded Meckel a venerable "Nestor of medicine" (1827, p. 88). When Serres formulated a version of the recapitulation principle in 1837—"Mollusks are the permanent embryos of the vertebrates and of man" (p. 370)—Darwin copied it in his notebook and took it to heart. But Darwin's own powerful use of the principle had first to be filtered through the analyses of its most influential antagonist, Karl Ernst von Baer.

Where anatomists such as Meckel, Treviranus, and Tiedemann assumed continuous progressive transitions from lowest organism to highest, von Baer sided with Georges Cuvier (1769–1832), maintaining that animal life revealed four fundamental arrangements of organic parts, four archetypes *(Haupttypen)* that remained distinct in nature. These archetypes, he granted, displayed different grades of formation *(Ausbildung),* depending on the heterogeneity and differentiation of the parts. In this respect a bee (a species type within the archetype of the articulata) would represent a higher degree of formation within its archetype than would most fish (vertebrates) in theirs. By so relativizing the idea of progressive scaling, the von Baerean framework frustrated a seemingly fundamental requirement of recapitulation theory.

Individual development, according to von Baer, proceeds from the general features of the archetype, during the early stages of embryological evolution, through higher grades of differentiation, until the particular features of the species are established. The vertebrate fetus does not pass through the permanent forms of the various articulata and mollusca. "The embryo of the vertebrate is already at the beginning a vertebrate" (1828, p. 220).

Although von Baer rejected the letter of recapitulation theory, he shared more than enough assumptions with its advocates that Ernst Haeckel (1834–1919) could reasonably call upon von Baer in defense of both species evolution and recapitulation theory (1866, vol. 2, pp. 6–12). Von Baer recognized, for instance, that the vertebrate archetype—because it required some organizational features common to all life—"is, as it were, composed of the earlier types" (1828, p. 212). More important, though, he admitted that within a type the organisms lowest in degree of formation, those least differentiated, would most closely embody the basic type (1828, p. 238). The archetype could, in a sense, be a real creature and not merely an ideal plan. In this respect, then, the more progressive creatures would indeed pass through some permanent types of those organisms lower in degree of development. These concessions had historical consequences. Though von Baer contended against recapitulation theory and the evolution of species, his own framework, having left his hands, underwent a development precisely toward these ends.

The evolution of species: Early theories. The idea that species have altered over time can be traced to ancient Greece, where Empedocles (fl. 444 B.C.) and Anaxagoras (534–462 B.C.) spun fables that Aristotle (384–322 B.C.) found incompatible with principles of natural law and stability of substantial structures. But even the Philosopher acknowledged spontaneous generation of insects and worms, belief in which during the Renaissance helped convince Francis Bacon (1561–1626) that while "the Transmutation of Species is, in the vulgar Philosophy, pronounced Impossible, . . . seeing there appear some manifest Instances of it, the Opinion of Impossibility is to bee rejected; and the Means thereof to bee found out" (1631, p. 132). Subsequently Benoit de Maillet (1656–1738), Julien Offray de LaMettrie (1709–1751), Pierre Louis de Maupertuis (1698–1759), and Denis Diderot (1713–1784) all spun gauzy notions about species change (see Glass, 1959). The great anatomist and founder of the Jardin du Roi, Georges Leclerc, Comte de Buffon (1707–1788), initially opposed the idea of species change, but breeding experiments and certain theoretical considerations brought him to the view (in his essay "De la dégénération des animaux," 1766) that the originally created kinds of animals (now the genera and families) have, owing to the influence of the environment, degenerated into the myriad species now populating this decidedly venerable earth—an assumption not far different from that of his enemy Linnaeus (1707–1778), who instead proposed the mechanism of hybridization to produce new species from original kinds. Buffon's and Linnaeus' quasi-evolutionary proposals became known to Erasmus Darwin (1731–1802), who advanced a more radical theory of species transmutation in his two-volume *Zoonomia* (1794–1796), according to which that original "living filament" created by God, through irritable response and acquired habit, became transmuted over eons into all the warm-blooded animals (vol. 1, pp. 138–139). Charles Darwin (1809–1882) first brushed against the hypothesis of species transmutation in reading his grandfather (1969, p. 49). But undoubtedly the more important early influence on Darwin, as well as on his contemporaries (especially Herbert Spencer, Robert Chambers, and Alfred Russel Wallace), was Jean Baptiste de Lamarck (1744–1829).

Darwin had learned of Lamarck's hypothesis in avocational study of invertebrates while at medical school in Edinburgh (1825–1827); and during the voyage on board H.M.S. *Beagle* (1831–1836), he had the leisure to examine Lamarck's *Histoire naturelle des animaux sans vertèbres* (1815–1822). In 1832, when sailing down the eastern coast of South America, Darwin received the second volume (1832) of Charles Lyell's *Principles of Geology* (1830–1833). It contained a sustained presentation and sharply negative critique (borrowed in large measure from Cuvier) of Lamarck's arguments "in favour of the fancied evolution of one species out of another" (vol. 2, p. 60). Lamarck had first suggested species change in 1800, but developed his theory more fully in later works, especially in his *Philosophie zoologique* (1809). It was this book that Lyell analyzed. Lamarck proposed that simple monadic life bubbled up from the muck, under the influence of the imponderable fluids of caloric and electricity. These forces, which continued to operate even now, caused the elementary vesicles to become more complex, eventually producing organisms that internalized those fluids. The incessant excavations and articulations by the internal fluid over generations led to greater perfection (i.e., complexity) of organisms, while the heritable effects of habit modified their parts to fit them into a changing environment (see Burkhardt, 1977; Richards, 1987; Corsi, 1988). Lamarck did not use the term "evolution" to describe this progressive transformation of animal species; rather he referred to the process, in obvious deference to his mentor Buffon, in a negative way, as the inverse of that "degradation" *(dégradation)* we observed as we cast our eye down the scale of life (1809, vol. 1, p. 220). Lyell gave the English usage of the term "evolution" currency by applying it to Lamarck's and Tiedemann's theories, but Lyell seems to have gotten the sterling for his coinage from Serres.

I have suggested that the principle of recapitulation transmitted to the idea of species transformation certain considerations born out of embryological theory. The term "evolution" indexes this transmission. The Germans, especially Tiedemann and Meckel, thought of the principle of recapitulation as demonstrating a unity of law that accounted for both individual development and species development. In his own consideration of recapitulation, which owed much to the Germans, Serres used the expression "théorie des évolutions" (1827, p. 83) ambiguously to refer to the "métamorphoses" of organic parts in the individual and the parallel changes one sees in moving from one family of animals to another and from one class to another.

Lyell cited Serres and referred to Tiedemann in evaluating the evidence that the principle of recapitulation provided for a real evolution of species (1830–33, vol. 2, pp. 62–64). Lyell thought the principle only demonstrated a Cuverian or Geoffroy-like unity of plan within the animal kingdom—transcendental links, but not real historical connections. Nonetheless, by use of the term "evolution," he suggested to his readers the intellectual connection between the theories of individual evolution and species evolution. One of those readers was Darwin.

Darwin's embryological theory of evolutionary progress. I believe that Darwin formulated and reformulated his ideas of species change in light provided by recapitulation theory and by the model of individual evolution. The principle of recapitulation and the embryological model allowed him to resolve the most pressing problem presented to him by his professional colleagues: how to account for the unity of type, the Cuverian *embranchements* of design, that almost every leading naturalist of the time recognized in the animal kingdom.

After returning from the *Beagle* voyage in December 1836, Darwin began arranging and cataloging his specimens. In March of the next year (see Herbert, 1980, pp. 11–12; Sulloway, 1982; Hodge, 1983), he set down in his "Red Notebook" brief speculations on species change. He further reflected on such changes in a series of notebooks beginning in late spring or early summer 1837, the so-called Transmutation Notebooks. In September 1838, he read Malthus, which, as he recalled, gave him a "theory by which to work" (1969, p. 120). The catalyst of Malthus precipitated out of Darwin's thought the essentials of his mechanism of natural selection. He further developed his ideas about species change in two essays, one in 1842 and its expansion in 1844. These became the spine for the more comprehensive expression of his theory, which he began drafting in 1856. The work on this book, which was to be called "Natural Selection," was interrupted in 1858 by a letter from Alfred Russel Wallace (1823–1913) that outlined a theory of transmutation very similar to the one Darwin had been laboring over for almost twenty years. In a white heat, he condensed his huge manuscript and added in smaller compass the further chapters he intended. He published *On the Origin of Species* in 1859. Through this long gestation, one can distinguish certain stages in the development of his theory (see Kohn, 1980; Hodge, 1983, 1985; Richards, 1987, pp. 83–105).

In the earliest stage (Darwin, 1987: Red Notebook, pp. 127–130; Notebook B, pp. 1–23), Darwin—reflecting ideas of his grandfather, Lyell (see Hodge, 1983), and Richard Owen (1804–1892) (see Sloan, 1986, and 1992)—considered species to be comparable to individuals. Under the influence of the environment, species changed over time; but, reaching the end of their allotted years, they gave birth to new species and then died off. In the second stage, Darwin retained the adapting mechanism of the heritable effects of environmental agents (taking many hints from Lamarck), but gave up the idea of species having a definite term of life. He supposed that there would be a branching of species: some would continue to "progress," "perfecting" through different forms; others, because they did not adapt fast enough, would go extinct (Notebook B, pp. 25–39). This, Darwin thought, was comparable to what occurred when individuals, having adapted to their circumstances, produced offspring with the new modifications (Notebook B, pp. 63–64). Like his grandfather and the German transcendentalists, Darwin understood the goal of species transformation to be higher creatures, ultimately man: "Progressive development gives final cause for enormous periods anterior to Man" (Notebook B, p. 49).

In the third stage of his reflections, Darwin realized that the mechanism of the direct effects of the environment would not easily adjust organisms to their surroundings. He then proposed that an animal would develop new habits to adapt to a shifting environment, and that these habits would, over time, be sifted, so that individual peculiarities would fall out. Finally, by dint of repetition over countless generations, such general habits would become instinctive. These finely articulated innate behaviors would, in their turn, slowly alter anatomical traits, changing the character of the species (Notebook B, p. 171). By postulating gradual modifications introduced by instinct, Darwin constructed his mechanism to Lyellian uniformitarian specifications, while rejecting what he took to be the Lamarckian attribution of conscious will-effort to animals (see Richards, 1987, pp. 85–98).

Malthus' *Essay on Population,* which Darwin read in late September 1838, enabled Darwin to appreciate the sort of pressure to adapt that would be produced by great fecundity. In this light he immediately understood the advantage a favorable trait would have, and how it might gradually alter species. In his reflections of September 28, Darwin's mechanism of natural selection came to birth. When it did, it slowly pushed the other devices of species change to the periphery of his theory of evolution. Though these "Lamarckian" instruments were no longer central to the production of new species, they were retained, and in various guises were preserved in the *Origin*.

Natural selection appears to modern eyes to foredoom any notions of evolutionary progress, especially those built into embryological models of species change. After all, Darwin's mechanism works on chance variations, which it selects to satisfy local requirements. Further, Darwin himself protested against what he took to be Lamarck's idea of an "innate tendency toward progression." Although Darwin did reject the hypothesis of an intrinsic cause buried in the interstices of organization that necessarily produced progressive evolutionary change, in the beginning he did insist, relying on the embryological model, that animals had an internal "tendency to change" (Notebook B, pp. 5, 16, 18, 20), which would be molded by the extrinsic agency of the biological environment. Because creatures would continually readapt to one another, each forcing the other to more complex alterations, natural selection would exert, as it were, an external pull, drawing most organisms to greater levels of complexity and perfection (see Ospovat, 1981, and Richards, 1988; see Hodge, 1983, for a different interpretation).

Darwin's notion of an innate tendency to change gradually faded in his theory; it came to be replaced by the supposition of environmental forces effecting variation (though he believed organisms might inherit a tendency to vary). He retained, however, his conception of a progressive dynamic; and in the *Origin* he augmented its power by attributing to natural selection the beneficent concern for the good of creatures that had been formerly expressed by the recently departed Deity. Artificial selection of animals was capricious and governed only by selfish desires of men; natural selection altruistically looked to the welfare of the creatures selected, as Darwin put it: "Man selects only for his own good; Nature only for that of the being which she tends" (1859a, p. 83). He concluded that natural selection therefore ensured progressive evolution: "as natural selection works solely by and for the good of each being, all corporeal and mental endowments will tend to progress towards perfection" (1859a, p. 489).

Darwin's theory of embryological recapitulation. The principle of recapitulation, almost from the beginning of Darwin's species theorizing, served as the aortic connection for three components of his early theory: the embryological model of evolution, the idea that the purpose of generation was progressive adaptation, and the assumption of common descent. Even after he had formulated the law of natural selection, Darwin continued to pivot his theory of evolution around the principle of recapitulation, as several passages in his notebooks suggest (e.g., Notebook E, pp. 83–

84). In the Essays and the *Origin,* Darwin's reflections on embryological recapitulation occur within his larger considerations of the doctrine of unity of types. He argued that the similarity of species within the class of vertebrata, for instance, could be explained by descent from a common ancestor. The generalized vertebrate archetype, Darwin suggested, was actually a primitive organism that had become more specialized during descent and that had given rise to the many genera and species it left as progeny. The embryo of a higher vertebrate, a human being for example, might then preserve a record of such descent in embryogenesis (see Gould, 1977a, pp. 70–74, and Ospovat, 1981, pp. 151–169, for a different interpretation).

In the Essay of 1844, Darwin indicated how his theory of recapitulation might explain the similarity of embryos of different species. He proposed that if selection operated, usually on mature organisms, then embryos of two breeds would resemble one another more than their adult forms would. Further, if these two breeds became the parent stocks of new species, then the embryos of those species would continue to resemble each other while selection gradually separated the adult forms. Thus these considerations showed "how the embryos and young of different species might come to remain less changed than their mature parents" ([1844] 1909, p. 225). In the several years before the publication of the *Origin,* Darwin conducted measurement experiments on a variety of breeds of dogs, horses, pigeons, and so on to demonstrate that resemblances were much closer among different breeds of neonates than among their parents.

Thus far we see Darwin preserving a set of facts that von Baer would not have denied, though of course Darwin framed these facts in terms of a theory of evolution that von Baer rejected in its German version and would continue to reject after he learned of the Englishman's new twist. But Darwin went further and wrote his initials on the very heart of the recapitulation principle. He drew from his several considerations what he thought to be a hypothesis "with much probability": that "in the earliest and simplest condition of things the parent and embryo must have resembled each other, and that the passage of any animal through embryonic states in its growth is entirely due to subsequent variations affecting *only* the more mature periods of life" ([1844] 1909, p. 230). Which is to say: if adults and their embryos were similar in earlier times and if embryos have changed little over evolutionary history, then embryogenesis of present-day organisms is a repetition of earlier *adult forms.* The archetype is the ancestor, and it is an adult. Darwin's discussion of embryology in the *Origin* resulted in this same conclusion.

At the end of the twentieth century, over 150 years after Darwin first formulated his theories of evolution, his conceptions have become, to paraphrase Auden's line about Freud, a whole climate of opinion. The debates today range over the tempo and mode of evolution: whether evolution is gradual and constant, or saltational and occasional; whether it is driven to great contingent branching by proliferating variability or more narrowly confined by reduction in variability; whether transformation is directed by natural selection alone or also by the structural *Bauplan* or blueprint of the organism; whether lonely genes are the focus of selection or whether the object of selection's desires are systems of greater comprehension—from individuals through kin groups, to populations and species (see Sober, 1984c). All of these different positions can easily be traced back through the nineteenth century, with Darwinians disputing among themselves then much as now. But neo-Darwinians seem to have reached general agreement that three older proposals—that species evolution should be modeled on individual evolution, that evolution is progressive, and that embryogenesis recapitulates phylogenesis—should be dismissed. These ideas, nonetheless, were the very seeds of Darwin's thought. And if one pries back the husks of rhetoric, these earlier stages in the formation of evolutionary theory can, with a little imagination, be yet discovered forming the fetal structures of the views of many contemporary neo-Darwinians.

EXTINCTION

John Damuth

IN MODERN usage, the word "extinction" refers to a terminal event in the history of a population, species, or higher taxon. "Extinction" is occasionally applied to other biological entities, such as alleles, gene lineages, or cell lines, but in these contexts the term is usually not being used in any technical sense that is different from the standard dictionary definition of the termination or death of a lineage. In contrast, the extinction of a population or taxon is often considered to be an example of a general biological phenomenon of considerable importance in biology and paleontology. Today it is a commonplace in evolutionary biology that most of the species that have ever lived are now extinct (Romer, 1949; Raup, 1984).

The application of the term "extinction" in the above sense both to taxa and to populations or demes, however, immediately causes a problem: the "lineage" that terminates must be defined differently for the two kinds of entities. Taxa, on the one hand, are chiefly the entities of history, and their existence is defined primarily by their phylogenetic relationships and by their extension in time; a species or genus is not extinct while any of its members are extant, wherever they may happen to be located. Populations, on the other hand, are functional ecological or genetic entities defined with reference to particular geographic locations. Thus, for taxa, the lineage in question is the traditional one, consisting of an ancestor and its descendants, without regard for their geographic distribution. For populations, the "lineage" consists of the individuals of a species found over a continuous time interval in a particular place. This includes immigrants to the population over this time period, but excludes emigrants.

Extinction is thus seen to occur across the widest of temporal and geographical scales, from an interval of a few generations within a local ecosystem to that spanning geologic time intervals on a global scale (Nitecki, 1984). These considerations lead one to doubt that the disparate phenomena embraced by the term extinction can all be related to a single set of unifying principles or common, natural causes and influences (Gould, 1985). Here, extinction of populations and of taxa will frequently have to be discussed separately.

Some brief remarks on the history of the concept of biological extinction will be more relevant than the history of the word itself, which has been used in essentially the modern sense since at least the beginning of the nineteenth century (e.g., Playfair, 1802). Awareness of the extinction of taxa long predates Charles Darwin and the general acceptance of evolution. Indeed, the disturbing implications of extinction, implying an imperfect Creation, were one of the initial impediments to the acceptance of fossils as remains of once-living creatures (Rudwick, 1976). It is usually agreed that Georges Cuvier (1812) demonstrated conclusively that the fossil record contained species not extant today and forced the general acceptance of extinction by the scientific community. Beginning about this time and continuing throughout the first half of the nineteenth century, geologists began to use the distinctive (and presumably extinct) fossil species of various rock units to correlate strata and to assist in the development of a geologic time scale. During this period, however, extinction was considered by many to be a fact opposed to evolution: the fossil species different from those observed today *either* represented extinct forms—with no species having undergone transformation—*or* all represented ancestors of the modern biota—with no lineage having become extinct (Rudwick, 1976).

Darwinism reconciled extinction and transformation by viewing the termination of lineages as merely a failure of certain poorly adapted lines to provide descendants. This view, combined with the recognition that descendants could be evolutionarily modified and may also have diversified into new lines, explained the differences between modern and fossil biotas. Extinction became a crucial part of the mechanism of evolution. For Darwin, evolutionary change by natural selection required the elimination of inferior varieties—the termination of their lineages. In explaining the dynamics of speciation and lineage transformation (e.g., Darwin, 1859b, pp. 116–126), he describes as *extinction* the disappearance of conspecific varieties resulting from intraspecific competition. Species were defined primarily on morphological grounds; there was no "speciation" process that was distinct from those processes that produced new varieties within a species. Thus if ancestral morphologies had disappeared within a lineage it was proper to say that those ancestral "species" (morphological types) had become extinct, even though there had not been any change in the number of lineages. Throughout what we would describe today as phyletic evolution within a species, intermediate forms were said to have become extinct. Darwin indiscriminately used the term "extinction" to apply to lineage-terminating events at all levels, including that of intraspecific varieties. This is also the usage of many neo-Darwinians, notably Rensch (1959).

By Darwin's day, students of paleontology had for some time been assigning formal Linnean names to fossil forms, and thereby they had established large numbers of extinct species. With the acceptance of evolution, it was recognized that some of these species could be connected in ancestor-descendant lineages. Because species were considered to be morphological types, however, this recognition of multiple successive "species" within a single lineage would not affect the status of these forms as true species, their status as being extinct, or the involvement of extinction processes in their transformation.

Thus in Darwinism and early neo-Darwinism (Huxley, 1942) the term "extinction," particularly when used to refer to species, came to denote both the termination of a species lineage without descendants and the disappearance of earlier (usually formally named) forms within the same species lineage, through evolution of that lineage. The latter usage, acceptable in the nineteenth century, was in conflict with the importance that the neo-Darwinian synthesis placed on speciation as a distinct process. Further, it confounded changes in diversity with changes in morphology. This incompatibility was clearly recognized by G. G. Simpson (1944, 1953) and others, notably Ernst Mayr (1949) and Sewall Wright (1949c). There has been a slow and steady movement toward disassociating the term "extinction" from within-lineage phenomena. The successive "species" within a single lineage are now termed *chronospecies,* to indicate that they grade into each other temporally and were not formed by splitting of ancestral lines (cladogenesis). Purists regard chronospecies as nothing more than arbitrarily named segments of a continuum. Likewise, the apparent extinction of one chronospecies through replacement by a successor has gone by a variety of terms, such as *phyletic extinction* (the converse of Simpson's *phyletic speciation*) (Raup and Stanley, 1971). Since the early 1970s, chronospecies "extinction" has been known as *pseudoextinction* (Van Valen, 1973; Raup and Stanley, 1978).

Extinction is forever . . . always? Extinction has always (at least) referred to an event that befalls taxa, yet the concept of a taxon has changed throughout the history of biology. Because of this long association of the term with taxa, many ambiguities in the way that "extinction" is used today are related to ambiguities in the definition of taxa and their roles in organismic biology.

Under idealistic views of nature, or views in which taxa are considered to be eternal classes (Hull, 1988), a taxon once extinct could reappear at a later time. The currently dominant view, however, is that taxa at all levels are spatiotemporally restricted individuals (Ghiselin, 1974b; Hull, 1976,

1988), and so extinction is permanent. Even should organisms evolve that are identical in every respect to members of an extinct species, the new organisms cannot be considered members of that species.

Consistency would appear to require a similar perspective on the extinction of populations: populations should be considered as spatiotemporally restricted individuals, and if members of a species become reestablished in an area where a population of that species has previously become extinct, they should be regarded as having formed a "new" population. Yet one can make a case for treating populations as temporally unrestricted classes, at least sometimes. Membership in a population is determined by presence in a particular location—not by genealogical descent. Because the location does not disappear, individuals inhabiting the area subsequent to local extinction and a period of nonoccupancy "belong" to the population in the same way that any immigrant does. Since eventual reestablishment by immigration from other populations is likely, it may make sense to allow populations to reach temporarily a size of 0 in a model of a natural system. Furthermore, strict application of a populations-as-individuals view may lead to occasional absurdities. Imagine that the individuals of a population leave their original area, disperse to different places, and then return. For long-lived species, the "new" population could then be composed of exactly the same individuals as the "old" population. In practice, it often makes no difference whether population extinction is regarded as terminal or as temporary. The usage intended is seldom explicitly stated and, if relevant, must usually be inferred from the context.

Global extinction is the complete extirpation of an entity—what we have been calling "extinction" up to this point, and what is usually meant if the term is not further qualified. In contrast, an entity is said to become *locally* extinct in a certain region when it no longer is found in that region. By definition, population extinction is always local—and usually simultaneously global. But taxa may become locally extinct as well. In this case, it is the local populations (or *avatars*—Damuth, 1985) of the relevant species that have become extinct. The species or higher taxa are not extinct globally, and may return to the region at a later time (e.g., Schankler, 1981; Johnson and Colville, 1982). In the special case that all of the populations of the taxon are found in the region in question, local extinction and global extinction are equivalent.

In paleontology, both local and global extinctions are important. Local extinctions can be used in regional biostratigraphy and can also indicate changes in paleoenvironment or paleocommunity structure. Global extinction is what is usually deemed to have more relevance in large-scale studies of changes in organic diversity. Because of its irreversibility, global extinction is also usually regarded as having the greater significance for macroevolution. But, because of imperfections in the fossil record, only local extinction is reliably observed for most taxa. Global extinction is inferred from absence in the later fossil record—though knowledge of the fossil record is always improving. The apparent "return from the dead" of taxa absent after a long interval (and whose disappearance would otherwise have been inferred to be a global extinction) has been called the "Lazarus effect," and taxa exhibiting such a fossil record are known as "Lazarus taxa" (Jablonski, 1986a).

Extinction in a hierarchical context. A given biological entity can be considered to belong to one or more of several nested hierarchies of natural entities. For example, a population is contained in a species and, in turn, the species is contained in higher taxa. Species and other taxa can belong to monophyletic groups (clades), whether or not these are represented in formal classifications; their nested nature provides many more inclusive groups to which an individual taxon may belong. A population may belong to a community and to other, even more inclusive, ecological entities such as regional biotas. It may also belong to an ecosystem and be part of a hierarchy of natural biotic/abiotic systems. A population or taxon can also be considered to occupy a place in a hierarchy of ever more inclusive geographic regions. All of these more inclusive entities can be said to change as the result of events happening to their constituents at all lower levels, and this results in a different usage of the term "extinction."

"Extinction" refers to a singular *event* in the history of a population or taxon. However, higher-level, more inclusive entities (consisting of aggregations of entities that experience extinction events) are said to suffer "extinction" as if it were a *process* or a *condition.* The higher-level entities change (their nature, their composition, their geographical extent, etc.) as a result of the irrevocable disappearance of some of their constituents. "Extinction" is thus a process of change that these entities undergo or experience, rather than a terminal event. Any historical entity (*sensu* Hull, 1988) that contains biological entities can suffer extinction as a process, whether or not it can suffer extinction as an event. For example, a continent or island can be said to suffer extinction as a process. Extinction *events* befalling an entity's members generate an extinction *process* that the entity experiences.

Extensions of the concept of extinction-as-a-process are fundamental to studies of taxonomic diversity over geologic time (e.g., Valentine, 1985; Van Valen, 1985). For taxa or for the biota as a whole, extinction *rates* are frequently calculated and given biological interpretations. The fates of taxa are often explained by the relative rates of competing processes, including origination and extinction.

In most of the literature both of these senses of "extinction" are employed without further qualification. There may be confusion if both senses are not recognized. For example, is "the extinction of a species" an *event* that befalls the species as an individual? Here, extinction may have the properties of an event, such as a probability and moment of occurrence, and it is clear that something has *happened* to the *species*. Or, is the "extinction of a species" merely the terminal event in a long, protracted series of events—a *process* of extinction (death) affecting the organisms of the species? Here, the death of the last member of the species—an ordinary event in itself—causes the extinction of the species as a side effect. From this viewpoint, there appears to be no specific event that affected the species as a unit.

Both ways of viewing "extinction" validly describe different aspects of extinction phenomena in a hierarchical context. The event and process usages are often clear from their context, yet conceptual difficulties have arisen when processes acting at more than one biological level have been sought (e.g., Vrba and Eldredge, 1984; Lloyd, 1988, pp. 108ff).

Mass and background extinction. Recent interest in extinction phenomena has intensified as a result of broad-scale statistical studies of the history of biotic diversity (see Nitecki, 1984; Valentine, 1985; Stanley, 1987; Larwood, 1988) and new explanations, involving extraterrestrial causes, for periods of global crisis (e.g., Alvarez et al., 1980; Russell, 1984). It has long been known that during certain periods of the earth's history unusually large numbers of species and higher taxa have become extinct. Such periods of time, where elevated extinction rates simultaneously affected many unrelated groups of organisms, are known as episodes of *mass extinction*. Mass extinctions can be distinguished from *background extinction*, which is extinction that occurs throughout the world or a region at the normal level of extinction rates experienced by the biota. During mass extinctions the number of extinctions rises—sometimes dramatically—above background levels. The factors that confer upon a species resistance to extinction during "normal" times may be different from those that enhance its ability to survive during mass extinction episodes (Jablonski, 1986b). The typical duration of mass extinction episodes is controversial, as are the criteria for recognizing them in the fossil record. The number of mass extinctions recognized varies from five over the last 600 million years to eight over the last 250 million years alone (Raup, 1986, 1988). A hotly debated topic is the suggestion that mass extinction episodes occur according to a regular periodicity (Raup and Sepkoski, 1984). Although much remains controversial in this area, there seems to be growing acceptance of a qualitative distinction between mass and background extinction. It has been suggested that human hunting played a part in the extinction of large terrestrial species during the Pleistocene (Martin and Klein, 1984). It has also been noted that the human activities of the past two hundred years have initiated an episode of mass extinction that may be as profound as any recorded in the fossil record (Myers, 1988).

FITNESS: HISTORICAL PERSPECTIVES

Diane Paul

"FITNESS" is perhaps the most contentious concept in evolutionary biology. How did the word enter the vocabulary of evolutionary biology, and how has its meaning changed over time?

Charles Darwin employed the verb "fit" synonymously with "adapt," and the adjective "fitted" with "adapted" or "suitable." The noun "fitness" appears only once in the first edition of *On the Origin of Species*: "Nor ought we marvel if all the contrivances in nature be not, as far as we can judge, absolutely perfect; and if some of them be abhorrent to our idea of fitness" (1859b, p. 472). But Darwin did, of course, ultimately use the expression "survival of the fittest," and this is the route through which the noun came into common use in biology. The actual phrase "survival of the fittest" first appeared in Herbert Spencer's *Principles of Biology* (1864), where it was employed simply as a synonym for Darwin's term "natural selection."

Notwithstanding his socialist sympathies, Alfred Russel Wallace was a great admirer of Herbert Spencer. Wallace was also concerned at the many misunderstandings that had arisen among Darwin's readers from the assumption that selection requires conscious thought and direction—that is, a selector. In 1866 he wrote to Darwin, arguing that much needless confusion had resulted from Darwin's choice of "natural selection" as well as his constant personification of nature as "selecting," "preferring," and so forth. In Wallace's view, "survival of the fittest" would both describe the theory more accurately and also avoid the misunderstandings that had plagued it from the beginning. As early as 1860 these difficulties had led Darwin to assert that, were he to start over, it would be with a different term. By 1866, however, he was reluctant to abandon "natural selection." He eventually agreed to compromise; although he would not give up his own phrase, he would work Spencer's into the *Variation of Plants and*

Animals under Domestication (1868) and the fifth edition of the *Origin* (1869). In his succeeding works, the terms are used interchangeably.

Darwin did not believe that he was taking a significant step in describing his theory as the "survival of the fittest." Fitness and its various cognates, however, had ordinary-language meanings that colored the ways in which it was read. In Victorian Britain and the post–Civil War United States, "survival of the fittest" came quickly to imply that the socially successful deserved the rewards of their position. To interfere with the status quo was to reverse the process of natural selection and thus ensure, in the words of the economist and sociologist William Graham Sumner, the "survival of the unfittest." And the unfit were already doing quite well; indeed, they appeared to be rapidly out-breeding their betters. "If the fittest do not serve as parents, the next generation will not inherit fitness" warned the biologist and eugenicist David Starr Jordan (1911, "Prefatory Note"). Thus it seemed necessary to adopt policies designed to increase the birth rate of the more prosperous groups and to limit that of the less prosperous. The "survival of the fittest" was invoked to support policies of economic laissez-faire and of reproductive intervention.

The development of population genetics in the 1920s and 1930s undermined the colloquial usage of fitness in evolutionary biology. In the work of J. B. S. Haldane, Sewall Wright, and R. A. Fisher, the gene was identified as the target of selection and selection itself was redefined as a change in gene frequencies. The measure of fitness became success in producing offspring, irrespective of the causes of that success. Moreover, what began as an indicator of fitness soon came to define its meaning. Haldane gave this new concept the (somewhat improbable) tag "Darwinian fitness" in his book *The Causes of Evolution* (1932).

During the 1930s and 1940s both old and new meanings of fitness coexisted. Thus Haldane himself could write that modern war destroys "the fittest members of both sides engaged in it," when it served his social purposes (1937b, p. 151). By the 1950s, however, biologists no longer employed the word in its vernacular sense, even in their popular writings. The growth of population genetics had played an important role in this development but so also had political events. With the rise of Nazism, evolutionary biologists outside of Germany struggled to disassociate their science from racial and class prejudice. Some tried to replace fitness with phrases whose past was less problematic, such as "adaptive value"; these efforts met with limited success. At the same time, politically liberal or leftist biologists such as Theodosius Dobzhansky, I. Michael Lerner, L. C. Dunn, and C. H. Waddington vigorously emphasized the concept's value-neutrality. A passage from the textbook *Principles of Genetics* by E. Sinnott, Dunn, and Dobzhansky illustrates the relationship between some biologists' social concerns and their insistence that fitness *is* reproductive success. After noting that nineteenth-century evolutionists had equated natural selection with a "struggle for existence" leading to the "survival of the fittest," the authors write: "These emotionally loaded phrases have been often misused for political propaganda purposes. A less spectacular but more accurate statement is that carriers of different genotypes transmit their genes to the succeeding generations at different rates . . . The 'fittest' is nothing more remarkable than the producer of the greatest number of children and grandchildren" (Sinnott et al., 1958, pp. 100–101).

In attempting to solve one problem, however, they had created another. If fitness is defined as success in surviving and reproducing, the statement that the fittest survive is apparently emptied of content. Thus was born the famous "tautology problem," which has bedeviled the field ever since.

Fitness: Theoretical Contexts

John Beatty

THE PRECISE meaning of "fitness" has yet to be settled, in spite of the fact—or perhaps because of the fact—that the term is so central to evolutionary thought. After all, evolutionary theory itself is still in flux.

An acceptable definition of the term "fitness" must be consistent with the role played by that term in evolutionary explanations. However fitness is defined, it must at the very least be positively correlated with evolutionary success, or in other words, with representation in future generations, for this is what higher fitness is invoked to explain.

One simple and apparently straightforward way of sustaining the connection between fitness and evolutionary success is by defining "fitness" as actual offspring contribution (see, e.g., Crow and Kimura, 1970, p. 5; Dobzhansky, 1975, pp. 101–104; see also Endler, 1986, pp. 27–51, for a discussion of a variety of different notions of fitness). But this construal has what are, to many, unappealingly counterintuitive consequences. Take the widely cited case of the identical twins, one of whom is, by chance, killed by lightning prior to its first reproductive encounter, the other of whom happens not to be in the path of the lightning and lives to contribute many offspring to the next generation (Scriven, 1959). Must we attribute different fitnesses to these otherwise indistinguishable twins?

One might suggest that the counterexample loses its force when the conception of fitness in question is suitably relativized. Fitness is offspring contribution *in a particular environment.* An organism that is highly fit in one type of environment—say, a strong and cunning lion in a savannah full of zebras—may be unfit in a different environment—say, a savannah full of lion hunters. Similarly, our imaginary identical twins have the same fitness only *in the same environment,* not when one is subjected to high voltage while the other is spared. The counterexample, it might be argued, does not apply to the twins' situation.

Yet in an important sense the twins do share the same environment, one in which the chance of an encounter with lightning is rare. The counter-example cannot be so easily circumvented.

One reason it seems so counterintuitive to suggest that the twins' different offspring contributions reflect differences in their fitness is that their different offspring contributions seem instead to be a matter of "chance." Indeed, evolutionary biologists distinguish those cases in which differential evolutionary success reflects differences in fitness and those cases in which it does not; the latter sort of case they refer to as evolution by "random drift."

The problem with the interpretation of fitness as actual offspring contribution is that it does not sustain this distinction. More specifically, the problem with the interpretation of fitness as actual offspring contribution is that it is not sufficiently probabilistic, because in order for the distinction in question to be sustained, there must be some chance that two organisms with the same fitness can leave different numbers of offspring. Interestingly, Darwin himself always wrote of those organisms that have the "best chance" of surviving and reproducing (see, e.g., Darwin, 1859b, pp. 61, 81).

Perhaps the simplest way of patching up the conception of fitness as actual offspring contribution is to construe fitness as the *average* number of offspring of a *type* of organism. Among organisms of a particular highly fit type, some will, by chance, do better, and some will, by chance, do worse, but the type as a whole will leave a relatively high number of offspring on average (this is a common conception; see, e.g., Emmel, 1973, p. 5).

Unfortunately, this more suitably probabilistic concept of fitness as average offspring contribution is still unsatisfactory. Our amended concept, like the original concept, is not explanatory. Offspring contribution and average offspring contribution are tallies, not causes, of the offspring contributions that lead to evolutionary success.

The offspring contribution of an organism of a particular type is much better explained in terms of particular traits of the organism in question: specific aspects of its physiology, anatomy, behavior, and so on, all of which contribute to its viability and fertility and ultimately to its overall *ability* to leave offspring in its particular environment. The most common construal of fitness in the philosophical literature today is one that identifies it with this ability and, ultimately, with the various properties that underlie this ability. This is the "propensity" interpretation of fitness (see, e.g., Brandon, 1978; Mills and Beatty, 1979; Burian, 1983; and Sober, 1984a).

The ability that is fitness bears a very special relationship to the abilities that underlie it. The manner in which any type of organism achieves high fitness is ultimately a matter of the physiological, anatomical, and behavioral traits that underlie its viability and fertility and in turn underlie its overall descendant contribution ability. And yet different types of organisms achieve high fitness (and low fitness) in very different physiological, anatomical, and behavioral ways. For instance, what one accomplishes in terms of increased viability, another may accomplish in terms of increased fertility. And what one accomplishes in terms of increased viability by way of possessing a particular camouflaging pigment, another may accomplish by way of possessing a particular form of enzyme that increases metabolic efficiency. Moreover, the particular physiological and anatomical traits that lead to evolutionary success when "placed" in combination with other specific traits and in particular environmental circumstances may, in combination with a different set of traits and in different environmental circumstances, lead to evolutionary failure. So while each and every manifestation of high fitness (or low fitness) can be explained in terms of underlying physiological, anatomical, and behavioral traits, there seems to be no particular function of traits in terms of which fitness can universally be expressed. Rosenberg speaks of the "supervenience" of fitness upon its underlying causal components in his discussions of this aspect of fitness (1985, pp. 112–117, 164–169).

It is important to note that the propensity interpretation also has a probabilistic component that is intended to accommodate the fact that there may be a *range* of offspring contributions (rather than just one single number) within the reproductive capabilities of an organism of a particular type in a particular environment. It may be that within a particular environment, an organism of a particular type will, because of the combination of traits it has, most likely leave many offspring. And yet, in that same environment, there may also be a small chance of being struck by lightning, and hence a small chance of leaving no offspring at all.

This raises the problem of how to compare fitnesses. Proponents of the propensity interpretation forsake identifying the fitness of an organism or a type with an entire probability distribution of offspring contributions, and identify it instead with one number, namely, the arithmetic mean, or "expected value" (e.g., Brandon, 1978; Mills and Beatty, 1979). This is simply the weighted sum of the values of the various possible reproductive outcomes, where the appropriate weights are the probabilities of the various outcomes.

One apparent difficulty facing our original definition of "fitness" that the propensity interpretation does *not* resolve has to do with the supposed problem of the circularity of the principle of natural selection. To be sure, the claim that "the fittest are most likely to leave the most offspring" is a tautology when "fittest" is defined in terms of actual offspring contribution. But the claim is no more empirical when "fittest" is defined as "best able to leave the most offspring."

The supposed tautology problem is really only a predicament, however, for those who believe that some such empirical principle is central to evolutionary theory. If some "principle of natural selection" were indeed the evolutionary equivalent of Newton's laws of motion, then one would expect to find it prominently placed in the standard textbooks; but it is not. Instead, one finds empirical formulae for predicting evolutionary outcomes based on assumptions about inheritance, with initial conditions covering fitness differences, population size, migration rate, mutation rate, and so on (basically, deductive consequences of the "Hardy-Weinberg law"). One formal axiomatization of evolutionary theory, by Mary Williams (1970), has a principle of natural selection as an axiom, though the principle is rendered empirical at the cost of treating "fitness" as an undefined primitive term.

There are a number of potentially more serious problems with the propensity interpretation of fitness. For instance, J. H. Gillespie (1972) has provided cogent reasons for suspecting that expected values of offspring contribution are not always appropriate for predicting and explaining evolutionary success. An analogy may help to make the point (Seger and Brockmann, 1987). A business may owe its long-term success to a costly insurance policy. The expense of the policy lowers the company's average yearly worth, but the policy itself considerably reduces the temporal variance in the company's worth and therefore helps prevent financial disaster. The business thus outlasts many competitors. In the case of evolution by natural selection, some types of organisms with relatively lower average numbers of offspring may prevail over types with relatively higher average numbers, as long as the former have smaller variances in offspring number and hence less chance of leaving no offspring at all—that is, less chance of evolutionary disaster. Gillespie argues that the evolutionary success of a type may represent a trade-off between the expected value and the variance of its probable offspring contributions (see also Gillespie, 1973, 1975, 1977). This means that the principle of natural selection, as stated above, with the expected-value version of the propensity interpretation of fitness inserted, is false, not circular. One way of construing the challenge now is to revise the definition of "fitness" along the line of Gillespie's suggestions, in order to maintain the circularity of the principle of natural selection!

Another problem with the propensity interpretation as usually elaborated is its emphasis on *offspring* contribution—its "shortsightedness," if you will. It is well known that, past a certain point, increased numbers of offspring can actually threaten the evolutionary success of a type, for example, by placing too great a demand on available resources or by minimizing the parental care that can be provided to each offspring. In these sorts of situations, more offspring may survive to leave more grand-offspring if there are fewer offspring in the first place. The literature on evolution of clutch size contains many such discussions (see, e.g., Lack, 1947b, 1954, 1966, 1968). To be sure, fitness as propensity to contribute offspring is positively correlated with very short-term evolutionary success—for example, with representation early in the life cycle of the next generation. But increased fitness, so construed, may be the very *cause* of decreased evolutionary success in the longer term.

The classic call for a long-term notion of fitness was by J. M. Thoday (1953). Some of his basic ideas have been recently recast in W. S. Cooper's (1984) interpretation of fitness as "expected time to extinction" (ETE). The ETE of a particular population or, more important, of a particular genotypic or phenotypic subpopulation, at a particular time and in a particularly specified environment, is just the probability-weighted sum of possible time intervals that might elapse before the (sub)population in question goes extinct. The most obvious problem with Thoday-Cooper notions of fitness is neatly recapped in the words of Keynes, who remarked that in the long run, "we are all dead." That is, what may concern us most is not *long*-term but, rather, more immediate evolutionary outcomes.

It might be possible to save the propensity interpretation by construing fitness as a "family" of propensities, including long-term and short-term descendant-contribution abilities, as summarized by a variety of different statistical parameters. It may be that different versions of the propensity interpretation will be appropriate under different circumstances. In this case, the concept will retain considerable intrinsic vagueness in spite of substantial technical elaboration.

FITNESS: REPRODUCTIVE AMBIGUITIES

Evelyn Fox Keller

A CHRONIC confusion persists in the literature of evolutionary biology between two definitions of individual fitness: one, the (average) net contribution of an individual of a particular genotype to the next generation, and the other, the geometric rate of increase of that particular genotype. The first refers to the contribution an individual makes to reproduction; the second refers to the rate of production of individuals. In other words, the first definition refers to the role of the individual as subject of reproduction, and the second to its role as object. The disparity between the two derives from the basic fact that, for sexually reproducing organisms, the rate at which individuals of a particular genotype are born is a fundamentally different quantity from the rate at which individuals of that genotype give birth—a distinction easily lost in a language that assigns the same term, "birth rate," to both processes.

Beginning in 1962, a number of authors have attempted to call attention to this confusion (Moran, 1962; Charlesworth, 1970; Pollak and Kempthorne, 1971; Denniston, 1978), agreeing that one definition—the contribution a particular genotype makes to the next generation's population—is both conventional and correct, and that the other (the rate at which individuals of a particular genotype are born) is not. Despite their efforts, however, the confusion persists (see Keller, 1987). In part this is because there remains a real question as to what "correct" means in this context or, more precisely, which definition is better suited to the needs the concept of fitness is intended to serve—in particular, the need to explain changes in the genotypic composition of populations. Given that need, we want to know not only which genotypes produce more but also the relative rate of increase of a particular genotype over the course of generations.

Perhaps not surprisingly, conflation of the two definitions of fitness is particularly likely to occur in attempts to establish a formal connection between the models of population genetics and those of mathematical ecology. But because the standard models for population growth all assume asexual reproduction, the two formalisms actually refer to two completely different kinds of populations: one of gametic pools and the other of asexually reproducing organisms. I suggest that in attempting to reconcile these two theories, such a conflation may in fact be required to finesse the logical gap between them. A more adequate reconciliation of the two formalisms would seem to require both the introduction of the dynamics of sexual reproduction into mathematical ecology and the introduction of a compatible representation of those dynamics into population genetics.

Perhaps counterintuitively, it is probably the second—the inclusion (in population genetics models) of fertility as a property of the mating type—that calls for the more substantive conceptual shifts. Over the last twenty years, we have witnessed the emergence of a considerable literature devoted to the analysis of fertility selection—leading at least some authors to the conclusion that the "classical concept of individual fitness is insufficient to account for the action of natural selection" (Christiansen, 1983, p. 75).

When fertility selection *is* included in natural selection, the fitness of a genotype, like the fitness of a gene (as argued by Sober and Lewontin, 1982), is generally seen to depend on the context in which it finds itself. Here, however, the context is one determined by the genotype of the mating partner rather than by the complementary allele. A casual reading of the literature on fertility selection might suggest that the mating pair would be a more appropriate unit of selection than the individual, but the fact is that mating pairs do not reproduce themselves any more than do individual genotypes. As E. Pollak has pointed out, "even if a superior mating produces offspring with a potential for entering a superior mating, the realization of this potential is dependent upon the structure of the population" (1978, p. 389). In other words, in computing the contribution of either a genotype or a mating pair to the next generation's population (of genotypes or mating pairs), it is necessary to take account of the contingency of mating in one's consideration. Such a factor, measuring the probability that any particular organism will actually mate, will incur a frequency dependence reflecting the dependence of mating on the genotypic composition and structure of the entire population. Given the theoretical connections between frequency dependence and higher-level selection processes (see, e.g., Uyenoyama and Feldman, 1980), the inclusion of a full account of reproduction in evolutionary theory may well necessitate the conclusion that natural selection operates simultaneously on many levels (gene, organism, mating pair, and group)—not just under special circumstances, as others have argued, but as a rule.

Gene: Historical Perspectives

Jane Maienschein

In his classic paper of 1917, "The Theory of the Gene," Thomas Hunt Morgan sought to explain what the "genetic factor" meant for biologists. He also intended to lay to rest the various objections that had been leveled against the gene theory. In his discussion, he revealed the profusion of different ways in which different people had used the concept of the gene, and the different aspects of gene theory that were regarded as important—or as objectionable (cf. Burian et al., 1988; Mayr, 1982a).

Some critics objected, Morgan explained, to the apparently static nature of the gene; truly scientific explanations must be more physiological or dynamic. Others pointed to the theoretical or symbolic nature of the gene, which as yet had no solid chemical nature to give it reality. And the gene theory seemed to some to involve merely juggling numbers and pretending to explain something about heredity, when in reality nothing of importance had been accomplished. Still others insisted that something fundamental about the organization of whole organisms was necessarily lost when researchers focused on parts, especially hidden genetic parts. Another line of criticism objected that the genetic parts appeared to be "fixed and stable in the same sense that atoms are stable," although in both cases, the real natural world showed a lack of such solidity and fixity (Morgan, 1917, p. 514).

Clearly the gene had come a long way in the few years since Wilhelm Johannsen had coined the term in 1909. At that time, Johannsen had proposed the word "gene" as a replacement for Darwin's "pangen" but intended that it serve the same purpose. It was a unit of heredity. But just what its nature was or how it functioned remained open to question. "The word 'gene' is completely free from any hypotheses;" Johannsen had insisted, "it expresses only the evident fact that, in any case, many characteristics of the organism are specified in the gametes by means of special conditions, foundations, and determiners which are present in unique, separate, and thereby independent ways—in short, precisely what we wish to call genes" (quoted in Carlson, 1966, pp. 20, 22). This lack of hypotheses about form or function obviously left the way open to confusion and competing alternative theories.

By 1917 Morgan and his group at Columbia University had begun to provide the missing hypotheses and had, in fact, developed a theory of the gene (Morgan et al., 1915; Morgan, 1917). In their view, the statistical data gathered from hybridization or breeding experiments provided virtually definitive evidence that some independent hereditary elements *must* exist in the germ plasm and must serve as the units of heredity. These factors, or genes, remain independent of genes responsible for other characters and they assort independently of each other, as Mendel had said. The only exception came when sets of genes were linked together, a fact that the group turned to its further advantage in support of the chromosomal basis for the gene theory more generally.

So genes must exist, but the breeding studies alone tell nothing about their nature or functioning. Such information comes from further studies, specifically on mutant strains in *Drosophila*. Mutation, it seemed, was a normal process. Evidence accumulated that mutations occur regularly, and not so rarely as to be useless. Indeed, normal mutations might occur sufficiently often to provide useful variations and thereby the necessary differences for selection to act on and to choose among. When Hermann J. Muller succeeded in 1927 in generating mutations in the lab using x-rays, mutation became an important research tool (Muller, 1922, 1927).

Mutant strains of *Drosophila* provided information about the effects of genes in this fly. In particular, mutants such as the white-eyed fly differ from the wild type in more than one respect. White-eyedness is associated with yellow body pigment, a lower productivity, and lower viability, for example. As a result, Morgan argued that it made sense to conclude that the factor in the germ plasm that produced the one character also produced the others. An individual gene could apparently have more than one effect. Other studies showed that more than one gene could contribute to each character. To complicate the picture further, different genes might even produce indistinguishable characters in some cases. All such conclusions pointed to the existence of something like genes as the units of heredity and to their complex role in effecting development. Separate pairs of genes, one from each parent, must exist in some form and remain independent in the germ plasm. With this sort of evidential support, Morgan's group felt, the gene theory had established itself as a leading scientific theory, despite the hypothetical nature of its central units and despite lack of evidence about their functioning.

Chromosome studies and linkage demonstrations lent further support and gave the gene theory a physical basis. William Bateson, Edmund B.

Wilson, Nettie Stevens, and others had begun to show the linkage of characters such as sex and eye color. Calvin Bridges (1916) had also demonstrated nondisjunction of the chromosomes, in which flies had one extra chromosome that presumably did not pair with any other in cell division; this also implied some sort of linkage of genes along the chromosome. Evidence of chromosomal crossing over and recombination during cell division added to this other work to support the idea of genes lined up like beads along a string. As Morgan put it, "While the linkage relations of genes do not *at present* have any immediate bearing on our conception of the nature of genes, they have a very important bearing on the problem of localization of genes in the germ plasm" (1917, p. 520). Because the Columbia group's gene theory encompassed more than the genes themselves, the questions of localization and function remained central. Though the distribution of genes during heredity remained a separate problem from the embryological questions about the genes' action during development, ultimately both were seen to form proper subjects for a full theory of the gene. Even in his role as geneticist, Morgan remained at least loyal in principle, if not in practice, to his embryological roots.

With the further elaboration of the theory by other members of the Columbia group, especially Muller, Morgan's theory of the gene achieved a quite obvious hegemony. Considerable resources, the Nobel Prize, admiring graduate students, and a host of other benefits accrued to the research program. And yet the theory underwent modification and refinement. The term "gene" had initially served for the (hypothetical) location in the germ plasm of the (hypothetical) hereditary unit. But it also referred to the specific occupant of the genetic locus. Not surprisingly, this caused confusion. Thus Morgan's group referred to the gene to indicate the relevant sort of hereditary unit. But each gene could have many alleles, or the specific genetic material that accounted for one version of a characteristic or another. There might, for example, be a gene for eye color with several alternative alleles—red, white, and whatever. Other modifications (e.g., crossing over) also called into question the static beads-on-a-string model of the genes-on-a-chromosome.

In addition, other views of "the" gene not only existed but represented reasonable lines of research responding to other basic commitments and other sets of concerns. In particular, William Castle's (1906, 1914) attempts to establish an alternative theory gained considerable notice. For Castle, unlike for the Morgan group, the gene was not inviolable and sacrosanct. Mendelian factors that determine characteristics could vary, Castle said. Furthermore, they were subject to the action of numerous "modifiers" as well. Castle rejected the position of the Morgan group, especially as put forth by Muller, that multiple genes may contribute to an individual character in the interest of simplicity. One variable gene made more sense to him than a host of cooperating genes. It provided more variation on which selection could act, for example. Citing the lack of any actual sighting of any gene, and to the very theoretical nature of such units, Castle pointed to the advantages of not rushing to judgment in favor of the Columbia group's theory. Persisting in his use of the term "unit character" as well as "gene" to refer to the hereditary unit, Castle showed his resistance to making the sort of genotype-phenotype distinction that the Morgan group felt was essential in order to make progress in understanding either heredity or development.

Castle felt that a return to Johannsen's more neutral view of the gene was more appropriate: he implied that the gene was a sort of black box or place holder, the "something" in the germ cell that gives rise to characteristics. To hypothesize that the gene must be stable and invariant, the assumption that formed the very basis of the Morgan group's theory of the gene, was to Castle unjustified speculation. Of course, Johannsen had not said anything to preclude further hypothesizing about the gene's nature and/or action, whenever relevant information came to light. And with time, evidence accumulated in favor of the Morgan group and against Castle.

Another critical attack came from the German biologist Richard Goldschmidt, who did not at all agree with the particular theory Morgan's group had put forth. In particular, he objected to the static nature of their gene. The organism as a whole is clearly dynamic and interactive, he insisted, and the attempt to explain its complexities in terms of stable hereditary units must fail. At first he suggested a model according to which the chromosomes are more or less hereditary place holders, with physiologically active genetic units moving in and out of the places along the chromosomes during cell division. Because he was primarily concerned with the way the genes act to effect proper development, he focused on the way that varying quantities of the gene elements act to produce characteristics.

Though sometimes modifying his views in light of new evidence, Goldschmidt (1928, 1938b) nonetheless consistently rejected the idea of the gene as a stable unit. The Morgan-Muller gene could not actually, physically exist, he felt, and it must remain a mere hypothetical construct with no reality and no function. Instead, the chromosome as a unit was what effects heredity and controls development for Goldschmidt. Rejected for his abstract "philosophical" position by the Morgan group, Goldschmidt attracted the attention of many others who developed their own rejections of the static gene and emphases on the functioning of genes within the whole, continually interactive, and dynamic organism.

Still another line of criticism came from those who advocated the importance of cytoplasmic inheritance. The Morgan group had stressed the chromosomal locations of genes, which therefore must reside in the nucleus. But as the historian Jan Sapp (1987) has explained, some researchers felt they had considerable evidence in favor of the role of cytoplasm in inheritance. Whether through cytoplasmic genes or through some other hereditary vehicle, these critics held that inheritance demanded more than nuclear lines of genes strung along a chromosomal string.

The 1940s brought modifications and additional views of the gene as geneticists moved beyond statistical breeding studies to embrace molecular biological and biochemical studies as well (Olby, 1974). Whereas initially the gene had remained a theoretical unit, with only indirect evidence even for its existence, by the 1940s a variety of studies had emerged to ground various more definite theories about the chemical nature of the gene. In particular, researchers had begun to make progress on understanding its physiological functioning. George Beadle, trained in the Morgan tradition by Morgan's student Alfred Henry Sturtevant, led one such effort. He had also worked with the French geneticist Boris Ephrussi, whose interest in showing how differentiation occurs in response to genetic action evidently influenced Beadle in his physiological emphasis. Beadle's evidence supported a theory that saw one gene correlated with one "primary" character and with one enzyme. Beadle and Edward Tatum (1941) developed such a theory, which they then continued to refine.

By that time, a host of researchers had entered the game of identifying the precise biochemical nature of the gene and its action. As a result, the 1940s and 1950s brought a host of alternative hypotheses. These culminated in the accepted structural model of the genetic material as a double helix of DNA, with various related functional theories.

While some gathered increasing support for the DNA nature and the double helical structure of the chromosomal material, others pursued work on the morphological nature of the gene. Seymour Benzer, for example, used new mapping techniques to suggest that research could soon come to recognize the units of recombination, of mutation, and of hereditary action. These might, in fact, not all be the same. The apparently simple gene might actually have several parts, or might operate in different ways for different purposes. Perhaps the gene locus actually contains more than one chemical part, each of which can nonetheless act separately for some purposes. In 1957, Benzer introduced the terms "recon," "muton," and "cistron" to correspond to his three different roles for the gene. Others have continued along such lines and have broken down gene action in other alternative ways, each of which has moved away from the simple concept of the gene as the straightforward hereditary unit. Different theories of the gene have emerged and have found support for one purpose or another.

The gene as a location along a chromosome, the gene as a particular type of biochemical material, the gene as a physiological unit directing development: are these all the same gene? At root, the dominant lines of research since the Morgan group's rise to power have assumed that there is one hereditary unit, the gene. That this unit has a location on the chromosome, has a particular biochemical nature, and acts in certain eventually specifiable ways has been the implicit assumption. Despite lack of direct evidence and despite Johannsen's emphasis on the lack of hypotheses concerning the nature and action of the gene, the dominant research groups have operated on the conviction that there is some underlying unit of heredity, recombination, mutation, and physiological action.

Others, for a wide range of epistemological and metaphysical reasons, have rejected that basic assumption. In the long run, however, no one of these alternatives has succeeded in establishing itself on equal ground with some version of gene theory. Yet at any given time, diverse alternatives have existed and have quite reasonably vied for attention. Some have called into question the continued value of using the term "gene" at all, because it has played so many different roles and has undergone such modification since its introduction. Others point out that, as with so many other basic concepts from "atom" to "species," the meaning may change but such change may form a continuous tradition. Through the changes the evolving concept plays an enduring and useful role, and the basic concept may remain useful and may provide a constant substratum to which to attach different interpretations. To date, biologists have found it useful to persist in the idea that there exists some underlying hereditary unit. Whatever its nature, structure, location, and action, they generally find it useful to persist in calling it the "gene" even while recognizing that the concept of the gene continues to undergo considerable revision and even fragmentation.

GENE: CURRENT USAGES

Philip Kitcher

WHAT'S IN A gene? Nucleic acid, usually DNA, but RNA in some prokaryotes. So it is easy to state a necessary condition for something to be a gene. Sufficiency is another matter, for we can segment the nucleic acid of the chromosomes (or the smaller chunks that inhabit the cytoplasm) according to many different principles of division. Which one is right?

At first it appears that the issue is relatively simple. Think of a gene as a functional unit, something that controls or affects the phenotype. Gregor Mendel, his rediscoverers, Thomas Hunt Morgan and his students, the early population geneticists: all picked out individual alleles by beginning with some aspect of an organism's phenotype. Take a trait—seed color, coat color, bristle number, what you will—and let the associated gene be that segment of nucleic acid that causes the trait to take the form it does. But what exactly does this mean?

The obvious suggestion is to look at differences. A segment of nucleic acid is causally relevant to a phenotypic trait just in case a modified form of the segment would yield a modified form of the trait. This preliminary analysis of the causal connection between individual genes and parts of the phenotype runs into immediate trouble when there are multiple genes associated with the same trait. (For further discussion of these problems, see Kitcher, 1982, and especially Rosenberg, 1985, chap. 4.) Suppose that the trait with which we begin is eye color in *Drosophila.* As aspiring geneticists quickly learn, there are numerous loci scattered over the *Drosophila* chromosomes at which mutations yield variants in eye color. So *the* gene for eye color in *Drosophila* is a disconnected assembly of chromosome segments? Better surely to suppose that this aggregate comprises *many* genes. Use the difference criterion to pick out the assemblage of genes relevant to a phenotypic trait and then divide that assemblage into maximally large connected segments, the individual genes that make up the assemblage.

This picture is still too simple. The form of a phenotypic trait may be affected by the nature of genes that act relatively early in epigenesis, genes that, intuitively, direct the formation of structures that make certain forms of phenotypic expression possible. The gene *Bicoid* controls the polarity of the *Drosophila* embryo. There are mutation sites within this gene, a scattered collection, not any connected subsegment, such that modifications at those sites produce nonfunctional gene products. Mutant *Bicoid* yields no polarity and death well before the organism develops eyes. So *Bicoid* includes (short) subsegments, differences within which change the form of the phenotypic trait of eye color—and, of course, many other phenotypic traits into the bargain. Is *Bicoid* a gene for just about everything?

Perhaps the difference approach can be revived by applying it in a more subtle way. After the advent of early molecular biology it became commonplace to distinguish units of function from units of recombination and mutation. Recombination can occur between any two nucleotides. Mutation can happen at any nucleotide through insertion, deletion, or substitution. But the units of function seem to be larger than one base segments of nucleic acid. Seymour Benzer (1957) characterized a *cistron* as a segment with the property that the phenotype of a double mutation within the segment differs according to whether the mutations occur in the *cis* arrangement or the *trans* arrangement.

Benzer's idea is ingenious, but it founders on the fact that many mutants are "leaky," allowing for partial, even full, enzymatic activity from the modified gene products and, in consequence, no differences between the phenotypes of the *cis* and *trans* arrangements. At least this difficulty will beset the proposal if we think of the organism's phenotype as comprising properties at some remove from the gene: traits such as bristle number or the ability to survive on some special medium. Another strategy for tackling the segmentation problem, proposed a decade before Benzer's refinement of the difference approach, is to push the phenotype inward. Why not focus on immediate gene action? Take a gene to be a maximal connected segment of nucleic acid that directs the formation of a polypeptide. (Regarding genes simply as segments that direct the formation of polypeptides allows smaller subsegments also to count as genes.)

This approach founders on the vicissitudes of gene regulation. The more we learn about transcription and translation the more tricky it is to elaborate it. For a segment of DNA to yield a polypeptide there has to be a binding site for an RNA polymerase, a molecule whose association with the DNA allows the nucleotide sequence to be read. The RNA that is transcribed is often post-transcriptionally modified, and, in any case, translation requires a complex ribosomal apparatus. How much of the nucleic acid should we reckon to the individual gene, and how much to the background facilitating conditions provided by the cell? Does the *lac* operon count as a separate gene or is it just a part of the *lac* gene? Should we extend the gene as far as the TATA box—a sequence of nucleotides that seems a regular precursor of the attachment sites of RNA polymerases? And what is to be said about nonfunctional genes, mutationally modified segments of nucleic acid, derived from segments that yield polypeptides but unable (for one reason or another) to generate polypeptides of their own?

Contemporary molecular biology does not worry much about these questions. Sequencing of individual genes goes on without too much fuss about boundaries. The successful sequencer offers the world a list of bases, a list that covers some region of interest. Theoretical discussions center on transcription of *DNA,* replication of *DNA,* repair of *DNA,* without worrying unduly about segmenting the nucleic acids into genes. Indeed, it is hard to see what would be lost by dropping talk of genes from molecular biology and simply discussing the properties of various interesting regions of nucleic acid.

But surely we need the notion of a gene for evolutionary studies? Why? Mathematical population genetics needs a criterion for segmenting nucleic acids no more than a principle for assorting organisms into populations. Just as we start with any of various idealized conceptions of what a population of organisms is, so too we can lay down idealizing conditions that genes are to meet. They are discrete units that are assorted into gametes according to specified rules, the rules of the Hardy-Weinberg-Fisher-Wright lotteries. To the extent that chromosomal segments, large or small, accord with these rules, the equations of mathematical population genetics will apply to them.

Evolutionary theorizing is full of references to genes, sometimes particular genes, sometimes genes in general. For the former, some one of the approaches that I have previously canvassed will suffice and residual ambiguities can be dealt with on an ad hoc basis. If you and I are discussing the effects of selection at a locus "for" some specified trait, we can agree on a genic region: some identified part of the assemblage of chromosomal segments, mutations within which make a difference to the form of the phenotype. Yet we may differ on whether some part of the regulatory apparatus in that region counts as part of the gene or as a separate gene. This is not likely to cause confusion for long, and, once it has become explicit, simple stipulations will enable us to communicate: if we continue to disagree we may present one another our sequences and assign labels to particular subregions.

The general discussion of genes is both more problematic and more interesting. Consider the thesis that the unit of selection is the gene. Doesn't this require some unambiguous general way of picking out genes?

Not really. As Richard Dawkins (1976, p. 35) quite insightfully notes, *The Selfish Gene* could have had the less euphonious but more accurate title *The slightly selfish big bit of chromosome and the even more selfish little bit of chromosome.*

Resolving the size of the gene is quite unnecessary for understanding genic selectionism.[1] One of Dawkins' best insights is pluralism about evolutionary modeling: for many evolutionary scenarios, there are alternative maximally adequate models, each of which takes selection to act at a distinct level; the genic viewpoint is privileged only because it is more widely available.[2] I want to go one step further. There are alternative versions of genic perspectives, versions that divide nucleic acids into shorter or longer segments. *Each* of the principles of segmentation that I have canvassed can be developed to yield an adequate genic perspective.

What's in a gene? Nucleic acid. How much? We don't need to say. There are many good ways to segment nucleic acid into genes—though not every way of segmenting nucleic acid is a useful way. Much of biology can be done without adopting any principle of segmentation at all. Where segmentation is needed, there are alternative principles of different utility in different situations. There is no need to seek the Holy Grail of the unique correct principle. It is enough to adopt one and to make one's choice clear.

A species, so the cynic says, is anything a competent taxonomist chooses to call a species. We can reach the same level of genuine insight and the same level of overstatement in the case at hand. A gene is anything a competent biologist chooses to call a gene.

1. Yet Dawkins sometimes seems to hanker after the idea that there is some privileged way to pick out genes, that there is an evolutionary unit, the "optimon," that strikes a compromise between power and fidelity. Longer bits of nucleic acid have a greater influence on the phenotype. Shorter ones are more likely to be preserved in the ravages of meiosis.

2. This position, defended in Dawkins (1982b, especially chaps. 1 and 14) is in fact highly controversial. Opposing views about the units of selection can be found in the writings of Richard Lewontin and Stephen Jay Gould. For recent discussions of debates about the issues involved here, see Sober (1984a), Lloyd (1988), Maynard Smith (1987a), Sterelny and Kitcher (1988), Sober (1990), and Waters (1991).

GENETIC LOAD

James F. Crow

THE TERM "LOAD" was introduced by Hermann J. Muller to dramatize the impact of recurrent mutation on human well-being. The word came into general usage following Muller's influential article "Our Load of Mutations," published in 1950. Although Muller's concern was for genes deleterious to human welfare, the term soon came to have a wider meaning. It was used to designate all fitness-reducing genes in any population, whether they are maintained by mutation or by other mechanisms. A load could thus be a consequence of normal genetic variability; the less favorable genes and genotypes constitute a load.

Much experimental emphasis has been on the *total* load, which includes both *hidden* and *expressed* loads (Morton, Crow, and Muller, 1956; Wallace, 1970). The hidden load is caused by those genes whose effect is not manifest—for example, recessives whose effect is concealed by heterozygosity with a favorable dominant—but which might be expressed at a later time or brought out by special techniques, such as inbreeding.

The deleterious genes that produce a genetic load may arise from mutation (mutation load), from genes that are favorable in heterozygotes but not in homozygotes (segregation load), from a mismatch between genes and environment (dysmetric load), or for other reasons. From the population standpoint, a load is not necessarily bad; it may provide the genetic variability needed to keep up with environmental changes and to allow for future evolution. Almost all new mutations are harmful, however, as are homozygotes for heterotic genes such as that causing sickle cell hemoglobin, so the load is deleterious to human welfare. Human geneticists tend to emphasize the latter aspect, and evolutionists the former. (For a general review of genetic load, see Wallace, 1970.)

Genetic load can be defined more precisely and treated quantitatively. The load, L, is defined as $L = (W_{max} - \overline{W})/W_{max}$, in which W_{max} is the fitness of the most fit genotype (perhaps hypothetical) and $\overline{W}$ is the mean fitness (Crow, 1958).

Consider first the *mutation load,* the decrease of fitness caused by recurrent mutation. The basic idea came from J. B. S. Haldane (1937a). He noted that at equilibrium the mean fitness of the population is reduced by a fraction equal to the total mutation rate per zygote (i.e., twice the rate per gamete), or half this value if the mutations are recessive. The Haldane principle is remarkable in showing that the impact of mutation on the population depends not on the severity of the individual mutations but on the mutation rate; if the impact is measured in terms of Darwinian fitness, the impact is equal to the mutation rate. This can be understood intuitively by noting that a mutation causing a lethal effect causes one death, which eliminates the mutant gene from the population. A mutation having one chance in 100 of causing death will persist in the population for an average of 100 generations before causing a death. With a system of mutation cost-accounting that equates 100 individuals exposed to a 1 percent risk of death to one individual exposed to a 100 percent risk of death, these are the same.

Haldane's principle depends on the assumption that mutations at different loci act independently. A formula permitting interactions was given by J. L. King (1967), who noted that the mutation load is approximately twice the mutation rate per gamete divided by the excess number of mutations in those individuals eliminated by natural selection. More exactly, the formula is $L = 2U/(\overline{z} - \overline{x} + 2U)$, in which U is the mean number of new mutations per gamete per generation, $\overline{x}$ is the mean number of mutations per individual before selection and $\overline{z}$ is the mean number in those individuals eliminated by selection (Kondrashov and Crow, 1988).

Haldane's interest was in the role of mutation in evolution; mutation was both a source of necessary new variability for future evolution and a source of reduced fitness currently. He regarded the mutation load as the price the species pays for the privilege of evolution; without mutation there would be no evolution.

The basic concept was discovered independently by Muller. Muller's interest was in the harmful effects of mutation and, therefore, of any agent increasing the human mutation rate. He used the mutation load as a way of measuring the impact of mutation on human welfare. This approach seems dubious to many, because it measures human welfare solely in terms of fitness (i.e., survival to adulthood and reproduction). Who would regard an early embryonic death as equivalent to the death of a young adult, or sterility as equivalent to death? Also, the principle does not take account of post-reproductive events or of suffering. Another difficulty is the assumption of equilibrium—clearly not true for the human species, for which the environmental conditions determining the equilibrium values are changing much faster than the rate of approach to a mutational equilibrium. No other measure, however, takes the whole mutational effect into account, and Muller regarded any partial estimate as ignoring the submerged part of an iceberg. A slightly different formulation of Muller's got around the equilibrium assumption, but at the price of not adequately considering future changes in population size. The issue became important in the mid-1950s because of the debate over nuclear testing (BEAR, 1956).

In view of these difficulties in its interpretation, the mutation load concept is no longer used as a way of assessing the impact of mutation or a change of mutation rate on the human population, although it still has a place in population genetics theory and in the study of natural populations. For example, the greater mutation load in an asexual than in a sexual population has been used as an argument for the ubiquity of sexual reproduction in nature (for a review, see Kondrashov, 1988).

Theodosius Dobzhansky (1955) was the first to extend Muller's use of the word "load" to include genetic impairment other than that caused by mutation. The mutation load is mainly the result of mutations that are deleterious when homozygous and deleterious or neutral in heterozygotes. In addition, there is a load caused by homozygotes at loci in which a heterozygote is more fit than any homozygote. Dobzhansky called this the *balanced load.* It is caused by segregation of less fit homozygotes from more fit heterozygotes. For this reason, and because the mutation load is also caused by a balance (between mutation and selection), *segregation load* is preferred by some. In analogy with Haldane's statement about the mutation load, the segregation load is the price the species pays for the privilege of Mendelian inheritance; in an asexual population there would be no segregation load. The segregation load per locus is ordinarily much larger than the mutation load (Crow, 1958; Kimura and Crow, 1964).

Any factor that leads to variability in fitness can create a genetic load. Others that have been discussed include *incompatibility load,* caused by maternal-fetal incompatibility, usually immunological; *recombination load,* caused by recombinational breakup of favorable linked epistatic gene combinations; *meiotic drive load,* caused by deleterious genes maintained by a meiotic or gametic advantage; *dysmetric load,* caused by less than optimum allocation of different genotypes to appropriate environments; *drift load,* caused by random gene frequency drift in small populations. (For a general discussion, see Crow and Kimura, 1970, pp. 297–312.)

The genetic load concept was extensively discussed during the 1950s and 1960s. The differing opinions were not about the definition but, rather, about the utility of a measure dependent on estimating fitnesses and fitness components, particularly those of a nonexistent optimum type. There were also great practical difficulties in measuring the relative values of the mutation and segregation loads. There were sharply divided opinions between those, especially Dobzhansky and Bruce Wallace, who thought that loci with superior heterozygotes were the rule, and others, especially Muller, who thought they were exceptional. If Dobzhansky and Wallace were correct, the typical individual would be heterozygous at the majority of its loci, and this would entail a large segregation load.

Realizing this, a number of geneticists suggested ways in which a large genetic load could be tolerated by a species with limited reproductive potential. It has long been known that the most efficient form of selection is truncation selection, by which the population is divided sharply into two groups, above and below a certain threshold, with those below the threshold removed by selection. In this way it is possible for several deleterious genes to be removed at once. Thus, to the extent that nature mimics truncation selection, the load is lessened, perhaps greatly.

This problem has not been resolved, but the application of molecular genetics in the 1960s has provided a direct answer concerning the amount of genetic variability in natural populations. The answer is complicated by the large amounts of highly variable DNA for which there is no known function, but if the question is confined to variability in protein gene products, the amount of heterozygosity in most diploid species is between 5 and 15 percent (Nevo et al., 1984). Nevertheless, this does not answer the question of how large a load is produced by this amount of heterozygosity, because the fitness effects of these molecularly detected alleles are largely unknown.

The relative fitness-reducing effects of the various loads are still largely unknown, although accumulating evidence from several sources has rendered generalized heterozygote superiority very unlikely. Questions concerning load are not as actively pursued as they were in the 1960s, not because they have been answered, but because other questions are more accessible to the methods of molecular biology. For example, is molecular evolution mainly driven by mutation and random drift rather than by selection, as argued by Kimura (1983)?

Finally it needs to be pointed out that the various genetic loads discussed above are static; that is, they refer to a population at equilibrium for the forces determining the load. Haldane (1957) introduced a similar idea related to population dynamics. He called it the *cost of natural selection*. This has sometimes been called the *substitution load*, but there is merit in following Haldane's precedent in using the word "load" for static measures and "cost" for dynamic. Haldane was attempting to quantify a problem, of which animal breeders are distressfully aware, that there is a limit to the number of traits that can be selected simultaneously.

Haldane's cost principle has the same elegance as his mutation load principle. He showed that the accumulated genetic load during the time a new favorable mutation is incorporated is a function of its initial frequency only (with minor modifications introduced by dominance and epistasis). Again the magnitude of effect of the mutation is irrelevant; a mildly beneficial mutation stretches weak selection over a long time, a strongly beneficial one has strong selection for a shorter time—the total amount of selection is the same.

The essence of the principle can be seen in a simple haploid model. Suppose that the new mutation increases the fitness of its bearer by a fraction s. Letting W be the fitness of the existing type and p be the proportion of the new mutant, the fitness of the mutant type is $W_{max} = W(1 + s)$, and the mean population fitness, $\overline{W}$, is $W(1 + ps)$. The load in any one generation then is $(W_{max} - \overline{W})/W_{max} = (1 - p)s/(1 + s)$. Assuming $s << 1$ and integrating this expression over the time required to carry out the gene replacement leads to approximately $-\log_e p_0$, where p_0 is the initial frequency of the mutant. For a diploid locus with semi-dominance (i.e., heterozygote halfway between the two corresponding homozygotes), the value is $-2 \log_e p_0$.

The cost of evolution has been much more problematical than genetic loads. How can there be a cost of substituting a beneficial new allele, when the population is steadily improved thereby? One might better ask what is the cost of not evolving! It has been difficult to use this principle in the actual study of evolution.

Yet a slight modification of Haldane's statement gives it a more transparent interpretation. If cost is measured each generation by $(W_{max} - \overline{W})/\overline{W}$, rather than by $(W_{max} - \overline{W})/W_{max}$, this is a measure of the reproductive excess required to keep the population from decreasing in size. Furthermore this interpretation makes Haldane's formula more nearly exact than his original interpretation (for details, see Crow and Kimura, 1970, pp. 244–251). This formulation helps place a limit on how rapidly gene substitutions can occur in a population with a limited reproductive excess.

As an example, using the Haldane formula, the amount of excess reproduction required to change a favorable semi-dominant mutation from a frequency of 0.0001 to 1 is $-2\log_e(0.0001)$, or about 18. If there were 10^5 genes evolving at a rate 10^{-6} per generation (roughly the observed rate for many proteins), the average cost per generation would be $18 \times 10^5 \times 10^{-6} = 1.8$. This means that the species would have to produce 2.8 times as many surviving offspring as the replacement number (i.e., 5.6 per pair of parents), to say nothing of offspring that die for reasons other than gene frequency change, in order to maintain the population size. This seems excessive for many species, especially large mammals. It was an argument of this kind that led Kimura (1968) to propose his "neutral theory" that molecular evolution is driven by mutation and random drift rather than by selection.

GENOTYPE AND PHENOTYPE

Richard C. Lewontin

THE DISTINCTION between the phenotype and the genotype of an organism was introduced into biology by Wilhelm Johannsen in 1911, to cope with a fundamental dichotomy in the Mendelian explanation of heredity. Gregor Mendel's scheme required the existence within organisms and their gametes of unobserved entities, hereditary factors (now called "genes"), whose state was not in a one-to-one correspondence with the appearance of the organisms themselves. So, in his experiments with tall and short pea plants, Mendel supposed two sorts of tall plants, one carrying two copies of the "tall factor" and one carrying a single copy of the "tall factor" and a single copy of the "short factor." Because of dominance of tall factors over short factors in plant development, these two sorts of tall plants were indistinguishable in appearance, but could be separated on the basis of their progeny in appropriate breeding experiments. In this explanatory scheme, it was the phenomenon of dominance that resulted in a many-one correspondence between the internal state of the organism and its outward appearance and demanded a dual system of classification of organisms. The "phenotype" of an organism is the class of which it is a member based upon the observable physical qualities of the organism, including its morphology, physiology, and behavior at all levels of description. The "genotype" of an organism is the class of which it is member based upon the postulated state of its internal hereditary factors, the genes.

Although the genotype-phenotype distinction was apparently introduced to cope with the many-one correspondence of postulated factors to actual appearance, the real source of the distinction is at a much more fundamental level. Even if heterozygotes carrying one copy of the "tall factor" and one copy of the "short factor" had been intermediate in height in Mendel's experiments, so that internal and external states were in a one-one correspondence, Mendel's system required a distinction between genotype and phenotype. The essential feature of Mendelism is a causal rupture between the processes of inheritance and the processes of development. What is inherited, according to Mendelism, is the set of internal factors, the genes, and the internal genetic state of any organism is a consequence of the dynamical laws of those entities as they pass from parent to offspring. Those laws of hereditary passage contain no reference to the appearance of the organism and do not depend upon it in any way. Mendel's two "laws," the law of segregation and the law of independent assortment (more honored in the breach than in the observance), are statements about the hereditary comportment of the internal factors, the genes. With respect to the causal processes specific to inheritance, the outward appearance of organisms is purely epiphenomenal. The latter is a consequence of different causal processes, so-called epigenetic processes of ontogeny that are dependent upon the state of the genes but not on the laws of their inheritance. Thus what is sometimes referred to as Mendel's third law, the law of dominance, is not a genetic law at all but an epigenetic one. Although it is August Weissman who is usually said to have introduced the distinction between the changeable soma and the underlying constant germ plasm—between the developmentally contingent phenotype and the hereditary genotype—it was in fact Mendel, twenty years before Weissman, who made this distinction fundamental to his theory. It is the existence of two separate causal pathways, one for the passage of the internal factors and one for the influence of those factors on the external appearance of organisms that requires two separate spaces of description, the genotypic space for the state of the internal factors and the phenotypic space for the manifest state of the organism.

In the Mendelian system of causal analysis genotype and phenotype appear asymmetrically. The genotypic status of the organism is causally prior to both its phenotypic state and the genotypic state of its offspring; the phenotype of the organism has no causal consequences. This asymmetry disappears in epistemological questions. The basic problematic of genetics is to explain the similarity and differences of phenotype between parents and offspring, or, more generally, between relatives of varying degree. Because of the Mendelian rupture between the processes of heredity and the processes of development, the explanation or prediction of the hereditary passage of *phenotype* requires a three-stage inferential and deductive process involving both phenotypic and genotypic spaces of description. Beginning with the phenotypic description of the parents, the first step is to infer their correct genotypic description. The second step is the deduction of the genotypes of the offspring, given the genotypes of the parents and the Mendelian laws of genetic inheritance. Finally, the genotypic description of the offspring must be mapped back into the space of phenotypic description by the forward application of the epigenetic rules.

The only unproblematic part of this inferential process is the second step of carrying parental genotypes into offspring genotypes. The development of genetics and cytogenetics since 1900 has provided a substantially complete description of this process, although surprises are always possible. There is, however, only rudimentary knowledge of the causal pathways of development, so the forward mapping of genotypic description into phenotypic description is not possible except in special cases. It is clear that the mapping is not one–one in general, and to the extent that a single genotypic class may correspond to multiple phenotypic classes, a determinate step from offspring genotypes to offspring phenotypes is not possible.

There is, however, considerable phenomenological evidence about the mapping of genotypic space into phenotypic spaces. Several distinctions are critical to an understanding of this evidence. First, we must recognize the type-token distinction. By "genotype" and "phenotype" we mean classes of organisms satisfying some genetic or phenetic criteria. Individuals belonging to those classes are tokens of those types. The actual physical set of inherited genes, both in the nucleus and in various cytoplasmic particles such as mitochondria and chloroplasts, make up the *genome* of an individual, and it is the description of this genome that determines the genotype of which the individual is a token. In like manner there is a physical *phenome,* the actual physical manifestation of the organism, including its morphology, physiology, and behavior. But it is immediately apparent that the phenome of an individual is in a constant state of flux so that the total phenotypic description of an individual has a temporal dimension. Not only do individuals change during development and aging, but physiological processes and behavior are, by their nature, temporal fluxes. This temporal dimension of the phenome is a major source of the difficulty in mapping genotype into phenotype.

Finally, it is necessary to distinguish between partial and total phenotypes and genotypes. If, by genotype or phenotype, we mean classes defined by total description of organisms, then all genotypes and phenotypes have at most one member. Given the known rates of mutation, the likelihood that two actually existing genomes are identical over their entireties is extremely low, even for those of identical twins or other clonally reproduced organisms. In practice genotypic and phenotypic descriptions refer to some part of the genome or phenome, the rest being regarded as irrelevant. Two individual white-eyed *Drosophila,* both of which are homozygous for the mutation *w* at the *white* locus on the X chromosome, are said to be genotypically and phenotypically identical. But of course they are not, and differences between them for other genetic and phenetic properties may be critical to their biology. Sometimes biologists act as if the rest of the genome were indeed constant, at other times they recognize the background heterogeneity but regard it as causally irrelevant, and at yet other times they explicitly "average over" the background, recognizing its causal efficacy, but treating it as random noise of no average effect. Underlying all these attitudes is a belief in the essential causal independence of different developmental processes so that the organism can be atomized into partial phenotypes and partial genotypes without any loss of essential information. It is seldom realized that the boundaries of partial genotype and phenotype are set by conceptual considerations that may have no relation to actual causal pathways. A study of, say, the evolution of the hand assumes that the hand is somehow a natural entity with genes "for" the hand acting on the hand independent of any other relevant aspect of the phenome.

The reverse mapping of phenotype space into genotype space requires not only that we know the epigenetic laws that convert genotype into phenotype, but that we be able to invert those laws to go from phenotype to genotype. Even if the forward laws of epigenesis are perfectly known, it does not follow that they can be uniquely inverted to provide unambiguous mapping of phenotypes into genotypes. Indeed, the simple phenomenon of dominance introduces an ambiguity in inferring genotypes from phenotypes, and although it is possible to resolve that ambiguity in very simple cases, as Mendel did, by further breeding tests, the complexity of developmental interactions between genes makes its resolution impossible in most cases in the absence of complete developmental information. Because inferences about genotypes depend entirely on observations of phenotypes, and because deductions of phenotypes depend upon the nature of the underlying genotypes, the entire structure of genetic prediction and explanation can become hopelessly circular. There are two routes open to break this circularity. One is to acquire detailed information on complex ontogenetic pathways, including the interactions of genes with each other and with the developmental environment. For the moment this seems utopian. The alternative is to solve part of the problem by inferring genotypes from phenotypic manifestations that do not have intervening developmental processes. That is the approach of molecular genetics, which reads the state of the genes from their molecular configuration.

The phenotype of an organism at any moment in its history is a consequence of the development of that individual from a zygote through a historical sequence. That sequence is a result at each moment of the previous state of the organism, of its genotype, and of the environment in which it is developing at that moment. Thus the law of transformation of genotype into phenotype is a law that must contain information not only about the genotype but also about the historical sequence of environments. For any given genotype there will be a developmental outcome specified by the environmental sequence. The function that maps the space of environmental sequences into the space of phenotypic outcomes for a given genotype was called by I. I. Schmalhausen (1949) the *norm of reaction* of the genotype. Of course, in practice, these are specified as the mapping of partial environment (e.g., temperature) into partial phenotype (e.g., body weight) for a partial genotype. Two genotypes are then recognized as different only if their norms of reaction differ for some aspect of the environment and some aspect of the phenotype.

In some cases the difference in norm of reaction between two genotypes is similar in all environments, either because the phenotypic manifestation has only a very weak contingency on the environment or because both genotypes map in a parallel fashion onto the phenotypic space. So on the one hand, for example, all *Drosophila* that are homozygous for the mutant gene *white* have white eyes, and all flies with the nonmutant form of the gene have red eyes, irrespective of the developmental temperature. On the other hand, as shown in the figure, eye size is reduced with increasing temperature, but flies carrying the mutation *Ultrabar* have smaller eyes than normal flies at all temperatures. Such norms of reaction are the exception, however, and in most cases the mapping of environment onto phenotype does not allow an unambiguous classification of genotypes in all environments. There is another mutant of *Drosophila, Infrabar* (see figure), whose eye size increases with temperature, so that at low temperatures it has smaller eyes than *Ultrabar,* but at high temperatures larger eyes.

The phenotype corresponding to a genotype is not completely specified even when the environmental history is given, because random developmental events at the level of cell division and migration cause differences in phenotype. There is random variation in bristle number between the left and right sides of a *Drosophila,* for example, even though both sides are genotypically identical and both have had the same environmental history. Moreover, because only partial genotypes are specified, the remainder of the genome, which will vary from individual to individual in the same partial genotypic class, will have effects that appear as random variation among individuals. Because of the contingency of development both on environmental sequence and on random developmental accidents, the mapping of genotypic space into phenotypic space is a many-many mapping. To each genotype there corresponds a characteristic distribution of phenotypes, and for each phenotype there is more than one corresponding genotype, even in a given environment.

The environmental contingency of the mapping of genotype onto phenotype depends on the developmental complexity that intervenes between genome and phenome, which is, in part, a function of the molecular details of the control for turning on and off the products of primary protein syn-

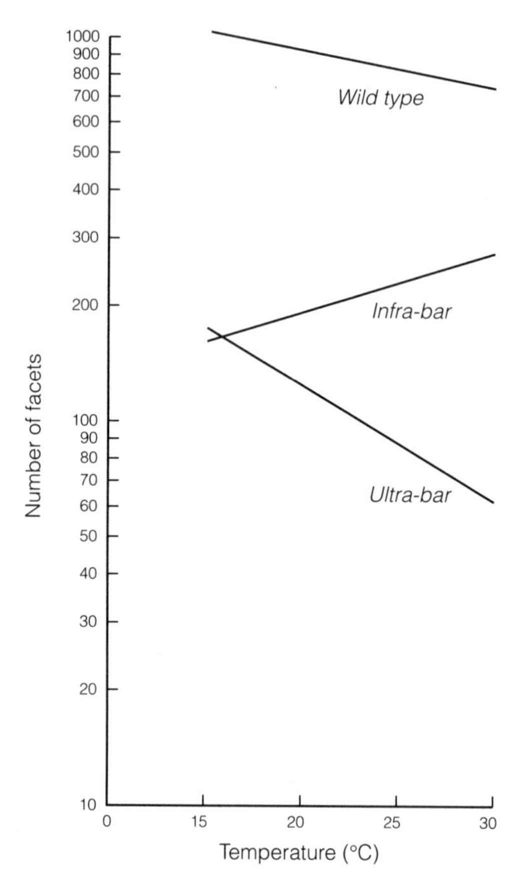

Norms of reaction to temperature for three different eye-size genotypes of *Drosophila melanogaster: wild type, infra-bar,* and *ultra-bar*. Eye size is measured by the number of facets in the eye.

thesis. The protein products specified by some genes are constantly being produced at levels that are only weakly influenced by the actual metabolic processes in which those proteins participate. Our fingernail protein is laid down at a fairly constant rate through most of our lives. Most of the enzymes specified by genes, however, are produced at rates that have a feedback control from metabolism itself. For example, the *lac* gene that specifies the structure of the enzyme beta-galactosidase in bacteria is transcribed only when an excess of the enzyme's substrate, glucose, is present.

In part, the degree of environmental contingency depends upon how many biosynthetic or developmental steps are involved in the final realization of the phenotype. There is a rough scale of environmental contingency as more and more development intervenes between gene and phenome. At the lowest level the DNA sequence of the genes itself is a phenotype, and a complete description of the DNA sequence is identical with a complete specification of the genotype. The discovery of this phenotypic level and the development of techniques for observing the phenotype have been revolutionary for genetic analysis, precisely because they solve the problem of inferring genotype from phenotype by eliminating development. All genotypes, irrespective of their influence on development, can be unambiguously discriminated at the molecular level of phenotype. The problem, of course, is to identify those genes that are related to higher developmental events.

The next most distal phenotypic level is the structure of the proteins specified by genes. There is a many-one relation even at this lowest biosynthetic level because of the degeneracy of the DNA code, so that many different DNA sequences are translated as the same protein. For DNA that is not itself coding sequence, but is involved in the rates of protein production, there is also a looseness of relation between gene sequence and function, but the redundancies are as yet not well understood. Yet other fractions of DNA such as introns and intervening sequences have yet looser and more poorly understood relations to the biosynthesis of proteins. There are, moreover, secondary modifications of proteins, both of their primary and tertiary structures, not under the influence of the genes that specified their original structure. Nevertheless, the qualitative structure of proteins is essentially independent of environmental contingency, although the amount of their production may be very strongly influenced by environment through feedback from substrate concentrations or directly by temperature effects. At the next level, that of cell differentiation, cell division, cell migration, and tissue differentiation there is yet greater contingency on environment. The relative growth rates of different parts of an organism are affected differentially by external environment or by the growth of yet other parts. Although no environment will map the genotypes of lions into the phenotypes of lambs, the environmental variation among organisms in morphology for a given genotype is of the same magnitude as the average variation between genotypes. Moreover, the norms of reaction, where they have been characterized, cross each other in an unsystematic way so that the phenotypes associated with specific genotypes cannot be easily characterized in general terms. For example, in the classical work of Clausen, Keck, and Heisey (1948) on growth of different genotypes of the plant *Achillea* in different environments, there was no correlation between the growth rates in the different environments, so that the tallest genotype in one environment might be the shortest in another and intermediate in yet another.

Finally we may expect that behavioral phenotypes will show an even greater environmental contingency, although very little evidence exists on the question. Species comparisons show a wide variation in the degree of genetic influence of behavior, even for what appears to be the same phenotypic trait. Among passerine birds some species, such as the Brown-Headed Thrasher, sing a species-specific song whose details are influenced only in minor details by experiencing the song before fledging. Other species, such as the Song Sparrow, can only sing a rudimentary form of their species-specific song unless they are exposed to it early. Yet others, such as the Mocking Bird, have only very general temporal outlines of a species-specific pattern and produce an essentially infinite variety of songs, both heard and invented. No generalization is possible, except to say that genotype-phenotype mappings of behavior, especially in organisms with very large and complex nervous systems, are likely to be such that the phenotypic variation within genotypes will be much greater than the average differences between genotypes.

GROUP SELECTION

David Sloan Wilson

NATURAL SELECTION produces adaptations—properties that cause organisms to survive and reproduce in their environments. Because adaptations can occur at a variety of levels, however, what is adaptive at one level may be maladaptive at another. It is adaptive for predators to capture prey, but if they are too successful the predator population can overexploit its prey and itself go extinct. It is adaptive for a group of monkeys to post sentries to watch for predators, but maladaptive for the individual sentry who must watch while its fellows eat.

These two examples illustrate that natural selection is a hierarchical process. If we focus on evolution within a single group, we find that it maximizes the relative fitness of genotypes within the group, with consequences that can be either positive or negative for the fitness of the group as a unit. To explain group-level adaptations per se, we must invoke a process of natural selection at the appropriate level—namely, that of many groups that vary in their genetic composition, some of which are more productive than others. This process is known as group selection.

The study of group selection has a turbulent history that has not yet subsided. Charles Darwin did not often address the issue, but he was remarkably perceptive when he did. He realized that self-sacrificial behavior in social insects (1859a, p. 258) and humans (1871, p. 113) must evolve by between-group selection. Other evolutionists, however, were not so perceptive and talked loosely of adaptations as "for the benefit of the species" without distinguishing between levels of selection. A few recognized the hierarchical nature of natural selection but assumed that group selection was so strong that group-level adaptations could be found everywhere (e.g., Emerson, 1960). This trend culminated in V. C. Wynne-Edwards' (1962) ambitious attempt, Darwinian in scope, to explain virtually all social systems as adaptations that prevent populations from overexploiting their resources. To Wynne-Edwards, group selection was "for everything concerning population dynamics, much more important than selection at the individual level" (pp. 19–20).

The reaction to Wynne-Edwards' book, especially from evolutionists well grounded in population genetics theory, was swift and devastating. Aside from a brief sketch by Sewall Wright in a book review (Wright, 1945), the process of group selection had never actually been modeled. Subsequent attempts uniformly concluded that group selection was a weak force relative to the selection of individuals within groups (see Wade, 1978, and D. S. Wilson, 1983a, for a review). Supported by such a flimsy theoretical foundation, the great mass of Wynne-Edwards' empirical observations collapsed (see Pollock, 1988, for a more positive assessment).

Although criticisms came from many sources, the mantle of the new individual selection movement fell on G. C. Williams, whose book *Adaptation and Natural Selection* (1966) is regarded as a modern classic. Williams adopted the extreme position that group selection, though possible in principle, could be ignored in practice and should never be invoked to explain an adaptation if a reasonable alternative based on individual selection could be found:

> It is universally conceded by those who have seriously concerned themselves with this problem that such group-related adaptations must be attributed to the natural selection of alternative *groups* of individuals and that the natural selection of alternative alleles within populations will be opposed to this development. I am in entire agreement with the reasoning behind this conclusion. Only by a theory of between-group selection could we achieve a scientific explanation of group-related adaptations. However, I would question one of the premises on which the reasoning is based. Chapters 5 to 8 will be primarily a defense of the thesis that group-related adaptations do not, in fact, exist. (p. 92)

The next decade was a dark age for the study of group selection. In an essay entitled "The Uses of Heresy," Stephen Jay Gould (1982b, p. xv) recalls with repugnance the "hooting dismissal of Wynne-Edwards and group selection in any form during the late 1960's and most of the 1970's." Nevertheless, during the 1970s a revised form of group selection arose that is accepted by many evolutionary biologists today.

Informally, most evolutionists would agree with Williams' definition that individual selection is evolution within single populations and that group selection is an analogous process at the group level. At the technical level, however, several definitions of group exist that in part account for the persistent nature of the controversy. Models that were built in response to Wynne-Edwards' book tended to define groups as discrete populations, isolated from each other except for a trickle of dispersers—populations that persist indefinitely unless driven extinct by the selfish genotype. (We might call such groups multi-generational demes.) These models tend to work poorly, in part because the "generation time" of extinction and colonization of groups is long relative to the generation time of births and deaths of individuals within groups (see Gilpin, 1975, Wade, 1978, and Boyd and Richerson, 1990, for a positive assessment of interdemic models).

A more recent class of models defines groups in reference to the traits that are being selected (reviewed by D. S. Wilson, 1983a; Wilson and Sober, 1989). For example, parasites within a single host interact with each other and are isolated from the parasites in different hosts. The population within one host can be defined as a group, and gene frequency change can be monitored within and between groups, even if only a fraction of the life cycle is spent within that host. Notice that, according to this definition of group (*trait group),* the generation time of a group need not be much longer than individual generation times and can even be shorter. For this and other reasons, between-group selection is an important component of models that define groups as sets of interacting individuals. This conception of groups also has the advantage of being comparable to the concept of an individual as a set of interacting genes and thus leads to a truly hierarchical theory of natural selection.

The importance of group selection depends largely on the way that groups are defined. The problem is aggravated by the fact that many models that were developed as *alternatives* to group selection according to the first definition of group (multi-generational deme) actually include a *component* of group selection according to the second definition (trait group). For example, Dawkins (1980, p. 360) states, "There is a common misconception that cooperation within a group at a given level of organization must come about through selection between groups . . . ESS [evolutionary stable strategy] theory provides a more parsimonious alternative." ESS theory is an application of game theory in which a large population of individuals is assumed to form randomly into groups of size N (usually $N = 2$). Fitness is determined during the grouped stage, after which the groups dissociate back into individuals and the process is reiterated. According to the first definition of groups, the entire population is a single group and ESS theory serves as an alternative to group selection. According to the second definition, a group is the set of N individuals that actually interact with each other, and the population consists of many such groups. In fact, the process of group formation, selection, and dissociation in ESS theory is mathematically identical to a standard population genetics model in which the "individual" is a gamete and the "group" is a diploid individual (Maynard Smith, 1987a). Rather than serving as a robust alternative to group selection, ESS theory now contains group selection as a vital component that, contrary to Dawkins' statement, is responsible for the evolution of cooperation.

At an even more technical level, proponents of the "new" group selection are not uniform in their definitions. Price (1970, 1972) advanced a method of partitioning gene frequency change into within- and between-group components that is often equated with individual and group selection (e.g., Hamilton, 1975; Wade, 1980). Several authors, however, have pointed out that individual selection appears in both the within-group and the between-group components of Price's method. See Heisler and Damuth (1987) for an alternative approach that avoids this problem.

Given the polemic nature of the controversy over group selection and the consequences of switching from one definition to the other, evolutionists can be expected to remain polarized for the foreseeable future.

HERITABILITY: HISTORICAL PERSPECTIVES

Michael J. Wade

THE DEGREE OF resemblance between offspring and parents is the most basic notion of heredity. Heredity is necessary for evolution by natural selection because, unless offspring resemble their parents to some degree, the effects of selection in one generation (the parents) will not be transmitted to the subsequent generation (the offspring). Heritability, the measure of this *degree* of resemblance, is a concept developed in biometrical or quantitative genetics from consideration of the biological facts that (a) genes are transmitted from parents to offspring in reproduction and (b) the variation and inheritance of continuous characters can best be understood as owing to many independent Mendelian genes.

R. A. Fisher (1918) and Sewall Wright (1921) investigated the theoretically expected degree of resemblance between parents and offspring using formal genetic models. They expressed heritability as a fraction of the total phenotypic variance in order to provide a means of estimating the relative importance of different causal agents to observed variations in the phenotype. In this way, the effects of heredity could be distinguished from, say, the effects of environment. Not all of the genetic differences in phenotype between individuals, however, are transmissible. This happens because parents in most diploid organisms transmit genes, rather than genotypes, to their offspring. Differently put, only a fraction of the genotypic differences among individuals are transmitted to the offspring or contribute to the resemblance between parent and offspring. Thus the transmissible fraction of the genetic variation among individuals is not equivalent in general to the genetic fraction of variation among individuals. It is the transmissible fraction also called the variance in breeding values that determines the response to selection.

In order to distinguish the transmissible fraction from the total genetic fraction, Wright identified three kinds of genetic variation. The additive genetic variance is the fraction of the total variance in phenotype among individuals owing to differences between these individuals in breeding value. Theoretical and empirical considerations indicate that this component of the total genetic variation is a useful measure of the degree of resemblance between parents and offspring. It permits prediction of the response to natural or artificial selection when the response is quantified as the change in the mean value of the phenotypic distribution. Because it describes the extent to which phenotypic change is possible by selection, it is the most frequently used definition of heritability.

The two other kinds of genetic variation, dominance (interaction between genes at the same locus) and epistasis (interaction between genes at different loci), can also affect the degree of genetic resemblance between parents and offspring. For this reason, heritability defined as the fraction of the total phenotypic variance owing only to additive genetic effects is often referred to as narrow-sense heritability (Lewontin, 1974a; Falconer, 1981) to distinguish it from the totality of genetic effects, called broad-sense heritability. (See also EPISTASIS and HERITABILITY by Feldman.)

The breeding value of each of a group of individuals can be estimated using controlled breeding experiments. The variance among individuals in breeding value, the transmissible variation, estimates the narrow-sense heritability. It is the mapping of a multilocus Mendelian genetic model onto an analysis of variance of the phenotypic values of individuals obtained from controlled breeding experiments that provides empirical estimates of the components of genetic variation. It is this mapping of the genetic model onto the statistical model that results in the frequent misconception that the existence of additive genetic variance indicates additive gene action. It is true that, if genes act additively in their effects on the phenotype (no dominance and no epistasis), then the existence of additive genetic variance is indicative of additive gene action. However, genes interact within and between loci in the development of most phenotypes, and the existence of these interactions prohibits a simple interpretation of the observed additive genetic variance in terms of additive gene action.

HERITABILITY: SOME THEORETICAL AMBIGUITIES

Marcus W. Feldman

IN COMMON parlance the word "heritable" describes a trait that is recognizably similar in parents and their offspring. In such a description, traits such as religious denomination, amount of real estate owned, language spoken, and hair color, all of which aggregate in families, would be regarded as heritable. Of these traits, only hair color is transmitted via the genes. Even hair color may exhibit complications in its mode of transmission, as two brown-haired parents may produce a red-haired offspring. The term "heritability" was introduced in order to quantify the level of predictability of passage of a biologically interesting phenotype from parent to offspring. It was in animal and plant breeding for agricultural purposes that accuracy of such prediction assumed economic importance, and it was for agricultural purposes that the mathematical expression of heritability was developed. (See also GENOTYPE AND PHENOTYPE.)

Consider a population in which there is variation among individuals in the numerical values of a trait that will be called the phenotype, P. This variation is measured by the variance, V_P. Suppose that different genotypic classes in the population have different mean phenotypes. Suppose further that the different environments, familial and extrafamilial, in which the individuals develop also contribute to the phenotypic variability. Then *heritability* is a number between zero and one which is intended to indicate that fraction of V_P which is the variance among genotypic means in the population.

To my knowledge the word "heritability" first appears in a book by J. L. Lush, *Animal Breeding Plans,* which was published in 1937. (I refer here to the third edition, 1945.) In Chapter 8 Lush divides V_P into two components that are nowadays usually written as V_G and V_E, where V_G is "that part of the variance caused by the heredity that different individuals have" (p. 91) and V_E is "that part of the variance caused by differences in the environments under which different individuals developed" (p. 91). Thus

$$V_P = V_G + V_E. \tag{1}$$

Heritability is then the "portion of the observed variance for which differences in heredity are responsible" (p. 91). Although Lush does not actually write

$$\text{Heritability} = \frac{V_G}{V_G + V_E} = \frac{V_G}{V_P}, \tag{2}$$

it is clear that this is the definition he intends.

To estimate the heritability defined in Equation (2), Lush describes experimental methods that involve correlations between phenotypes of animals with varying degrees of relatedness. In deriving these procedures, which Lush describes verbally, he was profoundly influenced by Sewall Wright's mathematical and empirical studies. Provine (1986b, p. 323) describes that relationship: "There can be no doubt that Lush relied heavily upon Wright, not only for help on quantitative questions, but even more for Wright's general formulation and approach to the basic problems of animal breeding."

Sewall Wright's interest in the relative importance of heredity and environment to phenotypic variability stemmed from his work on color patterns in guinea pigs. His statistical framework is now known as *path analysis,* and to this day it remains one of the central tools of correlational analyses in the social and behavioral sciences. Wright draws a series of lines or paths that describe the possible "causes" of an offspring's phenotype, and arrows indicate the assumed direction of causation. Such causes would normally include the parents' genotypes and some measure of the environment that contributes to the phenotype. Associated with each path is a number between zero and one that represents "the importance of a given path of influence from cause to effect" (Wright, 1921, p. 114). This number is equal to "the ratio of the standard deviation of the effect when all causes are constant except the one in question, the variability of which is kept unchanged, to the total standard deviation" (pp. 114–115). The accompanying figure, from Wright (1931), illustrates the conclusion from his application of this procedure to data collected by Burks (1928). The relationship between the path coefficients and partial correlations is described by Li (1975).

Wright then showed how to compute the expected correlations among the phenotypes of related individuals as functions of the path coefficients and of additional parameters such as the correlation among phenotypes of

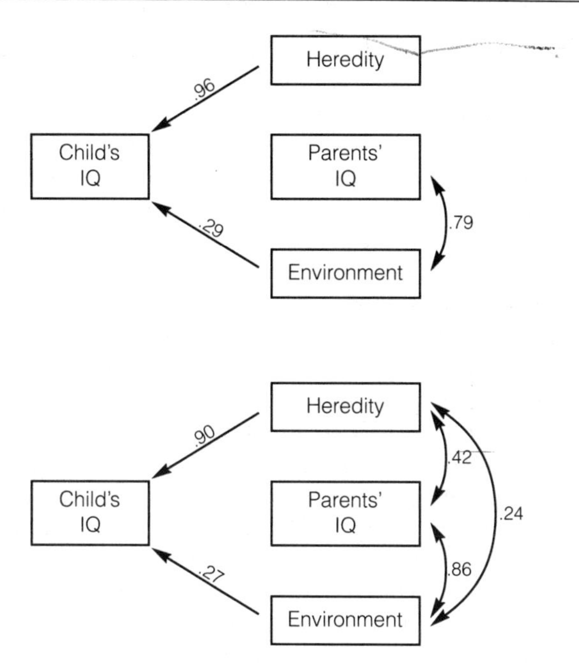

Path diagrams with estimates of the effect of heredity and environment on child's IQ. The upper diagram refers to adopted children and their natural and adoptive parents. The lower diagram refers to natural children and their parents. The numbers are obtained by solving sets of nonlinear equations given in Wright (1931, p. 160), from which this figure has been redrawn. The estimate of heritability (in the narrow sense defined below) is the square of the value representing the path from heredity to child's IQ, namely 0.92 in the upper diagram and 0.81 in the lower. Recent studies use more complicated diagrams and produce much lower estimates of heritability.

mates due, for example, to inbreeding or assortative mating. (Other parts of this now-famous series of papers in 1921 developed the theory of assortative mating and inbreeding in great detail.) From an observed set of correlations between relatives one can, in principle, work backwards and obtain estimates of the path coefficients. The square of the path coefficient between parental genotypes and offspring's phenotype represents the degree of genetic determination of the trait, h^2. In recent attempts to assess the contributions of heredity and environment to quantitative behavioral traits, this quantity is usually called the *genetic heritability* to distinguish it from the corresponding measure of the contribution from an environmental variable to the offspring's phenotype, which is called *cultural heritability*.

At the time that Wright developed his method for computing correlations between relatives and used these to assess the relative contribution of genetics to phenotypic variation, he was clearly unaware of a different approach that had been taken by R. A. Fisher (1918). In his influential paper, Fisher first considered the contributions of one locus with two alleles, say *A* and *a*, to the value of a phenotype. Assuming Hardy-Weinberg equilibrium for the genotypic frequencies, he computed the linear regression of the phenotype on the number of *A*'s in the genotype. The amount of phenotypic variance removed by this regression is called the *additive genetic variance*, V_A. If the substitution from *aa* to *Aa* produces the same average phenotypic change as that from *Aa* to *AA*, then all genetic variance for the *A* locus is additive. Fisher was able to show that, in general, V_G should include a component due to nonadditivity at one locus, which is called the *dominance* variance, V_D, as well as a term due to interaction between loci, which was called the *epistatic* component, V_I. By postulating that continuously varying phenotypes could be represented as sums of single-locus factors each obeying these rules, Fisher developed formulas for the phenotypic correlations between relatives that have been widely used in animal and plant breeding as well as in the description of genetic and behavioral variation in many species, including *Homo sapiens*.

The fraction of the total phenotypic variance that is the additive genetic variance, V_A/V_P, is what animal breeders usually call heritability and what in more recent texts is often called *narrow-sense heritability*. (By assuming that the nonadditive genetic components of variance are negligible, the *genetic heritability* of the preceding paragraph becomes identical to the narrow-sense heritability.) The ratio $V_G/V_P = (V_A + V_D + V_I)/V_P$ is called the *broad-sense heritability* or, less commonly, the degree of genetic determination (Falconer, 1960, 3rd ed. 1989, pp. 126–127). Wright (1921) also considered the nonadditive components of genotypic variance, but he did this outside the path-analytic framework that produced his results for the additive genetic components. Lush clearly meant Equation (2) above to mean broad-sense heritability. In his footnote (1937, 1945, p. 91) he states, "This definition includes as 'hereditary' the dominance and epistatic deviations since they result from differences between whole genotypes although they will not contribute to much of the likeness between relatives as the additive differences do." Lush goes on to claim that heritability in the broad sense is "useful in estimating the probable results of certain breeding systems in the next generation or two, but it tells nothing about the ultimate limits of the changes which might be made in that population either by breeding or by altering its environment" (p. 92). This statement is erroneous. It is only the narrow-sense heritability that is useful as an index of the amenability of a population to selective breeding.

Narrow-sense heritability *is* useful in agricultural applications where environmental conditions of growth and development can be controlled. A statistical estimate for the narrow-sense heritability pertains only to a particular population studied under a specific environmental regime. It cannot be regarded as valid for a different population or under different environments. There is nothing that can be inferred from such an estimate about the extent to which phenotypic differences between populations are due to genetic differences.

The most lucid account of Fisher's approach is to be found in Falconer (1989, especially chaps. 9 and 10). Falconer also describes practical ways by which narrow-sense heritability can be computed from the results of artificial selection as practiced by animal breeders. In artificial selection the *realized heritability* is the ratio of R, the mean deviation of the offspring of selected parents from the population mean, to S, the mean deviation of the selected parents from the population mean (Falconer, 1989, pp. 196–201). For reasons described by Falconer, R/S may not provide an adequate estimate of the narrow-sense heritability in the base population upon which artificial selection was carried out.

Lush (1945) makes no reference to Fisher (1918), although he does cite Fisher's famous (1930b) book and frequently uses the notion of additive genetic variance. Fisher (1930b) showed with his "Fundamental Theorem of Natural Selection" that under continued selection, V_A should decrease over time and approach zero, although the nonadditive genetic components of variance may not. Thus, under selection, narrow-sense heritability is not a constant parameter. Phenotypes such as calving interval in cattle, egg yield in poultry, or litter size in pigs have low narrow-sense heritabilities. This is usually explained as a consequence of the close relation between these traits and fitness, so that from Fisher's theorem their V_A's are low.

It is important to note that where the formulas from Wright's (1921) Table 2 and Fisher's (1918) paper can be compared, they agree. In particular, in the case where all genetic variance is additive, and even in the presence of assortative mating, correlations between relatives represented in terms of Wright's h^2 or Fisher's V_A/V_P are equivalent. Textbooks of population genetics, behavior genetics, human genetics, and quantitative inheritance tend to use Fisher's variance analysis in their treatments of phenotypic variation (see, e.g., Falconer, 1989; Cavalli-Sforza and Bodmer, 1971; Bulmer, 1980; Hay, 1985; Hartl and Clark, 1989), presumably because the authors wish to avoid having to develop the theory of partial correlation and regression. By contrast recent research in behavioral genetics, for example in the statistical analysis of IQ variation, has tended to use Wright's formulation.

The variance partition in Equation (1) entails the assumption that a given difference in the environment under which development occurs has the same effect on the different genotypes. If this is not the case, and for example genotype g_1 is better than genotype g_2 in environment e_1 but worse in environment e_2, then there is *genotype-environment interaction* and we must write

$$V_P = V_G + V_E + V_{GE}, \tag{3}$$

where V_{GE} is called the interaction variance. Falconer (1989, p. 135) writes, "Genotype-environment interaction becomes very important if individuals of a particular population are to be reared under different conditions"; Lush (1937; 1945, p. 91 footnote) in referring to the Equation (1), states, "In order not to confuse the argument, the nonadditive interactions of heredity and environment are neglected here. There is no reason to think that these effects will generally be small." The contribution of interaction to the total variance is difficult to assess, except under restricted experimental conditions that may be attainable in animal breeding but that are virtually impossible to meet in the study of human behavior, for example. A further complication to Equation (1) that is readily overcome in experimental populations results when some genotypes occur more often in specific environments than would be expected by chance; this is *genotype-environment correlation*. The effects of interaction and correlation should be separated from V_G and V_E, but this is very often difficult to do. In most applications of the study of humans, these effects become confounded with the purely genetic components of variance.

Each of the definitions of heritability is statistical in nature and does not involve a detailed specification of genetic or environmental transmission. It is rarely possible in the study of human quantitative variation—even for traits that appear to aggregate in families—to achieve the rigorous experimental design required to make inferences about genetic causality from the variance analysis (Lewontin, 1974c; Feldman and Lewontin, 1975). Indeed, each statistical procedure, whether it be analysis of variance or path analysis, involves a specific quantitative model of the way in which the phenotype is believed to be affected by genetics and environment. In addition, there are many ways to describe assortative mating according to how the correlation between mates is structured.

The importance of the model to the computation of heritability is evidenced by the changing estimates that characterize the history of familial studies of IQ. Prior to 1970, the accepted estimates for the broad-sense heritability of IQ were generally higher than 80%. In the early 1970s, specific forms of cultural transmission began to be included in the statistical models (Cavalli-Sforza and Feldman, 1973). In 1976, Rao et al.

claimed to have "resolved" the cultural and biological contributions to IQ with an estimate of 67% for the genetic heritability of IQ. More sophisticated models of assortative mating than that used by Rao et al. were suggested by a number of authors (Cavalli-Sforza and Feldman, 1978; Rice et al., 1978; Feldman and Cavalli-Sforza, 1979). The inclusion of these in the statistical analysis of U.S. IQ data by Cloninger et al. (1979) produced estimates of 32.6% and 27.2% for the genetic and cultural heritabilities, respectively, and an interaction component of 9.3% which should, in principle, be assigned to neither genetic nor cultural components. As is usual in such studies, nonadditive genetic factors are ignored. A reanalysis by Rao et al. (1982) resulted in estimates for the genetic heritability between 0.228 and 0.438 in children, and between 0.310 and 0.472 in adults and between 0.096 and 0.482 for the cultural heritability. These results are in good statistical agreement with those of Cloninger et al. It is interesting that despite these careful statistical studies that suggest roughly equal values for genetic and cultural heritabilities of about one third, the older estimates are still widely used. For example, Arthur Jensen (1989) claims, "The overall average of the best estimates we have for the broad sense heritability of IQ is between 0.60 and 0.70." In the popular press, Deborah Franklin (1989) reports in a *New York Times* article that "genes are 50 percent to 70 percent responsible for an individual's IQ." Not only are these figures different from those quoted above, but, as mentioned above, causation is incorrectly inferred from estimates of heritability.

An alternative to the use of a single summary statistic, heritability, is a descriptive representation of the relationship between phenotypic and genetic variation called the *norm of reaction* (see GENOTYPE AND PHENOTYPE). This term, originally used by I. I. Schmalhausen (1949), refers to a quantitative description of the phenotype of each genotype under study in an array of environments. Few such data have been obtained for normally heterozygous organisms. For inbred or homozygous strains of *Drosophila* and a number of plants, however, usually no strain is consistently above the others in all tested environments (Suzuki et al., 1989, pp. 655–658). Thus, even if the heritability in one environment were high, it might not be in another. By changing the environment, we find that both the amount and the pattern of phenotypic variation among genotypes may be changed. Although the norm of reaction is a difficult goal to achieve for most species, and some kind of summary description of variation is desirable, heritability, especially broad-sense heritability, may not be very informative.

Heterochrony

Stephen Jay Gould

Words and objects often, indeed usually, change their meaning when they enjoy a long history. But when, as sometimes happens, meanings turn into their opposites, then we feel that something extraordinary has occurred, something worth more than casual mention. The canonical examples are political—the old tale of the idealistic reformer turned wily despot, the pigs of Orwell's *Animal Farm* who begin their benevolent reign with the motto "four legs good, two legs bad," but end up as bipedal dictators (with the revised dictum "four legs good, two legs better"). But words can also turn into their opposites—and we encounter a fine example in the flagship of this volume, "evolution" itself. As a biological term, "evolution" had its first flowering as a name for the embryological theory of preformation—a denial of "evolution" in our modern sense, since all future generations are encapsulated as homunculi in the ovaries of Eve (or sperm of Adam). The turnaround in meaning, in this case, was not by subtle and insensible gradualistic continuity (itself called "evolution" in some circles!), but by substitution—as the embryological theory died and "evolution" was rerecruited by Herbert Spencer from the vernacular, and for a very different purpose.

"Heterochrony," used today as a general term for evolutionary changes in the timing of development, and a source of greatly expanded interest in the past decade (Gould, 1977a; Alberch et al., 1979; McKinney, 1988), has enjoyed (or, we might better say, suffered) an even more complex double shift in usage since its coinage by Ernst Haeckel in the 1860s. The first shift occurred in continuity, with the original definition retained—while the interpretation changed radically and became a source of interesting debate. The second shift, a mutational misunderstanding, was imposed by G. R. de Beer (1930) as a true discontinuity. But de Beer's redefinition has swept the field and become entrenched. Thus the odyssey of "heterochrony"—an epic in three chapters—becomes a pretty fair surrogate for processes of evolutionary change in biology. But why not? Ever since the Brothers Grimm put aside their fairy tales long enough to develop the laws of sound change in the diversification of Indo-European languages, scholars have found in the comparison of linguistic and biological change both a fruitful source of analogy and a quagmire of noncomparability!

Haeckel's original definition. Ambition seeks immortality in many ways. One surprisingly effective guarantee lies in the strategy of coining words with abandon and trusting to the cultural analog of natural selection for the survival of a few—the *r*-selective strategy for an enduring posterity. Haeckel followed this scattershot tactic with unparalleled zeal. Most of his terms are on the ash-heap of history, but a few—including heterochrony—have survived. Heterochrony may not have the glamorous familiarity of other Haeckelian creations—ecology, ontogeny, and phylogeny among them—but it holds a secure (and now growing) place in evolutionary terminology.

Haeckel gave heterochrony his best shot by investing this word as one of a half dozen or so key items within his major theoretical contribution—the doctrine of recapitulation, or the "biogenetic law": ontogeny recapitulates phylogeny. Haeckel wanted to use the biogenetic law as a chief guide to the reconstruction of evolutionary trees. He argued for a universal tendency to acceleration in developmental timing; adult characters of ancestors occur earlier and earlier in the ontogeny of descendants. Thus the younger an embryonic stage, the older the phyletic ancestry. Human embryos with gill slits are repeating the fish stage of phyletic history; later embryos with tails are passing through the reptilian or mammalian phase.

Haeckel hoped to use recapitulation as a guide to the reconstruction of phylogeny; he was strangely uninterested in its potential to provide insight into general mechanisms of evolution (Gould, 1977a). Thus Haeckel's concept of recapitulation included the vital premise that all parts and organs undergo developmental acceleration through time at the same rate, and in a fully coordinated way—so that the *entire* ancestral adult is transferred to an earlier ontogenetic stage where it can be recognized as a coherent ancestor.

Yet Haeckel was far too good a naturalist to argue, against the messy empirics of comparative embryology, that such a complete and coordinated recapitulation actually occurred. He knew perfectly well that other forces (juvenile adaptations, unequal rates of acceleration in different organs) impinged upon the embryo and prevented the idealized repetition of entire ancestral adults in embryonic stages. Rather, Haeckel, as a keen student of the uses of classification, developed a taxonomy of embryonic phenomena that exalted coordinated recapitulation as normal (and even virtuous) and relegated any departure to a class of disruptions and falsifications. Heterochrony, ironically, began its terminological life as a category of disruption to true recapitulation.

Haeckel established a dichotomous terminology of ontogenetic phenomena—good and bad—based on their utility in fostering the complete and coordinated recapitulation of ancestral adults in embryonic descendants. He designated as *palingenesis,* or epitomized history, the equal and coordinated acceleration of organs. All other phenomena he placed into his disruptive category of *cenogenesis,* or falsified history *(Fälschungsgeschichte).* The utility of recapitulation then resided in the power of palingenesis to overwhelm cenogenesis by relative frequency and clear signal. Haeckel's theory of recapitulation is not the naive claim that entire adults reside in descendant embryos, but the more subtle construction of a dichotomous classification accompanied by an empirical claim (later refuted by most naturalists) that palingenesis nearly always overwhelms cenogenesis. Haeckel wrote: "All of ontogeny falls into two main parts: first *palingenesis* or 'epitomized history' [*Auszugsgeschichte*], and second, *cenogenesis* or 'falsified history' [*Fälschungsgeschichte*]. The first is the true ontogenetic epitome or short recapitulation of previous phyletic history; the second is exactly the opposite: a new, foreign ingredient, a falsification or concealment of the epitome of phylogeny" (1875, p. 409).

Interestingly enough, Haeckel compared the recognition of palingenesis (and the casting aside of cenogenesis) to the purging of accumulated additions and interpolations from a text: "It is of the same importance to the student of evolution as the careful distinction between genuine and spurious texts in the works of an ancient writer, or the purging of the real text from interpolations and alterations, is for the student of philosophy . . . I regard it as the first condition for forming any just idea of the evolutionary process, and I believe that we must, in accordance with it, divide embryology into two sections—palingenesis, or the science of repetitive forms; and cenogenesis, or the science of supervening structures" (1905, p. 7).

Haeckel recognized two main categories of cenogenesis. The first, and most important, was *juvenile adaptation*—the interpolation of new features into early ontogenetic stages for their own immediate evolutionary utility. Adaptations of marine larvae for eating and staying afloat ranked highest in this category of regrettable disturbances to recapitulation.

The second category of cenogenesis included changes in timing and placement of one organ relative to others in the same body, rather than interpolation of new features. Haeckel devised two subcategories. First, and less important, was *heterotopy,* or change in place of differentiation (a term that has disappeared from use). As a primary example, Haeckel cited the current development of reproductive organs from mesoderm in most animals. Because mesoderm is, itself, a belated development in animals (for primitive creatures had but two layers of endoderm and ectoderm), and because ancestral animals with only two layers must have possessed reproductive tissues, these organs must have originated in endo- or ectoderm and later changed position to differentiate in mesoderm.

As a second, and far more important, subcategory, Haeckel identified changes in timing of development for one organ relative to others *in the same body*. Haeckel referred to these changes of timing as *heterochrony*—a good etymology, and the original source of the word. As a primary example, Haeckel cited the very early appearance of the heart at the beginning of vertebrate development—when the embryo is a scarcely differentiated tube and is therefore far before the "scheduled" appearance of the heart in phylogeny. By a more rapid acceleration in development (relative to other organs), the heart disrupts the true and full transfer of ancestral adults to earlier stages of ontogeny.

Thus, in summary, heterochrony began its life as a category of Haeckel's cenogenesis, or disruption and falsification of recapitulation. It referred to changes in the timing of development of one organ or part relative to others in the same creature—in particular, relative to those features obeying the regular pattern of palingenesis. Heterochronies are one of a class of embryonic adaptations (along with heterotopies and juvenile interpolations) that compromise the reading of phylogeny in ontogeny. Haeckel's final reformulation of the biogenetic law recognized this interplay of palingenesis and cenogenesis as primary: "The rapid and brief ontogeny is a condensed synopsis of the long and slow history of the stem (phylogeny): this synopsis is the more faithful and complete in proportion as palingenesis has been preserved by heredity and cenogenesis has not been introduced by adaptation" (1905, p. 415).

The first (and friendly) transformation. Haeckel's program of using the biogenetic law to search for entire ancestors in the embryonic stages of modern forms was pursued with much hope and fanfare, but led to few positive results and endless wranglings about untestable phylogenetic scenarios—all because the biogenetic law is basically false (see Gould, 1977a). By the closing years of the nineteenth century Haeckel's program had become a source of much ridicule, and the negative platform for launching other approaches to embryology—particularly the *Entwicklungsmechanik* of Wilhelm Roux and Hans Driesch.

Haeckel's downfall had many complex causes, but we may cite as one major reason the simple empirical conclusion that his exceptional category of cenogenesis—including heterochrony—turned out to be far more common than he had anticipated. So pervasive were embryonic and juvenile adaptations, and shifts in timing and placement of features relative to each other, that any potential signal of complete palingenetic repetition was completely swamped into imperceptibility.

In such a climate of trouble, supporters of a theory will try to preserve their basic scheme by making adjustments in the light of effective criticism. Haeckel's category of heterochrony became the source of a primary adjustment and therefore, in a major sense, the last battleground of important support for recapitulation.

Recapitulation requires a universal (or nearly universal) tendency for rates of development to speed up in phylogeny—so that ancestral adult stages appear earlier and earlier in the ontogeny of descendants. Haeckel had hoped for an even and coordinated acceleration of all organs—so that entire ancestors might be recapitulated. But he constructed the exceptional category of heterochrony to acknowledge that rates of change in timing could vary for different organs. As Haeckel's research program went forth and floundered, his supporters realized that heterochrony was a rule, not an exception. How might recapitulation be saved in the light of this fact?

Salvation can only come by redefintion—but the potential rescue is simplicity itself: redefine recapitulation as a process working separately on each organ. Entire organisms can never be recapitulated, only individual organs, each at a different rate. Recapitulation only requires a universal speeding up of development; it does not demand that the speeding up be equal for all organs (unless one wishes, as Haeckel did, to find entire ancestors in embryonic descendants). By giving up this chimerical hope of finding whole ancestors, recapitulation could be preserved on an organ by organ basis. Acceleration is universal. The embryonic phases of each organ will therefore represent a sequence of phyletic stages. But organs are accelerated at unequal rates, some rapidly (like the heart in Haeckel's canonical example of heterochrony), others more slowly. Still, all are accelerated, and recapitulation occurs in each organ. To save the biogenetic law, one must use Haeckel's maxim separately for each part.

This redefinition of recapitulation by organ rather than by creature had the curious effect of transforming Haeckel's "heterochrony" from a category of exceptions (to complete recapitulation) into a fundamental process of the new scheme. For heterochrony is the acknowledgment that organs are accelerated at unequal rates, and this observation became the linchpin of the revision. Without changing its definition, heterochrony made a transition from peripheral exception to central phenomenon.

The paleontologist Edward Drinker Cope, the most theoretically minded of American recapitulationists, spearheaded this revision by recognizing two categories for the biogenetic law: exact parallelism, when no heterochrony occurred and entire ancestral adults were pushed back into juvenile stages of descendants, and the far more common inexact parallelism, when heterochrony prevailed, organs changed their timing at different rates, and recapitulation occurred part by part (see also Weismann, 1881). Cope (1876, in 1887, p. 126) acknowledged the central role of heterochrony: "The phenomena of exact parallelism or palingenesis are quite as necessarily accounted for on the principle of acceleration or retardation as are those of inexact parallelism or cenogenesis. Were all parts of the organism accelerated or retarded at a like rate, the relation of exact parallelism would never be disturbed; while the inexactitude of the parallelism will depend on the number of variations in the rate of growth of different organs of the individual."

Heterochrony then became the chief focus of a major research program among German morphologists in the late nineteenth century (Oppel, 1891; Keibel, 1895; Mehnert, 1895, 1897). They attempted to construct *Normentafeln* ("normal tables") of expected rates of development for each organ in animals under study. They then tried to induce rules for varying intensities of heterochrony. These rarely proceeded much beyond the minimally enlightening claim that organs destined to be large or needing to function early usually showed the most intense speeding up of development relative to other organs. Franz Keibel wrote (1898, p. 786): "I seek a connection between the function of an organ and its time of appearance in ontogeny. I hold that an organ appears early in ontogeny if it begins to function early."

But these authors disagreed about the meaning of this pervasive heterochrony for Haeckel's faith in the biogenetic law. Ernst Mehnert found the law unblemished, even strengthened, provided that Cope's dictum of considering recapitulation organ by organ be followed. But Keibel, based on the same data, argued that Haeckel's law had been undermined, for he insisted on the original definition of complete repetition, with heterochrony as a disturbing exception.

Terminological wrangling of this sort often marks the death throes of a worldview. When the argument ceases to make sense, or gets too complex by backing and filling to meet valid criticism, or when new data cease to accumulate, one can always permute the meaning of terms and argue about their usages.

The second transformation. Recapitulation began with two simplistic notions: (1) that acceleration (or speeding up) of development might be universal as a mode of change in developmental timing; and (2) that all organs might be accelerated at the same rate, permitting a complete transfer of an ancestral adult to the juvenile stage of a descendant. The reform of the first transformation struck down the second claim, but did not annihilate recapitulation—for it permitted a redefinition in terms of organs rather than organisms.

The reform of the second transformation, by contrast, destroyed recapitulation while, ironically, making even more important its central theme of change in developmental timing as a mechanism of evolution. Both reforms represent a broadening of phenomenology—the first by allowing organs to change their timings at different rates, the second by permitting slowing down as well as speeding up of development.

But the two reforms had entirely different meanings for the fate of recapitulation. The first saved Haeckel's doctrine by reinterpretation for organs; the second disproved the biogenetic law by granting important, or even equal, weight to the anti-recapitulatory process of slowing down in developmental rates, leading to paedomorphosis, or retention of juvenile stages of ancestors in adults of descendants—the polar opposite to recapitulation (see also Gould, 1977a).

The validation of paedomorphosis undid the biogenetic law, but expanded the realm of phenomena in developmental timing with potential evolutionary significance. Thus when de Beer (1930) set out to write the short book that would form the basis for modern ideas on embryology and evolution, he faced a greatly expanded set of possibilities. Haeckel only thought in terms of acceleration—and for all organs at the same basic rate. De Beer had to consider both acceleration and retardation, and he had to affirm the mosaic character of organisms—with each organ undergoing its acceleration or retardation at a different rate in phylogeny.

De Beer therefore decided to make a new and more complex taxonomy of relationships between ontogeny and phylogeny. In so doing, he chose Haeckel's old term heterochrony as his general description for evolution by change in developmental timing—thus giving the venerable word a new life with greater scope and importance than it had ever possessed before. But he accomplished this by completely misunderstanding and changing Haeckel's meaning. I do not know (and the texts contain no internal evidence) whether he did this consciously, or by simple error. His action was, in any case, a creative confusion that has granted to the newly defined sense of heterochrony a central place in modern evolutionary theory.

De Beer first uses heterochrony (1930, pp. 9 and 24) in its proper Haeckelian definition to express a difference in the rate of timing for one organ *relative to other organs in the same body*. But then, starting on page 34 in a key chapter entitled "Heterochrony and Its Effect in Phylogeny," he shifts, and uses an entirely different definition of change in timing for an organ *relative to the same organ in an ancestor*. Instead of two organs in one body, we now have the same organ in two different bodies! He writes (p. 35): "Now, the principle of heterochrony will make it possible for any structure to appear later or earlier as well as at the same corresponding time, when compared with a previous ontogeny."

By this wrenching redefinition, de Beer made heterochrony the general term for the broader phenomenon that underlay both recapitulation and all the other phenomena linking ontogeny and phylogeny—phylogenetic change in developmental timing. One may claim that he did Haeckel a disservice in redefining the term. But one may also take the opposite position that this redefinition, after all, only made Haeckel's term the keyword of the large subject—the relationship between ontogeny and phylogeny—that Haeckel had tried to encompass (though too restrictively) with his biogenetic law. What better legacy could one want for a term? De Beer's books (1930, 1940, 1958) so dominated the field that his redefinition prevailed completely, and Haeckel's meaning became extinct.

In any case, de Beer then erected a prohibitively complex classification for the modes of heterochrony. By a series of redundancies and logical errors, de Beer had needlessly adumbrated a basically simple distinction between the two great heterochronic modes of speeding up and slowing down (Gould, 1977a). Some of the deficiencies of my own discussion were later resolved in Alberch et al. (1979), with its fundamental distinction of paedomorphosis (prolongation of early states to later times of descendant ontogenies) and peramorphosis (extension beyond an ancestral state at the same time in a descendant ontogeny) as two different modes of heterochrony. This simplification and forging of consistency in terminology has since been accepted by nearly all workers in the field. Today, the empirical study of heterochrony (McKinney, 1988) is a burgeoning activity in evolutionary biology, abetted by a widely shared feeling that the solution to many key riddles in evolution lies in the study of development and that new tools in genetic and embryological analysis now make resolution feasible and possible for the first time.

The *summum bonum* of Charles Darwin's world is survival and persistence, not progress or improvement. All species are engaged in a zero-sum game against a house with infinite resources, and all must eventually succumb. To remain at the table as long as possible, to have some fun and impact in the attempt, and to act with honor (if one happens to be a moral agent as well) can be the only benchmarks of success. If we might venture the same criteria for terms, heterochrony is an admirable creation indeed. By judicious "adaptation," involving both going with the flow and mutational saltation in meaning at an apposite time, heterochrony has not only survived; it has also enjoyed impact, and may ever hold the equivalent of moral virtue in its newest life as standard-bearer for such an important subject as the evolutionary significance of development.

HETEROSIS

Diane Paul

"HETEROSIS" HAS two meanings that are frequently confused—or simply fused. In one standard usage, it is employed as a synonym for "hybrid vigor"—the superiority of the offspring of a cross between two stocks to the better of the parents. The terms "hybrid vigor" and "heterosis" have been used interchangeably since at least the 1930s. Heterosis (or sometimes "single-locus heterosis"), however, is also employed as a synonym for "heterozygote advantage" or "overdominance," that is, a relation between alleles in which the heterozygote, AA′, is superior to either homozygote, AA or A′A′. Heterosis in the latter sense is often invoked to *explain* hybrid vigor.

In its original meaning, heterosis denoted a genetic-physiological process. The term was invented by George H. Shull in 1914 as a substitute for such phrases as "the stimulation due to differences in uniting gametes" and "the physiological stimulation of heterozygosity." In Shull's view and that of his collaborator Edward M. East, hybrid vigor was best explained by a beneficial stimulus to development produced (they knew not how) when different gametes united. It followed that "hybridity itself,—the union of unlike elements, the state of being heterozygous" was desirable (Shull, 1914, p. 126). Theirs was thus an early theory of per se heterozygote advantage.

The equation of heterosis with the view that vigor is a direct property of heterozygosity—and thus cannot be fixed—has survived the details of East and Shull's own theory and exerted enormous influence on agricultural practice. Hybrid breeding plans were based on the assumption that heterozygote advantage is an important source of vigor. This assumption has been challenged, however, both in Shull's day and in ours. In an alternative view, each parent contributes favorable dominant alleles that conceal the other's deleterious recessives. If this dominance explanation of hybrid vigor is correct, it should be possible to develop pure lines as good as hybrids—indeed, better, because deleterious "recessives" usually have slight phenotypic effects. (If the number of relevant loci is large and some are closely linked, it may take a long time to obtain high-quality inbred lines. The linkage point was made by D. F. Jones, 1917. G. N. Collins, 1921, demonstrated that, with a large number of factors, tight repulsion linkage was unnecessary.) That the same word serves as a synonym for one of these theories and the effect to be explained is the source of much confusion. How did this ambiguity arise?

Systematic research on hybrid vigor and its converse, inbreeding depression, dates at least from the eighteenth century. Charles Darwin conducted experiments comparing selfed and crossed plants from the same stocks of maize, and reported his results in *The Effects of Cross and Self-fertilization in the Vegetable Kingdom* (1876). By the turn of the century, it was generally understood that inbreeding produces many defective and sterile individuals and a general loss of "luxuriance" or "vigor." Cross-breeding reverses this process, and may even produce individuals superior in vigor to the original stocks. It was assumed that inbreeding results in deterioration because both parents are prone to have the same defects, which "intensify" in their progeny. Cross-breeding has the opposite effect because the parents possess different defects that "cancel out" in their offspring. After 1900, these insights were easily rephrased in Mendelian terms: in the course of selection for particular traits, breeders create strains that are homozygous for deleterious genes elsewhere in the genome. When these strains are crossed, the resulting hybrids acquire normal, dominant genes at most of their loci. This interpretation of hybrid vigor was first explicitly formulated by F. Keeble and C. Pellew, and independently by A. B. Bruce, in 1910.

A different interpretation, however, had already been advanced by East and Shull. Shull was interested in the effects of selfing and crossing on the expression of a quantitative character, and chose for his investigation the number of kernel rows in an ear of corn. He noticed that inbreeding the corn plants produced deterioration in such traits as the size and strength of the stalks and the number of ears (1909). This decline correlated with the increase in homozygosity and finally leveled off. Shull assumed that his inbreeding had resulted in the isolation of what Wilhelm Johannsen had termed "biotypes" (or pure lines). When these lines were crossed, the offspring were superior not only to their parents but in some cases to the original open-pollinated varieties. The source of this benefit, however, remained obscure. Why should the crossing of different biotypes prove beneficial? A paper by East (1909), published later that year, seemed to provide the answer.

East was convinced that Mendelism was insufficient to account for all the observed effects of inbreeding. He advanced an alternative theory based on a then common view of the functions of sexual reproduction: that it provides a physiological stimulation to development as well as a recombination of characters. East suggested that when "two strains differing in gametic structure" are crossed, this beneficial stimulation would increase (East, 1909, p. 177). He used the expression "physiological stimulation of heterosis" to describe the process.

Shull adopted, developed, and popularized East's idea. In the 1914 paper proposing the term "heterosis" (and its cognate "heterotic"), Shull elaborated the claim that heterozygosity per se is a virtue, and drew the logical conclusion for breeding practice. If "heterosis" explains hybrid vigor, it follows that pure lines cannot be as productive as hybrids. And because the quality of hybrids declines after the first generation, farmers must repurchase seed each year to exploit hybrid vigor fully.

In both East's paper and Shull's, however, the nature of the underlying process remains obscure. How is the physiological effect produced? East admitted that he had no idea of the mechanism through which such a stimulation could occur (East, 1909, p. 178). Thirty years later, he explained: "Genetic knowledge, at the time, was so meagre that it seemed necessary to assume that vigor is promoted when the genes at certain loci are unlike,—an assumption for which there was no proof, and which was not illuminating as a dynamic interpretation" (East, 1936, p. 375). Perhaps for this reason, the idea of heterosis was soon supplanted by the dominance interpretation of hybrid vigor.

But the East-Shull theory was only dormant, not dead. In a 1922 paper, "On the Dominance Ratio," and again in his 1930 work *The Genetical Theory of Natural Selection,* R. A. Fisher demonstrated that when the heterozygote has a selective advantage, a stable equilibrium will result. Fisher's was a purely theoretical discussion. In the 1940s, however, Fred Hull argued on the basis of experimental work that heterozygote advantage (now called "overdominance") was responsible for some of the actual hybrid vigor in corn. Hull's views, in turn, were popularized by Jay Lush, author of a leading text on breeding. (Later experimental evidence, however, indicates that overdominance, at least in corn, is unimportant.)

Work in medical genetics also played an important role in reviving the concept of heterozygote advantage. In 1954 Anthony Allison provided a convincing explanation for the high frequency, in much of Africa, of the gene for S hemoglobin. Individuals who are homozygous for this gene suffer from a severe disease, "sickle-cell anemia"; heterozygotes express a mild form of anemia, "sickle-cell trait." At the time of Allison's research, as much as 3% of the population in highly malarial regions was afflicted with the disease, and 30% with the trait. A mutation rate high enough to account for these gene frequencies was unthinkable. But Allison demonstrated that the heterozygotes were much more resistant to infection by the malarial parasite *Plasmodium falciparum.* In areas where *falciparum* malaria was common, the heterozygotes were adaptively superior even to the normal homozygotes. This model seemed to explain the high frequency of some other hemoglobin disorders as well.

In the 1950s and 1960s, the significance of heterosis/overdominance became the central issue in a controversy dominated by H. J. Muller and Theodosius Dobzhansky. None of the parties to what came to be called the "classical-balance" controversy denied the existence of heterozygote advantage. Muller agreed that heterosis explained the high frequency of sickle-cell anemia and a few other diseases, but he considered such cases rare. In his view, heterozygote superiority was a trivial fact about the world. Dobzhansky, however, considered it of great importance. Indeed, he came to believe that it accounted for most of the genetic variability in natural populations. Dobzhansky's friend and collaborator I. Michael Lerner tried to develop a mechanics of heterozygote superiority in his 1954 book, *Genetic Homeostatis.* Lerner argued that because heterozygotes made two kinds of gene products, they were better buffered than homozygotes against environmental changes. Thus in variable and changing environments (which he thought typical), they would be fitter. But neither Lerner's nor later efforts have succeeded in elucidating the mechanics of gene action underlying heterosis/overdominance. Whether there *are* many cases of overdominance remains a matter of dispute.

In 1944 W. Gordon Whaley protested the growing use of "heterosis" as a synonym for "hybrid vigor." He noted Shull's insistence, in his 1914 paper, that more than gene action might be involved in the stimulus of heterozygosity. "By the original definition 'heterosis' refers to the developmental stimulation resulting, from whatever mechanism, from the union of different gametes" he wrote. "'Hybrid vigor' denotes the manifest effects of heterosis" (Whaley, 1944, p. 463). Whaley advised keeping these terms separate. Discussions of heterosis in recent texts, especially on plant and animal breeding, appear to confirm the wisdom of this advice.

Whaley's protest was generally disregarded. It did, however, elicit a reply from George Shull, who insisted that his term was perfectly neutral in respect to cause. "The word 'heterosis' was chosen in the same spirit as Johannsen's word 'gene,' namely that it should be *free from every hypothesis.* It represented a group of observable phenomena for which any subsequent student was free to propose his own explanation without thereby being obliged to abandon the word 'heterosis'" (Shull, 1948, p. 440; emphasis his). As Shull wished, the word was not abandoned. But his claim in respect to its meaning was simply untrue. Heterosis was from the beginning—and remains today—the most theory-laden of terms.

HOMOLOGY

Michael J. Donoghue

HOMOLOGY is widely considered to be among the most important principles in comparative biology (Bock, 1974; Riedl, 1978). As Colin Patterson (1987, p. 18) put it, "all useful comparisons in biology depend on the relation of homology." Yet there are still significant differences in the meaning attributed to the term (and related words such as analogy and homoplasy), even among authors professing the same general outlook. Advances in phylogenetic systematics and in developmental and molecular biology have renewed interest in homology, and recently there have been several reviews that focus attention on alternative concepts. The papers of Patterson (1982, 1987, 1988), Roth (1984, 1988), and Wagner (1989a, b) are especially valuable.

Homology and evolution. The term homology is generally attributed to the British anatomist Richard Owen, who defined "homologue" in 1843 as "the same organ in different animals under every variety of form and function." Later its meaning was extended to include what he called "homotypy" (the "same" organ *within* an organism) and "general homology" (structural resemblance to an archetype) (Owen, 1848). Owen did not take credit for the concept, citing Etienne Geoffroy Saint-Hilaire, who noted its presence in even earlier German literature (see references in Sattler, 1984). Indeed, as Wagner (1989b) points out, the concept was already developed in Goethe's writing. At the very least, however, Owen provided "precision and currency" (Lankester, 1870).

The first and most fundamental shift in the meaning of homology accompanied the rise of evolutionary thought. Owen's definition was more or less retained by Charles Darwin (1859a) and his contemporaries—homology was morphological correspondence as determined primarily by relative position and connection. But the associated explanation was radically different. Metaphysical archetypes and "essential" similarity were replaced by material ancestors that could evolve. This connection having been made, the observation of homology could then be used as evidence for evolution and phylogenetic relationship.

Although ancestry was at first viewed only as an explanation for homology, it soon was incorporated into the definition (Haas and Simpson, 1946). E. R. Lankester's paper of 1870 played an important role, as evidenced in later editions of the *Origin*. He was critical of essential similarity and instead connected homology and related terms to common ancestry. Furthermore, he suggested that homology be subdivided. For homologous similarity due to inheritance from a common ancestor, Lankester coined the term "homogeny," and for such similarity resulting from independent evolution he introduced "homoplasy." "Analogy" then referred to similarities that would not be accepted as homologous based on standard criteria. The term "homogeny" was never widely adopted, and "homology" quickly became associated with similarity that could be traced to a common ancestor. This kind of transition—from explanation to definition—is probably commonplace (see Donoghue, 1985, on "species").

In retrospect, the connection between homology and evolution created at least two difficulties that appear to be common sources of semantic confusion (Donoghue, 1985). First, two potentially dissociable elements were combined in one definition (Boyden, 1943, 1973). If there are cases (as we suppose there are) in which similar structures originate independently, or cases in which very different structures originate through evolutionary transformation, one is forced to choose which element—similarity or ancestry—is to be given primacy. Second, homology was now expected to account simultaneously for the maintenance of similarity and for the transformation of form—for both constancy and change.

Homology as similarity. One response to these difficulties was to continue to associate homology with similarity alone. This is sometimes defended on the grounds of historical precedent, but the argument has generally been ineffective, as the meaning attached to words often changes as associated theories are modified. Alternatively, equating homology with similarity has been endorsed on the grounds that similarity is independent of, and logically precedes, the inference of common ancestry. However, precedence in an inferential process hardly implies greater importance, nor does it determine what a particular term is to signify (de Queiroz and Donoghue, 1990). Indeed, it could be argued that ancestry is fundamental, because it provides an underlying rationale for choosing among possible similarity criteria.

Others have argued that evolutionary homology is nonoperational and that it is circular to define homology in terms of ancestry if it is then to be used as evidence of ancestry (e.g., Woodger, 1945; Sokal and Sneath, 1963). These arguments are countered by Ghiselin (1966), Hull (1967, 1968), and others (see Hennig, 1966, p. 94). Here it is important to recognize that when homology is defined in terms of ancestry, particular instances are recognized by using a set of similarity criteria and by testing the congruence of presumed homologous traits with other such traits (see below). In this formulation, similarity criteria are a means of helping to identify homology, and particular similarity criteria can be evaluated on the basis of how well they do. But when homology is defined in terms of similarity, one wonders how one set of similarity criteria can be justified as better than another.

Three criteria discussed by Remane (1952) are widely cited both by those who define homology in terms of similarity (e.g., Kaplan, 1984) and by advocates of evolutionary homology (e.g, Hennig, 1966): similarity in relative position, similarity in structural detail, and the presence of transitional forms, including intermediates in ontogeny. If there are reasons for *defining* homology in terms of Remane's criteria, these have not been clearly stated. And if the ultimate intention is to discern which structures were derived from a common ancestor, then why not apply the word homology to traits that are actually believed to be historically unique, rather than to structures that merely conform to Remane's criteria (Sanderson, 1989)?

Even if one wished to equate homology with Remane's criteria, there are obvious difficulties when some but not all of the criteria are met. Sattler (e.g., 1966, 1984; see also van der Klaauw, 1966, and Meyen, 1973) notes that structures sometimes combine the features of two or more different organs. For example, the "phylloclades" of some monocots are borne in the position of a branch but are similar to leaves in terms of development, symmetry, and internal anatomy, perhaps as a result of homeotic mutations that cause developmental processes to occur in new locations (Sattler, 1984, 1988). To accommodate such cases, Sattler proposed that a semi-quantitative index be used to reflect the "partial homology" of the structure to a leaf and to a stem, thereby circumventing the need to assert that it is either "essentially" a leaf or a stem. Partial homology is incompatible with standard evolutionary views, according to which structures are either homologous or not (e.g., Patterson, 1987).

The so-called operational view of homology developed by pheneticists (e.g., Jardine, 1967; Key, 1967; Jardine and Sibson, 1971; Sneath and Sokal, 1973) has similar consequences. According to Sokal and Sneath (1963) two characters are operationally homologous if they are "very much alike in general and in particular." This, too, opens the way for partial homology, because the extent to which structures are similar may vary continuously. It is noteworthy that operational homology has resurfaced in the literature on morphometric techniques, wherein homology is associated with points or "landmarks" on a structure, which can then be used in calculating a series of measurements (Bookstein et al., 1985).

The equation of similarity with homology, and consequently the acceptance of partial homology, is also widespread among molecular biologists (e.g., Winter et al., 1968), despite arguments against such usage (e.g., Reeck et al., 1987) and at least one journal that specifically prohibits it (Patterson, 1988, in reference to *Molecular Biology and Evolution*). Here, when it is said that two polypeptides or segments of DNA are 75% homologous, it means that they can be aligned in such a way that the same amino acids or nucleotides are present at 75% of the sites. In this context, the issue is complicated by the argument that sequences which are sufficiently similar must also be homologous in the evolutionary sense, simply because the chances of independently acquiring a high percentage similarity are vanishingly small (for examples of such calculations, see Fitch, 1966; Doolittle, 1981; Dayhoff et al., 1983; Karlin et al., 1983).

Transformational homology. A more popular view is that structures are homologous if they can be traced to a particular condition that originated once in a common ancestor (Simpson, 1961; Hennig, 1966; Mayr, 1982a). This includes retention in a more or less unchanged form or transformation, directly or sequentially (and possibly radically), from an ancestral condition. Similar features that originated independently from states in different ancestors are not homologous. Such cases, whether involving reversal, convergence, or parallelism, are generally covered by the term "homoplasy" (Sanderson and Donoghue, 1989; but see below). "Analogy" also commonly means the opposite of "homology" (e.g., Haas and Simpson, 1946; Riedl, 1978), but it is restricted by some authors to cases of convergent similarity shaped by selection, or even to similar function regardless of ancestry (Owen's original definition; see Boyden, 1943).

One set of difficulties with the transformational view of homology concerns the range over which the term is to be applied. According to Riedl (1978, p. 41) the lower limit of applicability—the "homonomy limit"—is the level at which one encounters a large number of virtually indistinguishable building blocks (e.g., bone cells). At the other end of the spectrum, Riedl identifies organisms, considering it inappropriate to refer to whole organisms as homologous with one another. But these ideas are not universally accepted. Rieger and Tyler (1979) extend homology to ultrastructural features, while Patterson (1988) and Nelson (1989) would use homology in reference to organisms, and even species or "higher" taxa. There is also disagreement about whether homology should be limited to structures or extended to encompass functions and behaviors (see Hubbs, 1944; Haas and Simpson, 1946; Cracraft, 1981; Lauder, 1986, 1990).

A related difficulty concerns the distinction between homology and nonhomology. Some evolutionary taxonomists (e.g., Hecht and Edwards, 1977) have promoted the view that homology excludes convergence but includes parallelism, where the latter is interpreted as independent derivation from a condition in a single common ancestor (see also Vavilov, 1922; Mayr, 1974). Although this view is rejected by many authors (e.g., Simpson, 1961; Hennig, 1966; Holmes, 1980), some use the term "homoiology" to refer to instances of independent derivation in closely related taxa or within "narrow kinship groups" (Hennig, 1966, p. 117). However, homoiology has also been used to mean "analogies on a homologous base" (Riedl, 1978; Patterson, 1982) and has been applied more broadly, even to the independent derivation of structures such as the "fins" of fish and whales. Clearly, parallelism and convergence grade into one another if the distinction is based solely on some estimate of the relative distance of ancestors.

Perhaps the greatest tension in the literature on evolutionary homology revolves around the role of similarity. Earlier definitions explicitly refer to structural similarity, but the tendency has been to drop such references (Patterson, 1982), on the view that transformation is logically independent of similarity. After all, two structures may be effectively identical yet derived independently, or they may be derived from the same precursor but so highly transformed that they are not at all similar. Bock (1963) made the latter argument in reference to the ear ossicles of mammals (see also Cracraft, 1967; Mayr, 1969). The counterargument (e.g., Ghiselin, 1969b; Inglis, 1970) is that even though homologous structures may be very different, "without some similarity, we should not even dream of homology" (Stevens, 1984, p. 403). In fact, similarity is evident in the conditional phrase (often implicit) that specifies what exactly is being compared. Thus Patterson (1982, p. 24) maintains, in reference to Bock's example, that the stapes in mammals is similar to the hyomandibular of sharks "as that element of the gnathostome hyoid arch which articulates with the braincase."

Taxic homology. One response to difficulties with the transformational view is to equate homology with the term "synapomorphy" (Patterson, 1982), where the latter is taken to mean a derived trait characterizing a monophyletic group. For Patterson the underlying motivation is to tie homology, which he believes "exists only in the human mind" (Patterson, 1982, p. 59; Nelson, 1970), to something that exists in the world, namely, monophyletic groups. Homology becomes "the relation through which we discover" monophyly, and the result is what Patterson called the "taxic" view of homology. Much the same view is evident in earlier cladistic literature (e.g., Eldredge and Cracraft, 1980; Rieppel, 1980; Wiley, 1981; Nelson and Platnick, 1981) and has since been adopted by other systematists (e.g., Stevens, 1984; see also Rieppel, 1988).

The taxic view squares nicely with the traditional evolutionary view if monophyly is defined in terms of common ancestry, as it is by phylogenetic systematists *sensu* Hennig (1966). And when it is appreciated that homology includes both synapomorphy and symplesiomorphy (because the latter are synapomorphies that circumscribe more inclusive monophyletic groups), this view is also consistent with the observation that homologies are arranged in a hierarchy corresponding to the hierarchy of taxonomic groups (Bock, 1963, 1974, 1977; Riedl, 1978).

The main departure from the standard evolutionary view concerns the emphasis placed on identifying transformations. Evolutionary definitions focus on how attributes are related in the sense that one was transformed into another during the course of evolution. Under the taxic view this becomes irrelevant. Instead, it suffices to identify which characters mark monophyletic groups, regardless of exactly where these traits came from. Patterson illustrates this difference in analyzing controversies over the derivation of mammal ear ossicles: all that matters under the taxic view is whether the incus and malleus are in fact synapomorphies of mammals. Curiously, Patterson (1982, p. 36) concedes that "the transformational approach to homology may be more informative, and a lot more interesting, than the taxic approach." Indeed, the transformational approach forces us to try to connect characters to one another—to formulate bolder hypotheses.

A corollary of the taxic view is that nonmonophyletic groups are characterized only by nonhomologous features. This seems clear enough in the case of polyphyly, but Patterson means to include paraphyletic groups as well. Thus, although the scales in lizards and crocodiles are homologous, scales are not a homology of "reptiles" but of all amniotes, when the derived (transformed!) states of birds and mammals are also considered. It is even more tempting to think of homologies of paraphyletic groups in cases of secondary loss; for example, the loss of wings in some lines of insects (compare Patterson, 1982, and Hennig, 1966, p. 95). The taxic view contrasts in this regard with standard evolutionary definitions, which do not directly connect the status of groups and the status of character transformation series. The question, then, is whether there is any value in maintaining some terms to reflect the phylogenetic relationships of groups (mono-, para-, and polyphyly) and their characters (synapomorphy and symplesiomorphy), and other terms (homology) to reflect a relationship among character states, irrespective of group membership (Sanderson, 1989). Patterson's taxic view ties these together, as though there were only one legitimate concern. In contrast, Hennig (1966, pp. 95, 120) recognized related but separate concerns (see also de Queiroz, 1985, p. 294; Sober, 1988, p. 117).

Patterson's perspective is especially useful in focusing attention on tests of homology, which include *similarity* in position, structure, and development; *conjunction* (or coexistence) of presumed homologues in a single organism; and *congruence* with other presumed homologies. Of these, he singles out congruence as the most valuable (at least for morphological data; see Patterson, 1988). The basic idea is to hypothesize that states are homologous based initially on similarity criteria and then to test this by including the character in a cladistic analysis to see whether it actually marks a monophyletic group (see also Wiley, 1975; Rieppel, 1980; Stevens, 1984). Notice that this relegates Remane's primary criteria to a role in initial evaluation and elevates congruence (one of Remane's auxiliary principles, which was meant to apply only to simple characters) to the ultimate role in establishing homology.

Patterson's analysis also allows a rather clean distinction among different forms of homoplasy. Convergences fail both the similarity and the congruence tests (although in practice such "characters" are usually rejected before cladistic analysis). Parallelisms, however, pass the similarity test but fail the congruence test. Homonomy (repeated units within an individual organism, such as hair; Riedl, 1978) or serial homology (repeated units arranged along the body axis, such as leaves on a plant) is indicated by failure of the conjunction test. Patterson (1987, 1988) pointed out that paralogous genes (Fitch, 1970) also belong in the last category, at least when organism phylogeny is being considered (e.g., the globin gene family in vertebrates).

Development and homology. The taxic view of homology elevates the ancestry component of traditional evolutionary definitions to supreme importance, while deemphasizing the retention and transformation of characters. Another response takes the opposite tack, playing down the unique origin of traits while emphasizing mechanisms responsible for character retention and transformation, especially the developmental basis of constraint. This approach, developed to some extent in G. R. de Beer's *Homology, an Unsolved Problem* (1971b), finds perhaps its fullest expression in Rupert Riedl's *Order in Living Organisms* (1978). More recently, it has been taken up by Louise Roth (1984, 1988) and, independently, by a former student of Riedl, Gunter Wagner (1989a,b). As evidenced by the titles of their papers, both wish to formulate a so-called biological concept of homology—one that encompasses homology concepts of all sorts and accounts for the underlying observations in developmental terms. In contrast to Patterson's view, the emphasis is squarely on causes, and little attention is paid to testing hypotheses of homology.

Iterative homology, including homonomy and serial homology, has been a central concern in this literature. Although this is often not admitted as true homology (e.g., Boyden, 1943; Mayr, 1982a), which is generally said to deal with comparisons among organisms or species, proponents of the biological concept insist that it be considered alongside standard homology (Moment, 1945; Ghiselin, 1976; Van Valen, 1982; Roth, 1984, 1988; Wagner, 1989a,b). As Riedl (1978, p. 38) put it, "in the last analysis we are dealing with the same mechanism which is of the same fundamental importance . . . whether such identical individualities become separated from each other to occur in different individuals or whether they replicate within the same individual."

In her first paper on homology, Roth (1984, p. 27) explored the view that it is "based on the sharing of pathways of development which are controlled by genealogically related genes." Several conclusions follow from this. First, homology cannot be viewed as an all-or-none phenomenon (cf. "partial homology," discussed above). Instead, Roth (1984, p. 18) argues that "it is important to recognize *degrees* of homology," with the strength determined by the stage at which developmental paths diverge. She proposes, for example, that petals are more strongly homologous to leaves than sepals are. Second, she concludes that "at some level, distinguishing homology from parallelism will neither be possible or useful" (1984, p. 23), and that insisting on doing so would render the concept nonoperational because identical mutations can arise independently within interbreeding populations.

In her paper of 1988 Roth opts instead for Van Valen's (1982) definition of homology: "a correspondence between two or more characteristics of organisms that is caused by continuity of information." This, she maintains, has the virtue of being the most comprehensive, ideologically neutral, and flexible definition (although Ghiselin's [1976] definition—"parts that arise from the same source"—may be equally flexible). It also accommodates the recognition that conservatism at one level—genetic, developmental, or morphological—is not necessarily mirrored at other levels. Indeed, Roth emphasizes ways in which these levels can be dissociated, including changes in embryological source tissue, changes in inductive relationships, and what she calls "genetic piracy." The last refers to cases in which genes are "'deputized' in evolution; that is, brought in to control a previously unrelated developmental process" (Roth, 1988, p. 7).

Wagner (1989a,b) covers many of the same points in his papers, but from a somewhat different angle. For example, he emphasizes that morphological characters are not replicators, that is, that they are "not directly inherited but are built anew in each generation" (Wagner, 1989b, p. 54; see also Sattler, 1984). However, Wagner is not satisfied with Van Valen's definition of homology. As he says, "its charm as well as its weakness lies in the term 'information,' which does not imply a particular mechanism" (Wagner, 1989b, p. 60). He prefers a developmental definition, but one focused specifically on aspects of development that constrain modification: "structures from two individuals or from the same individual are homologous if they share a set of developmental constraints, caused by locally acting self-regulatory mechanisms of organ differentiation" (Wagner, 1989b, p. 62). This obviously applies to iterated structures and cases of differentiation between the sexes (e.g., penis and clitoris), but it is much more restricted than other views. It is difficult, for example, to accommodate such things as nucleotide substitutions and color variants, and it is unclear whether it applies to behavioral traits. Furthermore, Wagner's definition says nothing of ancestry, and it is compatible with phylogenetic definitions only insofar as structures that qualify as homologous under Wagner's view are also likely to be historically unique.

Wagner's treatment is especially valuable in focusing attention on the individuation of parts. As he points out, traditional comparisons have sometimes focused on structures that are not individuated (even though they appear to be), such as tarsal elements in vertebrates (Goodwin and Trainor, 1983) or the cones of mammalian teeth (Van Valen, 1982). By way of explaining the origin and maintenance of individuality Wagner develops the idea of "epigenetic traps." These include hierarchical ontogenetic networks, with the property that some characters have greater "burden" (dependent traits or downstream effects; Riedl, 1978) and therefore are likely to be more highly constrained (see also, Wimsatt, 1986; Arthur, 1988; Donoghue, 1989). Cyclical networks, wherein traits are mutually coupled through inductive relationships, can result in even greater constraint (Wagner, 1989a).

Some developmental biologists have taken a more extreme view of homology than have the authors discussed here. For example, Goodwin (1982) believes that attention to the generative dynamics underlying morphogenesis will allow a return to "the original definition of homology": "an equivalence relation on a set of structures, partitioning the set into classes whose members share certain invariant internal relationships and are transformable one into the other while preserving the invariance" (Goodwin, 1982, p. 51). In this context, "transformation" is not to be interpreted in historical terms, but instead refers to the possibility of deriving one structure from another in a formal, atemporal sense. If not for the difference in motivation, Goodwin could easily be allied with those who hold a classical or phenetic view of homology.

The history of the word homology can be interpreted as a series of responses to challenges brought on by underlying conceptual changes. The rise of evolutionary thought forced a reconsideration of the meaning of homology and related terms. In turn, the recognition that similarity and common ancestry do not necessarily coincide resulted in an important split, some biologists preferring to maintain the connection between homology and similarity and others opting to associate homology with ancestry. Those who chose similarity faced the problem of justifying the selection of similarity criteria. This led, along several paths, to the notion of partial homology. Those who pursued ancestry faced other difficulties, aside from the usual demands for an operational definition. The problem of identifying transformations suggested the possibility of shifting the focus to monophyly—Patterson's taxic view. In contrast, the indirect inheritance and iteration of morphological traits suggested a definition based on developmental mechanisms that account for the individuality and conservation of structures—the "biological" view of Roth and Wagner.

The homology problem focuses attention on issues of even greater generality. The choice of definitions is dictated to a large extent by how narrow or how broad one wishes to be. Patterson is satisfied to equate homology with synapomorphy, which has as one of its benefits the ability to specify critical tests. Roth apparently wants to avoid associating homology with any one theory, seeing virtue in a definition that is all-encompassing. Achieving consistency with every version of homology may yield a definition that is of little use to anyone, however. Wagner's definition, in contrast, may be overly specialized, as it excludes structures and events that are of considerable interest.

The choice of a definition is, at least in part, a means of forcing other scientists to pay closer attention to whatever one thinks is most important. It is hardly surprising, therefore, that morphologists such as Kaplan and Sattler favor definitions of homology that focus attention on position, structure, and development, for these are the aspects they find most interesting. Patterson's taxic view focuses attention on monophyly and away from identifying individual transformations, which he presumably hopes will result in more interest in cladistics. Roth and Wagner focus on retention and transformation almost to the exclusion of ancestry, hoping to stimulate more studies of the genetic and developmental mechanisms underlying constraint. For better or for worse, the choice of definitions helps determine whose agenda will attract the most attention. Fortunately, however, attention to the variety of legitimate concerns associated with the term homology is not entirely dependent on the choice of a definition.

INDIVIDUAL

David L. Hull

INDIVIDUALITY CAN be treated from a commonsense perspective or as a technical notion in either philosophy or science. Examples of ordinary individuals are natural objects such as rocks, islands, trees, and termites. Artifacts such as tables and chairs also belong on the list. In addition, scientists posit all sorts of peculiar individuals, from quasars and electrons to genes and genets. Philosophical individuals such as Leibniz's monads, Wittgenstein's (1923) objects, and the most infamous of all, bare particulars (Ryle, 1935), are stranger still. Here I begin with commonsense conceptions of what counts as an individual and then turn to the role of individuality in evolutionary biology.

As might be expected, there turns out to be no such thing as "the" scientific conception of "the" individual. Instead there are several different conceptions, depending on current theories. If nothing else, I hope to show the advantages of applying the techniques of conceptual analysis to detailed examples in the practice of science in contrast to the sketchy, fictional examples preferred by analytic philosophers. Although my examples are all drawn from evolutionary biology, I am committed to the view that comparable investigations in other areas of science would produce very similar results.

Although I do not discuss any of the technical notions of individuals produced by professional philosophers, I do introduce the sorts of distinctions and considerations that generations of philosophers have found important (e.g., Hobbes' ship of Theseus). The result is a combination of philosophy and science. As Julian Huxley (1912, p. 1) remarked when he turned his attention to such questions, the "idea of individuality is dealt with of necessity both by Science and by Philosophy, and in such a difficult subject it would be a mistake to reject any sources of help."

Commonsense individuals. One reason that we have not exposed the notion of individuals (or particulars) to the same intense scrutiny that we have lavished on the polar notions of classes (or kinds) is that, initially at least, ordinary individuals seem to fulfill the needs of both philosophical and scientific analyses. Socrates is, after all, an individual human being, and lead balls are clearly individual material bodies. However, as usual, appearances are deceptive. Atoms may fulfill all the traditional criteria of individuals, but electrons are very peculiar individuals indeed. Conversely, one of the most paradigmatic examples of classes or kinds, biological species, turns out to be more problematic than one might expect.

One of the most obvious instances of individuals are spatiotemporally localized material bodies that either remain unchanged through time or else undergo relatively continuous change. Of these, the two examples that are most commonly used to illustrate individuals are organisms, in particular vertebrates, and such artifacts as automobiles. Both exemplars have their weaknesses. What if one were gradually to replace the parts of an automobile as they wore out until all of them had been replaced. Would this be the same individual automobile or a new one? What if one kept all the old parts and reassembled them? Would this individual be identical to the original even though for a while it was nothing but a heap of parts? What is one took apart a car, let the parts sit for a few years, and then reassembled it? What if one gradually modified a car into a van or a truck? Would it still be the same individual? And on and on. Such examples are frustrating because they highlight problems without providing any way of resolving them other than intuitions, and intuitions are as variable as societies and people.

Use of organisms as instances of individuals is a step in the right direction, but unfortunately "organism" in such contexts is not a technical term in biology but another instance of ordinary usage. For instance, in his investigations into the notion of individuals, Strawson (1959, p. 42) claims that "*This is an animal* entails *There is some birth which is the birth of this.*" I don't know about the ordinary concept of animal, but many animals are not in any recognizable sense of the term "born." By "animal" Strawson probably means "vertebrate." Julian Huxley's discussion of individuality provides another example of the damage that our vertebrate bias can do. According to Huxley (1912, p. 9), inorganic systems form particulars but not individuals. "Cause half a mountain to be removed and cast into the sea: what remains is still a mountain, though a different one. Take away a planet, and the Solar System still works: its working is different, but, as we can see, only different, not less perfect." But as Huxley well knew, similar observations hold for many living organisms, especially those that exhibit modular organization (Jackson, Buss, and Cook, 1986). Cause half a poriferan colony to be removed and cast into the sea. What remains is still a poriferan colony. It still works. It is no less perfect. In fact, depending on the species, what remains may be two colonies, not just one. Once we direct our attention away from Vertebrata to other phyla, the notion of an organism becomes much richer and more diverse than any science fiction writer has yet to imagine.

Three traditional criteria for individuality in material bodies are retention of substance, retention of structure, and continuous existence through time (genidentity). If organisms are to count as individuals, then the first two criteria are much too restrictive. In point of fact, many organisms totally exchange their substance several times over while they retain their individuality. Others undergo massive metamorphosis as well, changing their structure markedly. If organisms are paradigm individuals, then retention of neither substance nor structure is either necessary or sufficient for continued identity in material bodies. The idea that comes closest to capturing individuality in organisms and possibly individuals as such is genidentity. As its name implies, this criterion allows for change just as long as it is sufficiently continuous. The overall organization of any entity can change but it cannot be disrupted too abruptly. To this criterion is often added the requirement that an individual have relatively discrete boundaries in space and time. As reference to "relatively" discrete boundaries and to change not being "too" abrupt and disruptive indicates, numerous borderline cases are bound to occur. Furthermore, which entities count as individuals depends on frame of reference. From one perspective, boundaries may be sharp and change continuous. From other perspectives, a sharp boundary can become fuzzy and continuous change becomes abrupt.

"Common sense" is strongly biased by our relative size, duration, and perceptual abilities. Given our human perspective, vertebrates have discrete boundaries in space and time and none of the changes that they undergo are all that abrupt or disruptive. But if we direct our attention to other organisms, common sense begins to crumble. For example, is a patch of crab grass one individual or several? Because the runners connecting the various tufts of grass are buried, we are likely to treat each little tuft as a separate individual. Because the runners in strawberry patches lie above the ground, we might be tempted to treat all the interconnected plants as a single individual. Slime molds present even more puzzling problems. At certain stages in their life cycles, they form multicellular slugs that crawl around until they find an appropriate spot, plant themselves, and turn into a treelike structure that buds off free-living cells that fan out over the substrate, only to reassemble later into another slug. If these examples do not shake one's faith in the power of common sense, an exploration of modular organisms might. In sexually reproducing organisms, growth and reproduction are easily distinguished; in asexually reproducing, modular organisms, they are not (Jackson, Buss, and Cook, 1985).

Expanding our attention from a small percentage of extremely peculiar organisms, such as vertebrates, to all organisms erodes our confidence in organisms as clear examples of individuals. If we change our frame of reference to that of an atom, organisms as well as all macroscopic objects become mainly empty space. From this perspective, clouds of atoms seem no more individuals than does the Milky Way seen from the human perspective. Perhaps distant galaxies may seem like individuals, but not our own, seen from the inside. Expanding our perspective to that of a planet might well result in species taking on the appearance of individuals as they expand and contract their ranges like giant amoebae. If common sense is wedded to the human perspective, then the commonsense distinction between individuals and crowds of individuals becomes a matter of scale and vantage point.

Perhaps spatiotemporally localized material bodies are paradigm commonsense individuals, but people seem happy to extend the idea of individuality to things such as nations, in which spatiotemporal contiguity is not required. Identity in these contexts is determined by common laws, economic relations, and the like. The United States of America did not become a new nation when Alaska and Hawaii became states. Nations can also split in two, as in civil wars, only to merge again later. As a result of conquests, treaties, and the redrawing of political boundaries, nations can go out of existence, but sometimes, the "same" nation can reappear after a period of time. In such cases, the only continuity is in the loyalties of the people involved. Are these same individuals, different individuals, or individuals at all?

Individuals in evolutionary biology. A continuing problem in philosophy is to find some principled way to distinguish between all the welter of classes that clutter our conceptual landscapes and some set of privileged classes, commonly termed natural kinds. A parallel problem exists for individuals. Out of the welter of individuals that clutter our conceptual landscapes, how are we to pick out "natural" individuals? I would suggest that "natural" individuals are those entities to which laws of nature apply. If gold is a genuine natural kind, then an atom of gold is a genuine natural individual. Ontological status is thus theory dependent. If you want to discern natural individuals, you must understand the processes in which they function, and this understanding cannot be casual, commonsense understanding. It must be detailed and professional. Casual references to genes as beads on a string will not do. Even in the context of Mendelian transmission genetics, genes are much more complicated than this. In the context of molecular biology, the complexities only multiply as we learn about introns, exons, regulatory genes, and so on (see GENE; see also Kitcher, 1982; Rosenberg, 1985).

In the middle of the nineteenth century, biologists debated the status of siphonophorans such as the Portuguese man-of-war. T. H. Huxley (1852) concluded that the large float and numerous tentacles together form a single organism, while Louis Agassiz (1857–1866) came down on the side of their being a colony of distinct organisms. Ernst Haeckel (1862–1868) tried to dissolve the dispute by distinguishing between morphological individuals (morphonts) and physiological individuals (bionts). Morphologically a Portuguese man-of-war consists of numerous organisms; physiologically it forms a single organism (see Jeuken, 1952; Winsor, 1976; Rinard, 1981; Gould, 1984a).

The trouble with Haeckel's solution to the problem of biological individuals is that morphology and physiology do not provide sufficiently well articulated theoretical contexts. Biologists have been engaged in the study of anatomy and physiology for centuries, but no "theories" of morphology and physiology have materialized in the same sense that evolutionary theory is a "theory." In order to see the dependence of individuality on theories, one must investigate more highly articulated areas such as evolutionary biology.

Which entities function in the evolutionary process? What are these functions? From the beginning the units of selection controversy has hinged on the notion of individuality (see UNIT OF SELECTION). If selection is subdivided into the two interrelated processes of replication and interaction, the problem then becomes what sorts of things are replicators and interactors? The characteristics usually attributed to replicators and interactors are those commonly possessed by individuals. For example, those who oppose group selection begin by arguing that groups do not possess the characteristics necessary to count as individuals, while those who advocate selection at higher levels of organization argue to the contrary that they do.

In his drive to break the hold that the organismic perspective has had on evolutionary biology, Richard Dawkins (1982b) argues that phenotypes are necessary for selection to occur but that these phenotypic characters need not come packaged in cohesive wholes such as organisms. They need not, but in point of fact, they usually do. One possible explanation for the prevalence of spatiotemporal individuals at all levels of organization of the living world is that individuality is inherent in selection processes, not incidental to them. If so, then understanding selection processes requires us to understand individuality. According to Dawkins (1988), one of the most fundamental distinctions in understanding selection is between those organisms that periodically go through a cellular bottleneck and those that do not. In this connection he describes two hypothetical species of seaweed that exhibit these two extremes of organization, bottle-wrack and splurge-weed. Dawkins (1988, p. 263) concludes, "In bottle-wrack, the individual plant will be a unit with genetic identity, will deserve the name individual. Plants of splurge-weed will have less genetic identity, will be less entitled to the name 'individual' than their opposite numbers in bottle-wrack."

The issue of individuality is relevant to the evolutionary process in a second way as well. Relative success (fitness at the genotypic level) is determined by counting. Regardless of the entities involved, principles of individuation have to be provided to determine which entities to count. In the context of interaction, one must be able to determine which entities are causally interacting with which. As Lloyd (1988, p. 64) argues, the "theoretical delineation and empirical detection of interactors" is "*the* central point of debate among those participating in the units of selection controversies."

On the simplest perspective of all, biological evolution is analyzed initially as changes in allelic frequencies at a single locus. More complicated phenomena must be explained by means of combinations of these minimal units. But how big of a stretch of DNA is to be considered a single allele, and must all parts of this allele be contiguous at all times? If interaction is central to selection processes and organisms are important interactors, then we need criteria for individuating them. Unfortunately, criteria that work so well for vertebrates break down dramatically for many invertebrates and virtually all plants. Janzen (1977) was not just being cute when he asked, "What are dandelions and aphids?" Instead he intended to point out that dandelion and aphid clones, even though their "parts" are not spatiotemporally contiguous, are the operative evolutionary "individuals" because they are the entities that exhibit "reproductive fitness."

For too long evolutionary theorists have concentrated on organisms in which physiological individuals (ramets) coincide with genetic individuals (genets). Issues such as the evolution of sex and senescence take on quite a different character when they are investigated in clonal organisms. As Jackson (1986, p. 5) remarks, "One of the most remarkable features of modular organisms is that iterated growth has the effect that particular genes or gene combinations are repeatedly expressed, repeatedly hazarded to a diversity of environments and selective forces." Buss (1983) in turn distinguishes between those organisms that "sequester" their germ cells and those that do not. The usual sequence in organisms that sequester their germ cells is mutation, propagation, and selection, but in clonal organisms selection can occur after mutation but before propagation. The point of these examples and others is that the sorts of entities that typically come to mind when we think of organisms are not in fact typical. Depending on how one counts, they are relatively rare. Ideally, an adequate theory of biological evolution must apply to all organisms—plants as well as animals, clonal and aclonal forms alike.

As if lower levels of organization did not afford sufficient conundrums for any general analysis of individuality, levels of organization more inclusive than organisms are even more puzzling. The existence of entities more inclusive than single organisms is central to the controversies surrounding "group selection." The sorts of groups discussed by V. C. Wynne-Edwards, Sewall Wright, and David Sloan Wilson are representative. Wynne-Edwards (1963) is interested primarily in groups that possess adaptations which characterize groups as entities in their own right, rather than their members individually. E. Vrba and N. Eldredge (1984) further investigate this strong sense of group selection, arguing that if groups are to be selected in the same sense that organisms are, then they must exhibit "emergent" group-level adaptations. Not only must these adaptations be more than simple functions of the organisms that make up the group, but also they must exhibit heritability.

For over half a century, Wright (1931, 1980) developed his shifting balance theory of evolution in which species are subdivided into partially isolated demes. On Wright's theory, diffusion from demes that have acquired superior coadaptive combinations of genes is more prevalent than from other demes. As a result, new gene complexes can be established more readily than under total panmixia. For our purposes, Wright's demes possess fewer characteristics of individuals than do Wynne-Edwards' groups. D. S. Wilson's (1980) trait groups are even less like individuals. In Wilson's model, populations are periodically subdivided into temporary subgroups that vary in their genetic composition. Interactions among the individuals within each of these ephemeral subgroups results in a differential contribution of certain subgroups to the gene pool as a whole. These various sorts of group selection raise a variety of questions, but the one that concerns us is the extent to which these groups actually count as individuals and the effects that this status has on the evolutionary process.

Thus far, I have discussed only replication and interaction, arguing that entities that have the characteristics of individuals most highly developed function best as either replicators or interactors. I have not discussed the *results* of selection processes: the evolution of lineages. One of the most controversial topics in evolutionary biology during the past couple of decades is species selection, the suggestion that possibly entire species are selected in the same sense that organisms are, that is, that they function as interactors (Eldredge and Gould, 1972; Stanley, 1975). But regardless of whether species can function *in* selection processes, they certainly result *from* them.

Because of the heritability requirement, selection processes are going to produce lineages, entities connected by the ancestor-descendant relationship. Traditionally these entities have been considered species. Species may lack other characteristics of individuals, but at least they are spatiotemporally restricted. Because spatiotemporal restrictedness and unrestrictedness is so central to our understanding of nature, this conclusion is not trivial. Numerically the same species cannot pop up here and there at different times and places. In early versions of Charles Darwin's theory of evolution, he treated species and organisms as if they were the same sort of thing, replete with life cycles, but by the time he published *On the Origin of Species,* he had abandoned the notion of species having life cycles (Hodge, 1983). Even so he continued to insist on genealogical connections. Darwin found the genealogical character of species so important that he was prepared to end his Big Book with the following quotation:

> Hence I conclude that it has not as yet been absolutely proved that the same species has ever appeared, independently of migration, on two separate points of the earth's surface: if this were proved or rendered highly probable, the whole of this volume would be useless, & we should be compelled to admit the truth of the common view of actual creation; & that organic beings are not exclusively produced by ordinary generation, with or without modification. (in Stauffer, 1975, p. 566)

Darwin is exaggerating. Although admitting the independent appearance of the same species more than once would have required him to rework the conceptual foundations of his theory extensively, he is unlikely to have abandoned it entirely. The quotation, however, belongs on a very short list of phenomena that Darwin considered to threaten the entire fabric of his theory. Although the notion of individuality has not played a prominent role in the literature of evolutionary biology, it has lain just below the surface, rising to the level of explicit discussion only rarely. If nothing else, the discussion here has surely shown that analyses of such metaphysical notions as individuality conducted in a specific scientific context are vastly more productive than those that take ordinary usage as their subject matter.

Lamarckism

Peter J. Bowler

THE TERM "Lamarckism" conventionally refers to the evolutionary mechanism of the inheritance of acquired characteristics, popularized through the influence of the French biologist Jean Baptiste de Lamarck (1744–1829). In fact, a number of distinct biological processes have been included within the general framework of Lamarckism, although all depend upon the assumption that characters acquired by the adult organism can somehow be transmitted to the offspring and can ultimately become incorporated into the species' hereditary constitution. All of these processes are now rejected as incompatible with the central dogma of molecular biology, which forbids the transferral of information from the somatic tissue to the DNA of the reproductive cells. Before the emergence of genetics, however, Lamarckism was popular with many biologists and was for a time seen as a major rival to Darwinism. The popularity of Lamarckism arose largely from the fact that it was a natural extension of a pregenetical way of looking at reproduction, in which growth and inheritance were seen as aspects of a single integrated phenomenon. The separation of growth and development from transmission by modern genetics has robbed Lamarckism of its basic plausibility, and efforts to "update" the theory have not been successful. Yet support for the theory has erupted from time to time even in the twentieth century, mainly because it has become associated with a particular philosophical position centered on rejection of Darwinian materialism and opposition to conservative social policies. But history reveals that some Lamarckians have been materialists; others have adopted very conservative positions. The theory comes with no guarantee of ideological purity.

Popular sentiment has always inclined to the belief that acquired characters such as the strong arms developed by a blacksmith might have some tendency to be inherited. Many naturalists built this folk belief into their theories before Lamarck's name became associated with the process (Zirkle, 1946). But it was Lamarck who first incorporated the idea into a systematic theory explaining the development of life on earth (Burkhardt, 1977; Hodge, 1971; Mayr, 1972a; Jordanova, 1984). His best-known work was the *Philosophie zoologique* of 1809 (trans. Lamarck, 1914). The inheritance of acquired characters was not the chief mechanism of development in this theory. Lamarck postulated an automatic progressive trend that would produce a linear sequence of animal forms (and a parallel sequence for the plants). The inheritance of acquired characters was only introduced as an adaptive mechanism that would distort the linear progressive trend. Lamarck insisted that the process worked differently for plants and animals. Plants experienced a direct effect when placed in a new environment; they simply grew different structures adapted to the new conditions. These structures would be passed on to the next generation and would eventually become fixed in the species. Lamarck insisted that animals only experience an *indirect* effect of changed conditions. When placed in a new environment, the individual animals feel new needs and adopt new habits; they use their bodies in a different way, some organs now being used more than before, others less. The more active organs naturally increase in size; the others waste away. Lamarck insisted that these individual effects were to some extent inherited by the offspring and over many generations would accumulate to produce a permanent change in the species. A classic example is the neck of the giraffe, which grew as a result of the new habit of feeding from trees (1914, p. 122). In the introduction to his *Histoire naturelle des animaux sans vertèbres* (1815–1822) Lamarck defended the same theory, but put less emphasis on the existence of an autonomous progressive trend.

It has traditionally been assumed that, thanks to the efforts of his great rival, Georges Cuvier, Lamarck's theory was ignored. But we now know that more radical naturalists did take him seriously (Appel, 1987; Corsi, 1988). Even in Britain, the radical Scots anatomist Robert Grant supported Lamarck—although he was gradually marginalized as a result (Desmond, 1984, 1989). Charles Darwin's earliest evolutionary speculations were based on the inheritance of acquired characters, although his experiences with animal breeders soon persuaded him that variation was random rather than directed, and he went on to develop the theory of natural selection. Darwin never denied the possibility that the inheritance of acquired characters might act alongside natural selection in the evolutionary process. After the publication of *On the Origin of Species* many so-called Darwinists, including Herbert Spencer and Ernst Haeckel, placed great emphasis on the inheritance of acquired characters. Some biologists became convinced that the effect was more important than selection; they founded a "neo-Lamarckian" movement that was particularly influential during the period of Darwinism's eclipse in the last decades of the nineteenth century (Bowler, 1983).

A number of paleontologists supported Lamarckism because they thought the fossil record revealed linear evolutionary trends so regular that they could not have been produced by the selection of random variation. The American school of neo-Lamarckism, led by Edward Drinker Cope and Alpheus Hyatt, supposed that new habits defined trends of specialization that would affect the whole future evolution of the species. They saw evolution as a process in which the growth of the individual was gradually and systematically extended in a particular direction. As a consequence they—along with Haeckel—were leading supporters of the claim that ontogeny recapitulates phylogeny (Gould, 1977a). Hyatt also supposed that growth could be directed along nonadaptive lines, leading to racial senility and extinction. For many paleontologists, Lamarckism was thus associated strongly with orthogenesis: they believed that the direction of evolution was determined by both adaptive and nonadaptive modifications to the process of individual growth.

At the same time many naturalists thought they could provide both experimental and indirect evidence that modern species were being modified through the inheritance of acquired characters. In the United States, Alpheus Packard (1888) explained the origin of blind cave animals by invoking the inherited effects of disuse. Other studies involved what Lamarck had called the direct modification of organisms through exposure to new conditions (for details, see Bowler, 1983). There were some attempts to demonstrate the inheritance of mutilations. Immense controversy surrounded the efforts of Paul Kammerer (e.g., his "midwife toad" experiments) to demonstrate the inheritance of characters acquired by animals forced to adopt new habits (Kammerer, 1924; Koestler, 1971).

Much of the plausibility of Lamarckism rested on the assumption that inheritance and growth were inextricably linked. Even Darwin had not challenged this view: his theory of pangenesis was a classic product of the old developmental viewpoint (Hodge, 1985). Pangenesis typified an approach to inheritance in which it seemed only natural to suppose that acquired characters were inherited. The parent's body actually manufactured the material responsible for transmitting the information of heredity to its offspring. Heredity became, in effect, a process that *remembered* the pattern of parental growth, and evolution occurred when the memory-pattern was extended to include new characters acquired by the parents. It was widely believed that the supposed fact of embryonic recapitulation proved Lamarckism to be the chief mechanism of evolution. This way of thinking only began to break down with the emergence of the hereditarian attitude that was to become typical of Mendelian genetics (Bowler, 1989).

Once characters were visualized as units transmitted independently of growth, it became difficult to see how newly acquired modifications could be "remembered"; evolution came to be seen as the accumulation of mutations that were produced independently of the growth process.

The first step in this conceptual revolution took place when August Weismann developed his concept of the "germ plasm" during the 1880s. Anticipating the central dogma of molecular biology, Weismann declared that the germ plasm, the material of heredity contained in the chromosomes, was totally isolated from what happened in the rest of the body. In the short term, Weismann's dogmatism backfired and led to a temporary increase in support for Lamarckism among the biologists he had alienated. But once Weismann's concept of the role played by the germ plasm was incorporated into Mendelian genetics, the days of Lamarckism were numbered. By the time Kammerer performed his experiments with the midwife toad in the early years of the twentieth century, Lamarckism was already being dismissed as incompatible with the new understanding of heredity. When an apparent fraud was detected in his experiments, Kammerer committed suicide and, at least in Britain and the United States, geneticists succeeded in purging Lamarckism from scientific biology. In France and Germany, however, biologists continued to adopt a more flexible view of heredity in which there remained some possibility of Lamarckian effects (Sapp, 1987). More seriously, the growing influence of T. D. Lysenko in the Soviet Union allowed a form of Lamarckism to displace genetics altogether for several decades (Medvedev, 1969; Joravsky, 1970; Roll-Hansen, 1985).

Even in the English-speaking world, many paleontologists continued to give credence to Lamarckism well into the 1930s. In 1927 the psychologist William MacDougall reported experiments that were supposed to demonstrate the inheritance of learned habits in rats. Experiments were also performed in an effort to show that some chemicals could "damage" the genes (e.g., Harrison and Garrett, 1925–26)—although this was a far cry from the positive effect that the old Lamarckians had postulated. In more recent times, the most active effort to revive a form of Lamarckism has been made by Ted Steele (1979). Significantly, Steele proposed a mechanism that was compatible with the central dogma. Accepting that a selection process took place within the immune system to produce cells capable of dealing with toxins, he supposed that the resulting genetic information was transmitted to the DNA of the germ cells through infection by viruses. Although geneticists admitted that this effect could work in principle, Steele's results on the inheritance of immunity have not been replicated and his views are now largely dismissed.

If Lamarckism has been abandoned by science, the outside world retains an interest in the theory because of its alleged philosophical and ideological implications. Literary figures from Samuel Butler to George Bernard Shaw and, more recently, Arthur Koestler have tried to convince us that Lamarckism offers a more humane alternative to the depressing materialism of the Darwinian theory. Butler's *Evolution Old and New* (1879) launched a personal attack on Darwin and hailed the inheritance of acquired characters as a mechanism that would allow us to believe that living things were something more than puppets in the hands of material forces. If Lamarckism were a valid evolutionary mechanism, it would be possible to suppose that evolution was directed into new channels by the voluntary choice of new habits made by animals when confronted by new conditions. Mind thus played an active role in evolution and could be seen as an expression of the Creator's power built into material nature. Although ostracized by the Darwinians, Butler's views were shared by American neo-Lamarckians such as Cope and played a significant role in encouraging support for Lamarckism in the late nineteenth century. Similar views were expressed in the Preface to Shaw's *Back to Methuselah* (1924) and in the writings of Koestler (1967, 1971). In the early years of the twentieth century the Darwinians tried to defuse this argument against their theory by claiming that natural selection could also be directed by consciously chosen habits. This was the so-called Baldwin effect, also known as "organic selection" (Baldwin, 1902).

Many Lamarckians supposed that their theory provided a better foundation than Darwinism for the belief that evolution was necessarily progressive. Because mind was so important for allowing animals to adjust to changes in their environment, the inheritance of acquired characters would naturally promote the development of the mental powers in every generation. Lamarckism was also given an ideological dimension by writers who insisted that it offered the prospect of improving the human race through social reform (Stocking, 1962). Whereas Darwinism promoted the view that the unfit must be eliminated, Lamarckism suggested that improved conditions would produce better people—and that the improvements would become built into the future constitution of the human race. This view was adopted by Kammerer (1924) and was still capable of attracting considerable attention in the popular press. The Lysenko affair also illustrates the possibility of constructing a link between anti-Darwinian, anti-hereditarian biological theories and socialism.

Lamarckism has no intrinsic links to these philosophical and ideological positions, however. Lamarckians have been able to use their theory to justify a whole range of beliefs. In the early nineteenth century, Lamarckism was seen as a component of a materialistic way of thought that was challenging the traditional view of the human soul. A little later on Herbert Spencer built the inheritance of acquired characters firmly into his evolutionary philosophy, thus linking it with his support for laissez-faire capitalism as the force that would drive social progress. Although Spencer has been called a "social Darwinist," his support for the free-enterprise system was in fact based on the view that struggle was the best way of stimulating all individuals to improve themselves and hence to improve the race (Bowler, 1988). This was a Lamarckian view of social development, which Spencer backed up by his support for the theory in biology. Spencer and the socialists merely differed over the nature of the force that would guide people toward better habits, Spencer preferring natural pressures while the reformers opted for state control. Spencer was also hostile to the almost idealist philosophy of "mind in nature" that Butler used to support Lamarckism. It is also obvious that Lysenko and his Marxist followers would not have followed Butler in this respect.

The Lamarckians of the late nineteenth century were deeply involved in the construction of scientific justifications for racism. Both Haeckel and the American neo-Lamarckians used the link between Lamarckism and the recapitulation theory to support the view that African races were immature or underdeveloped forms of humanity. Lamarckians such as Spencer also used the inheritance of acquired characters to argue that races that had evolved in the less stimulating climate of the tropics would not have been driven so far up the ladder of progress (Bowler, 1986). One of the last British Lamarckians was the embryologist E. W. MacBride, who attempted to defend Kammerer against the geneticists during the "case of the midwife toad." Koestler (1971) called MacBride the "Irishman with a heart of gold"—yet MacBride was a vicious racist who wrote openly in support of the German Nazis and argued that the Irish component of the British population should be eliminated by compulsory sterilization (Bowler, 1984a). That a humane social thinker such as Kammerer and a racist such as MacBride could find themselves on the same side in the defense of Lamarckism illustrates the impossibility of constructing a direct conceptual link between the idea of the inheritance of acquired characters and any particular philosophy of human nature.

Macromutation

Michael R. Dietrich

Macromutations constitute a special subclass of mutations usually invoked in discussions pertaining to macroevolution and speciation. In general, the term "macromutation" refers to a heritable change in the genetic material that produces a large effect usually associated with speciation. The size of a mutation, according to Ernst Mayr, can refer to effects on three levels: "the level of the gene (amount of reorganization of the DNA), the level of the phenotype, or the level of the resulting fitness" (Mayr, 1960, p. 356; for a brief history of the concept of mutation, see Mayr, 1941, pp. 65–67; Dobzhansky, 1951, pp. 25–31). If macromutations are simply thought of as "big" mutations on any one of Mayr's levels, then the difference between macromutations and micromutations might be thought of as simply a matter of degree. "Macromutations," however, are also thought of as producing sudden speciation events or saltations. If macromutations can produce speciation and micromutations cannot, then the differences between macromutations and micromutations would also be a matter of kind.

Historically, the various arguments for the existence and importance of macromutations came in three waves. The first "macromutationists" were actually "mutationists" rallying around William Bateson's theory of discontinuous variation (1894) and, more important, Hugo de Vries' mutation theory (1910). The second wave of macromutationists reached their acme in the 1940s with the publication of *The Material Basis of Evolution* by Richard Goldschmidt. Goldschmidt's advocacy of the evolutionary importance of macromutations was notably shared by the plant biogeographer J. C. Willis (1922) and the paleontologist Otto Schindewolf (1936, 1950). The third wave of macromutationists includes a wide range of contemporary scientists such as Guy Bush, Richard Dawkins, T. H. Frazzetta, Stephen Jay Gould, Thomas Kauffman, Rudolf Raff, and Steven Stanley. Even though these contemporary advocates of macromutation may be the intellectual heirs of Richard Goldschmidt in some ways, their concepts and advocacy of macromutations diverge from Goldschmidt's and from each other's.

Goldschmidt may have been the first to use the term "macromutation," but the concept of a macromutation as a speciation event is usually traced back to the mutation theory of de Vries. De Vries' mutations were heritable, discontinuous changes in the elementary characters (pangenes) of an organism. Because he understood the difference between real species to be that of a difference in elementary characters, mutations (production of a new elementary character or unit) by definition produced new species. In his words, if one assumes that the "attributes of organisms are composed of definite units that are sharply distinct from each other [then] the occurrence of a new unit signifies a mutation" (de Vries, 1910, I, v). Furthermore, "each new unit, forming a fresh step in this process [of mutation], sharply and completely separates the new form as an independent species from that from which it sprang. The new species appears all at once; it originates from the parent species without visible preparation, and without any obvious series of transitional forms" (I, 3). De Vries developed his mutation theory further by distinguishing between mutations of different kinds and fluctuating variations of phenotypic characters, but his central thesis remained that mutations are changes in and among material units that produce new forms or species (as de Vries defined them) in a single discernible step (see Allen, 1969). So, although de Vriesian mutations were thought to have some material basis, they were recognized and defined in terms of their phenotypic and evolutionary effects. This advocacy of mutation as a cause of sudden speciation is what makes de Vries a macromutationist from a current perspective.

De Vries' concept of mutation, however, is recognizable as a concept of macromutation only in retrospect. De Vries is labeled a macromutationist in light of the reinterpretation of his mutation theory by later genetic theorists, such as Thomas Hunt Morgan. Morgan's conception of mutations (1932, chaps. 1, 6) distinguished between mutations that were in fact changes in chromosome number and gene or point mutations. A gene mutation or a point mutation was a "change in only a single element in a chromosome" (p. 32). Morgan made this distinction because, by 1932, de Vries' observed mutations in *Oenothera* were known to be the result of changes in chromosome number. Morgan's conception, therefore, included de Vriesian mutations, but his emphasis was on genes and point mutations. (Steven Stanley [1981, p. 66] includes Morgan as a macromutationist, but Morgan's early enthusiasm [1908] for de Vries' mutation theory was later [1912] replaced by his advocacy of both large- and small-scale mutations without the sudden creation of new species [Allen 1968, p. 131].) As de Vries' theory lost supporters, Morgan's theory of the gene and, with it, the concept of mutation as point mutations gained support (Allen, 1968, pp. 130–132). Only after Goldschmidt popularized the distinctions between macroevolution and microevolution and between macromutation and micromutation, however, did de Vriesian mutation get relabeled as a form of macromutation and point mutation come to be seen as a form of micromutation (Goldschmidt, 1940, p. 8).

According to Goldschmidt, a mutation acts as a macromutational agent when it produces "suddenly, a huge effect upon a series of developmental processes leading at once to a new and stable form, widely diverging from the former" (Goldschmidt, 1952, p. 96). Goldschmidt thought that macromutational events of this sort were necessary in order for there to be evolution at and above the species level; slow accumulation of micromutations (i.e., point mutations) could produce variation within a species but never a new species. But in order to understand Goldschmidt's theory of mutations, his own somewhat radical theory of genetics needs to be taken into account.

In 1938 Goldschmidt called for the rejection of the concept of the particulate gene (Goldschmidt, 1938a, pp. 309–316). In its place he offered a theory that treated the chromosome or large parts of it as a unit. Instead of thinking of genes on a chromosome as "beads on a string," Goldschmidt suggested that the chromosome be thought of as a long chain molecule involved in a number of interrelated reactions. The chromosome was therefore tied into a coordinated system of reactions such that a change or changes in the molecular chain would alter the reaction system. Not all changes in the system of reactions would have phenotypic effects; only those changes that produced reactions strong enough to exceed a threshold would be expressed. A phenotypically expressed change that was based on an alteration in the chromosome and its reaction system constituted a mutation. As Goldschmidt noted in 1946, this model of chromosome action did not prevent localizable changes in chromosome structure; sections of chromosome of any size could be active units at different times (1946, pp. 250–253). Gone, however, were qualitative changes in the genetic material at specific points on the chromosome (point mutations); all mutations were instead understood as rearrangements or repatternings of different sizes (p. 253). Just as the particulate gene was replaced by the chromosome with its associated system of reactions, mutations as changes in the material constituents of genes were replaced by mutations as repatternings of the molecule chain that made up the chromosome.

This model of the chromosome as genetic unit lay behind Goldschmidt's elaboration of his theory of evolution in *The Material Basis of Evolution* (1940). His principle thesis is that microevolution (i.e., evolution below the species level) and macroevolution are distinct phenomena, and that the slow and gradual accumulation of micromutations characteristic of microevolution is not a sufficient mechanism to bridge the gaps between species. What Goldschmidt claimed was needed was a mechanism for bridging these gaps: a mechanism for macromutation. The answer for Goldschmidt was the complete repatterning of the chromosome. In his words: "A complete repatterning might produce a new chemical system which as such, i.e., as a unit, has a definite and completely divergent action upon development, an action which can be conceived as surpassing the combined actions of numerous individual changes by establishing a completely new chemical system" (1940, p. 203). The production of a completely new chemical system or reaction system by repatterning could bridge the gap and form a new species; such a repatterning was called by Goldschmidt a *systemic mutation*.

A systemic mutation consists of a change in the pattern (serial order) within a chromosome that leads to a new species (Goldschmidt 1940, p. 206). According to Goldschmidt, "A complicated change of intrachromosomal pattern may occur instantaneously or in a few consecutive steps, and, if it leads at all to a stable condition, may at once produce the new reaction system, the species" (p. 244). But some chromosomal repatternings, such as inversions and translocations, were known to occur without noticeable effect. In light of this evidence, Goldschmidt allows that systemic mutations may produce new patterns that are not above a "threshold of pattern action": "Only when by chance a pattern, viable in homozygous condition and above the threshold, has been reached; i.e., such as the patterns actually found when comparing species, does the new system of reaction suddenly emerge, though prepared by subliminal steps" (p. 246). The qualifications in Goldschmidt's statement are important. Systemic mutations had to affect complete developmental processes in such a way that a new system of coordinated reactions could be reached rather suddenly, if they were to function as macromutational events. Moreover, systemic mutations would have to occur in small populations and they would have to be preadapted to available niches, if they were to survive in a population.

In his first article concerning macroevolution (1933), using Morgan's classical theory of the gene and its concept of point mutations, Goldschmidt advocated "hopeful monsters" resulting from mutations in genes controlling development. But by 1940 the classical theory of the gene was replaced by a chromosomal model, and mutation was understood as chromosomal repatterning with resulting changes in coordinated reactions. Systemic mutations gave Goldschmidt a hypothetical mechanism for speciation that agreed with his theory of genetics and allowed for radical changes in developmental patterns that reflected their potentialities. Goldschmidt's detailed treatment of the potentialities of development and of mutations with large effects on development in *The Material Basis of*

Evolution is meant to provide plausibility for systemic mutations. He writes, "I think that only the linking of such an idea [the idea of evolution in large steps based on changes in embryonic development] with the facts of genetics and physiological genetics can raise it from the status of a hint to that of a theory" (1940, p. 310). Goldschmidt's hopeful monsters, the results of mutations in genes affecting development, were just such a hint that was later developed, using his theory of chromosomal repatterning, into the formation of new species by systemic mutations. *The Material Basis of Evolution* and later work (1952) indicate that Goldschmidt thought that simple macromutations depending on genes controlling development were possible, but unlike systemic mutations, they lacked an explanation of how the genetic material functioned. It was thus Goldschmidt's concern with gene function and physiology that prevented him from advocating the existence of particulate genes or macromutations based on them. Most contemporary macromutationists, however, are fonder of Goldschmidt's hints than of his more fully developed theory.

In correspondence with Goldschmidt in 1949, Sewall Wright goes far in helping to sort out Goldschmidt's theory of mutations. Wright distinguished three major classes of mutation: "(1) mutations (such as your 'chromosome repatterning') with more or less complete isolating effects (2) normally mendelizing mutations with major effects on characters (e.g. homeotic mutations [mutational substitution of one tissue type for another]) but no direct isolating effects and (3) your micromutations (contributions to quantitative variability, modifiers, isoalleles, etc.)" (1949b, p. 1). Wright points out that Goldschmidt tends "to group together classes (1) and (2) as macromutations (giving rise to higher categories) in opposition to (3) micromutation" (p. 2). Wright accepted all three classes of mutation, but thought that micromutations ought to include mutations with both major and minor effects on phenotypic characters, classes (2) and (3), without isolating effects. Mutations with isolating effects (class 1 mutations) were macromutations and were important for either "creating an isolation which permits character differentiation later, or, more often, in clinching a species difference that has already arisen" (p. 2). What Wright picked out as crucially important for macromutations were their isolating effects. Goldschmidt too thought these were crucial (without such effects macromutations could not produce a new species), but he also wanted to argue for a specific genetic mechanism (repatterning) and for the idea that the mutations with large phenotypic effects were probably the kind of mutation usually involved in speciation, especially rapid speciation.

This apparent conflation of mutations with isolating effects and mutations with large phenotypic effects was furthered in 1951 by G. S. Carter's redefinition of macro and micromutations in terms of the degree of their phenotypic effect. According to Carter, macromutations were the "large, easily recognizable mutations of genetic experiments" (1951, p. 95) and micromutations were "mutations of smaller phenotypic effect due to changes in modifying genes and to position effects" (p. 131). Carter admits that new species can occasionally form by "sudden evolution of genetic isolation . . . as the result of macromutation or chromosome re-arrangement" (p. 235) and that these sudden changes require macromutations (p. 268), but throughout his work *Animal Evolution,* macromutations are strongly associated with large phenotypic effects and only weakly associated with speciation. As G. G. Simpson notes in *The Major Features of Evolution,* Carter's use of the terms "macromutation" and "micromutation" to represent different degrees of phenotypic effects confuses the terminology, because Goldschmidt meant for the terms to represent a difference in kind on a cytological and evolutionary level (Simpson, 1953, p. 92). Simpson was careful to distinguish between phenotypic effects, on the one hand, and, on the other, cytological (genotypic) differences underlying different mutations and the potential for producing genetic isolation from changes in chromosome number. The only real resemblance Carter's macromutations bear to Goldschmidt's is that what each calls macromutation provides the only means for sudden genetic isolation and speciation.

The concept of macromutation most commonly appearing in contemporary literature is at least in part built on Goldschmidt's concept of macromutations and their role in developmental systems. Occasionally, macromutations are associated with the theory of punctuated equilibrium. Stephen Jay Gould advocates both punctuated equilibrium and macromutation, but notes emphatically that "punctuated equilibrium is not a theory of macromutation" (1982a, p. 88). Gould's advocacy of macromutation is limited to what he calls "legitimate" forms of macromutation. He writes: "Illegitimate forms of macromutation include the sudden origin of new species with all their multifarious adaptations *ab initio,* and origin by drastic and sudden reorganization of entire genomes. Legitimate forms include saltatory origin of key features (around which subsequent adaptations may be molded) and marked phenotypic shifts caused by small genetic changes that affect rates of development in early ontogeny with cascading effects thereafter" (pp. 88–89).

Four kinds of macromutations are listed by Gould in this passage. The first type are fully adapted macromutations, where a new fully adapted species arises from a single mutational step. The second type are reorganizing macromutations, which are distinguished by their genetic mechanism of rapid reorganization of the genome. The third type are partly adapted macromutations, where new key features arise in a single mutational step but are not perfectly adapted. The fourth type are developmental macromutations, based on changes in genes controlling early development. Gould's underlying interests are admittedly in the developmental potentialities and the change in phenotypes that can result, so it is no surprise that he favors partly adapted and developmental macromutations (1982a, p. 89). Gould does not speak of his legitimate macromutations as necessarily leading to new species, however. Indeed he distinguishes between discussions of "morphological shifts (legitimate macromutation), speciation (various theories for rapid attainment of reproductive isolation), and general morphological pattern in geological time (punctuated equilibrium)" (1982a, p. 90). These different levels of discussion, he writes, are "not logically interrelated" (p. 90). Fully adapted macromutations are the only form of macromutation that he explicitly links with speciation, and they are "illegitimate." It seems, therefore, that Gould wants to define macromutations not in terms of isolating effects and speciation but in terms of morphological shifts.

Richard Dawkins also distinguishes between conceivable and inconceivable forms of macromutation. The sudden origin of a new complex feature requires a sudden increase in new information that Dawkins finds inconceivable (1983, pp. 414–415). On these grounds he would probably find Gould's fully and partly adapted macromutations objectionable. Dawkins and Gould would agree, however, that developmental macromutations are conceivable and legitimate, because "legitimate" macromutations only require extensions of current systems; they are morphological shifts. Homeotic mutants (mutational substitution of one tissue type for another) are a good case of "legitimate" macromutations because there is a large morphological shift, but there is not a large increase in information—the wing growing where halteres normally grow in *Drosophila* is not a new kind of wing, a new piece of information, or a complex new feature; it is a regular wing in a new location. Like Gould, Dawkins focuses on the origin of complexity, not necessarily speciation; both focus on morphological shifts and developmental opportunities and constraints, while rejecting Goldschmidt's emphasis on the genetic mechanism of repatterning and on the exclusive role of macromutations in speciation.

Goldschmidt's emphasis on repatterning and large-scale genetic changes has not been lost on all contemporary scientists. Darryl Reanney has suggested a reexamination of their role in mutation. He distinguishes between (1) entropic mutations, i.e., random point mutations; (2) reciprocal recombination, such as crossing over; and (3) quantal recombination, such as sequence inversion, duplications, and translocations (1977, p. 249). According to Reanney, quantal recombinations are the "prime source of usable variation in evolution" (1976, p. 556). These kinds of changes requiring recombinational enzyme systems are taken to provide a powerful adaptive mode for the creation of genetic novelty. However, Reanney does not explicitly call quantal recombinations macromutations and he does not link them with speciation events. Scientists such as M. J. D. White, Hampton Carson, and Guy Bush do, however, link chromosomal repatternings with speciation in their theories of chromosomal or stasipatric speciation (see White, 1978; Atchley and Woodruff, 1981).

Guy Bush's version of stasipatric speciation is speciation without complete geographic isolation, initiated by chromosome rearrangements that establish a degree of reproductive isolation and initiate novel adaptive change. According to Bush, a macromutation is "any mutation that results in a major adaptive shift, a change in the way of life that opens a new adaptive zone" (1982, p. 125). Bush's macromutations would not necessarily be expressed as visible, large morphological changes associated with some other forms of macromutation. Indeed many large morphological changes, such as the effects of homeotic mutants in *Drosophila,* would not qualify as macromutations for Bush unless they produce a significant adaptive shift and "enhance the prospects for the development of reproductive isolation" (p. 125). Bush's macromutations, then, are defined only in terms of the large increase in fitness that they produce, although he clearly leans toward chromosomal rearrangements as the genetic basis for speciation.

In general, macromutations have traditionally been thought of as heritable changes that directly lead to or cause rapid speciation; they have some isolating effect. Carter, Gould, and Dawkins diverge from this general rule in emphasizing morphological shifts and minimizing the connection of isolating effects and speciation with macromutations. Gould and Dawkins concur with Goldschmidt's interests in developmental potentialities in relation to macromutations; Bush and others share more of Goldschmidt's interest in underlying genetic mechanisms for producing genetic isolation. The end result is a lack of consensus concerning the connection of macromutations to speciation events and its relative importance as a possible mode of speciation.

MONOPHYLY

Elliott Sober

A COLLECTION OF species is said to form a monophyletic group when the species in it are related to each other in certain ways. Usage is far from uniform regarding what those relationships must be, however. And the variation in usage is not at all random, in that evolutionary taxonomists such as Mayr (1969), Simpson (1961), and Ashlock (1979) endorse one usage, whereas cladists such as Eldredge and Cracraft (1980), Nelson and Platnick (1981), and Wiley (1981) follow Hennig (1966) in endorsing another.

Ernst Mayr (1969, p. 75) describes the cladistic usage as "completely contradicted by common sense" and "in opposition to the phenomenon of evolutionary divergence." Cladists have reciprocated, claiming that the evolutionary taxonomists' idea fails to pick out real units in the phylogenetic hierarchy. Mayr (1982a, p. 228) also complains that Hennigians expropriated a word whose meaning had been stable since Ernst Haeckel, on which they then imposed their own idiosyncratic meaning.

The cladistic usage, though of more recent vintage, is simpler to understand. Consider a branching process in which earlier objects give rise to one or more later objects, but in which an object has precisely one immediate ancestor. Species that are formed by the splitting of earlier species count as an example, as do asexual organisms that reproduce by fission. Species that form by hybridizing and organisms that reproduce sexually do not generate a branching process in this sense. Rather, they are said to produce a *reticulate* genealogical pattern.

A diagram of a branching process might have time moving from the bottom of the page to the top, with nodes representing objects and lines between nodes representing the ancestor/descendant relationship. The branches form a tree in which nodes split but never join. In Figure 1, we see the genealogical relationship produced by a single species (Species 1)

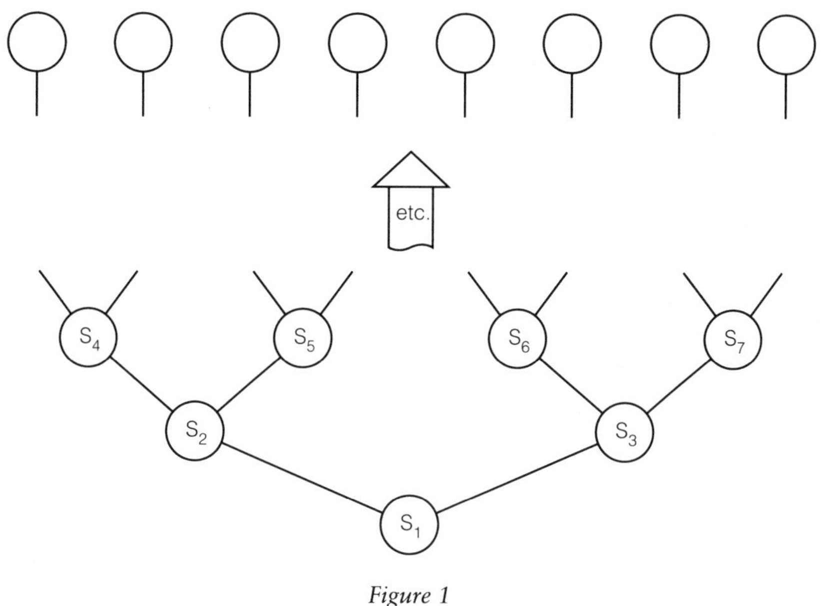

Figure 1

that gives rise to two descendants, which each produce two more, and so on.

Some collections of these species form monophyletic groups; others do not. Willi Hennig (1966, p. 73) formulates the cladistic concept of monophyly as follows: "a monophyletic group is a group of species descended from a single ('stem') species, and which includes all species descended from this stem species." Hennig's idea is easily illustrated by means of the "cut method" (Sober, 1988). Draw a line through any branch in Figure 1. All species above that cut comprise a monophyletic group.

Several consequences of Hennig's definition are worth noting. First, the complement of a monophyletic group will not itself be monophyletic. Cladistic philosophy maintains that the only "real" taxonomic units are the monophyletic ones; this entails that complements of monophyletic groups should not be given systematic recognition. This policy has led to fundamental revisions in traditional systematic practice. For example, such categories as the "Acrania, Agnatha, Pisces, perhaps the Amphibia, and certainly the Reptilia are groups of unrelated taxa," and so should not be accorded taxonomic status (Eldredge and Cracraft, 1980, p. 164).

Second, to ask whether two or more species belong to a monophyletic group is to pose an incomplete question. On the one hand, in a fairly trivial sense, any two terrestrial species belong to some monophyletic group or other, because all terrestrial life is related. On the other hand, if we relativize our question about the two species to some third species (an "outgroup"), the question becomes more interesting. Human beings and dogs belong to a monophyletic group to which trout do not belong. However, it is false that human beings and dogs belong to a monophyletic group that excludes chimpanzees. Note that in saying that human beings and dogs *belong to* a monophyletic group, one is not saying that there is a monophyletic group *composed just of* human beings and dogs. In taxonomic practice, questions about monophyly concern, not two species, but at least three: given three species, which two of them belong to a monophyletic group that fails to include the third?

The third point is that the cladistic requirement of monophyly applies only to superspecific taxa, not to species themselves. Species can be ancestors of other species. This means that the cut method cannot isolate an ancestor species from its descendants. Monophyly is a way of describing *sets* of nodes in a branching process. The nature of those nodes must be specified by other means (Sober, 1988).

Fourth, notice that the branching structure depicted in Figure 1, together with the cladistic definition of monophyly, completely settles whether a collection of species forms a monophyletic group. This point is important, because it is not a feature possessed by the concept of monophyly deployed by evolutionary taxonomists.

To illustrate how the usage of evolutionary taxonomists differs from that of cladists, consider the relationship of lizards, crocodiles, and birds, illustrated in Figure 2. Notice that cladistic usage implies that lizards and

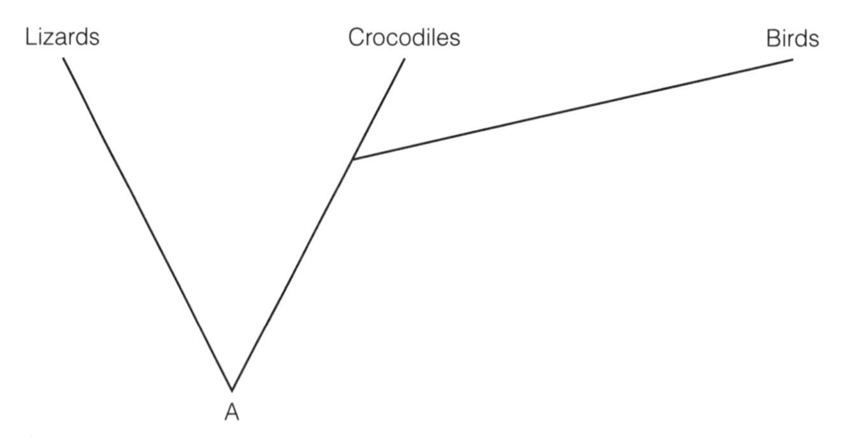

Figure 2

crocodiles do not form a monophyletic group apart from birds. Yet evolutionary taxonomists believe that lizards and crocodiles share a number of adaptive similarities not possessed by birds. According to the evolutionary taxonomists' usage, there is a monophyletic group that includes lizards and crocodiles but excludes birds.

The ensemble of characteristics possessed by lizards and crocodiles allows both to be called "reptiles." Moreover, evolutionists believe that the ancestor of lizards, crocodiles, and birds (denoted by "A" in Figure 2) was also a reptile. As a result *Reptilia* counts as monophyletic in the evolutionary taxonomists' sense of the word, but does not do so according to cladistic usage.

How should monophyly be defined? G. G. Simpson (1961, p. 124) says that "monophyly is the derivation of a taxon through one or more lineages from one immediately ancestral taxon of the same or lower rank." Mayr (1969) accepts this definition. The mention of "rank" indicates that whether a collection of species comprises a monophyletic group in this sense is not settled by the ancestor/descendant relationships depicted in Figure 1. This means that the evolutionary taxonomists' concept of monophyly requires further spelling out.

Let T be any set of species. Let "$x \in T$" express the proposition that species x is a member of T. Let "$x\mathrm{A}z$" express the proposition that species x is an ancestor of species z. The evolutionary taxonomist's concept of monophyly ("monophyly$_e$") may be stated as follows:

> T is monophyletic$_e$ if and only if there exists a species $s \in T$ and for any other species d, if $d \in T$, then $s\mathrm{A}d$.

Reptilia is monophyletic in this sense, because there is a (stem) species that is itself a reptile such that all other reptiles are descended from it.

The cladistic concept of monophyly ("monophyly$_c$") may also be defined in this format:

> T is monophyletic$_c$ if and only if there exists a species $s \in T$ and for any other species d, if $s\mathrm{A}d$, then $d \in T$.

Reptilia fails to be monophyletic according to this definition, because *Reptilia* does not include *all* the descendants of the stem species.

Evolutionary taxonomists regard *Reptilia* as monophyletic because they are willing to call the ancestor of lizards and crocodiles a "reptile." But what principles govern this practice? Without further restrictions, a liberal policy of introducing taxonomic names would allow any motley collection of species to count as monophyletic.

The evolutionary taxonomist places great weight on the difference between divergent and convergent evolution. As already noted, the cladist can identify the monophyletic groups depicted in Figure 1 without knowing which characteristics are possessed by any of the species there represented. The evolutionary taxonomist cannot do this, and requires information about the character states and evolutionary processes involved.

In the case of lizards, crocodiles, and birds, the first two taxa retained ancestral characteristics (the ones that count as "reptilean"); the birds diverged from this ancestral form by evolving novel characters. Figure 3(a) represents the kind of evolutionary process at work here; "0" denotes the ancestral form and "1" the derived form. Because the stem species and species *P* and *Q* exhibit the ancestral form, one is entitled to introduce a taxonomic term that encompasses the three, but that excludes species *R*; hence the term "Reptilia."

Species may be similar, however, because of convergence—that is, because of the independent origination of derived characters. Figure 3(b) illustrates this pattern. Here *X* and *Y* are similar, but because their common ancestor does not also exhibit the similarities, the evolutionary taxonomist (not to mention the cladist) will decline to say that *X*, *Y*, and the stem species comprise a monophyletic group that excludes species *Z*.

So a collection of species is monophyletic$_e$ when one member of that collection is an ancestor of all the rest and when the ancestral character states are present in all members of the collection. It is the difference between the conservation of ancestral forms and the convergent origination of derived forms that makes all the difference for the evolutionary taxonomists' concept of monophyly.

Not surprisingly, each systematic school has produced terminology to

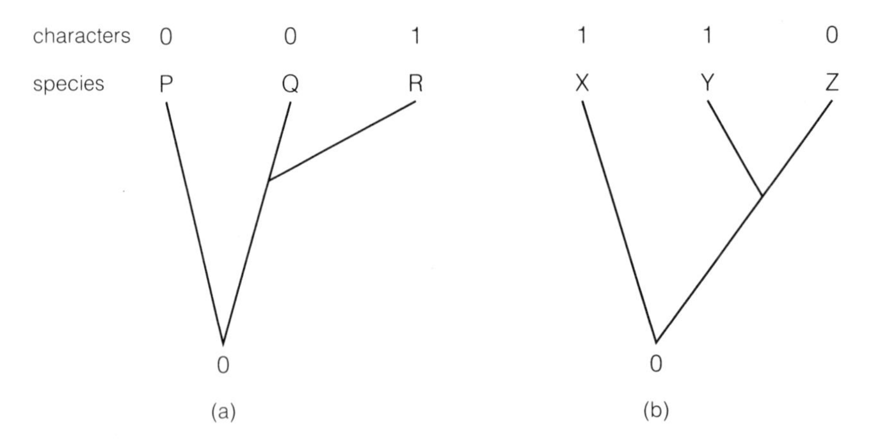

Figure 3

capture the distinctions drawn by the other. An evolutionary taxonomist will use the term "holophyly" to capture what a cladist means by "monophyly" (Ashlock, 1971). A cladist will use the term "paraphyletic" to describe the group that includes *P* and *Q* but not *R* in Figure 3(a). And the group that includes *X* and *Y* but not *Z* in Figure 3(b), which both schools regard as beyond the pale, is called "polyphyletic."

MUTUALISM AND COOPERATION

Douglas H. Boucher

MUTUALISM AND cooperation are the generally used terms for positive interactions between organisms, that is, for interactions that are beneficial to both participants. "Mutualism" is applied to interactions between organisms of different species, and a convenient shorthand for the three fundamental kinds of interactions between species is as follows: *mutualism* is represented as a +/+ interaction, *predation* as +/−, and *competition* as −/−, where the plus and minus signs indicate the direction of the effect one organism has on the other. "Cooperation" is somewhat more vaguely defined, but is coming to be used by many as the analogous term for positive interactions between organisms of the same species. Terms related to mutualism and cooperation, often used with partial or total overlap in meaning, are *symbiosis, protocooperation, obligacy, facilitation,* and *altruism.*

Because the definition of mutualism involves effects on both participants, a different term needs to be introduced to denote a unidirectional positive effect of one species on another. This "sound-of-one-hand-clapping" kind of mutualism, called *facilitation* of species 1 by species 2, denotes positive effect of species 2 on species 1, without any reference to the effect of species 1 on species 2.

Both mutualism and cooperation came into biology from socioeconomic discourse. Each is derived from a Latin verb (mutualism: to change; cooperation: to work together) and was part of ordinary speech long before being borrowed by biologists. Both also have lent their names to economic institutions that join people together to work for their common good, such as cooperatives (as in "Rural Electric . . ."), the French Mutualité movement, and insurance companies. Colloquially, mutualism denotes a relationship between two persons or groups in which each helps the other.

Biologists began to apply the terms to nature in the latter part of the 1800s. Mutualism was coined as a scientific term by the Belgian zoologist

Pierre Van Beneden in 1873, in a lecture entitled "A Word on the Social Life of Lower Animals" (Van Beneden, 1873): "There is mutual aid in many species, with services being repaid with good behaviour or in kind, and *mutualism* can well take its place beside commensalism."

Around the same time, it became common to contrast cooperation with competition as a second, and equally important, force in evolution. With the development of Social Darwinism, this contrast became a major point of debate. The most prominent proponent of the cooperation side was the Russian geographer and anarchist Peter Kropotkin, whose series of articles in reply to Thomas Huxley's "The Struggle for Existence in Human Society" were collected in the best-selling book *Mutual Aid* (Kropotkin, 1902). Kropotkin traced cooperation from animal societies through the development of humanity to modern times, arguing that it had become progressively more important in the course of evolution. Despite his title, most of his examples are of positive interactions *within* species, and thus would be considered cooperation rather than mutualism by our present definitions.

Another example of the terminological confusion that quickly engulfed the subject is the conflation of mutualism with *symbiosis* ("living together"), coined by Anton DeBary to designate the close association of two species, whether mutually beneficial or not (DeBary, 1879). By 1893, the botanist Roscoe Pound had already noted that "symbiosis in the strict sense and mutualism are often confounded" (Pound, 1893). The distinction between them—mutualism denoting mutually beneficial interactions between species, whether closely associated or not—has only recently begun to be made clearly.

In yet a third example of confusion, a major ecology textbook (Odum, 1971) applies the term mutualism only to those mutually beneficial relations between different species that are *obligate* to both. That is, neither species can survive in the absence of the other. If both can survive alone, that is, if the relationship is *facultative* for both, Odum calls it *protocooperation* rather than mutualism. It is unclear which term applies if the interaction is obligate for one species and facultative for the other.

Further imprecision comes from the failure to distinguish exactly who are the "both species" to whom a mutualism is beneficial—individuals or populations? The distinction is important for two reasons. First, the definition of "beneficial" should differ accordingly, involving increases in fitness in the case of individuals but increases in population growth rate in the case of populations. Second, some population-definition mutualisms, such as the dispersal of nuts by mammals, can involve the death of most of the individuals of one of the species—squirrels can eat 99% of the acorn crop, and bury 1%, but these survivors will have a high probability of germination. This is assuredly not beneficial to the acorns that are eaten, but may be to the acorn population's rate of establishment.

Because of these confusions, the present definition is not really adequate for rigorous studies of evolution. For this purpose, mutualisms (and all other interactions) should be denoted in terms of gene frequency changes within populations (Templeton and Gilbert, 1985). Nonetheless, this ideal is seldom even sought, let alone reached, in contemporary ecological studies.

Ecologists, evolutionary and otherwise, in practice consider an interaction to be mutualistic if they can show that fitness decreases in its absence. The simplest sort of demonstration of mutualism is experimental: exclude one species and observe that the second species has reduced fitness. Then exclude the second species, and observe that the first has reduced fitness. Though conceptually simple, this is sometimes impossible in practice, either because exclusion is physically difficult or because it requires drastic modifications of the ecosystem that, for conservation reasons, we are unwilling to do.

As a substitute, such variables as the fecundity, survivorship, growth rate, seed set, or seed establishment of each population can be shown to vary directly with the population of the other species.

A third, and weakest, way to indicate mutualism is to demonstrate transfers of resources (e.g., photosynthate, protein, minerals) that are known or can be assumed to be limiting (see RESOURCE).

In the absence of genetic information, all three of these techniques are adequate for ecology but inadequate for evolutionary studies. For example, cooperation is present in all species, in the obvious sense that mates, and parents and offspring, increase each other's fitnesses. But biologists generally use the term only in cases in which cooperation is unexpected, because of low genetic relatedness. Individuals sharing few genes are expected to be competitors, so that such phenomena as cooperative nesting by anis are considered noteworthy, warranting a special research effort and a separate term.

These issues are crucial to consideration of the altruism problem that has dominated animal behavior studies in the years since W. D. Hamilton's papers on kin selection (1964) (see ALTRUISM). They are also relevant to mutualism, for as G. C. Williams (1966) pointed out, participation in a mutualism with an organism of another species amounts to enhancing the fitness of organisms that can help an organism's conspecifics. It should thus be selected against except under conditions in which the direct benefit received by an organism outweighs the indirect benefit to competitors.

Williams felt that "the really good examples of mutualism are relatively rare, and it must be that these necessary preconditions seldom arise" (1966, p. 247). Although Williams' assessment that mutualisms are rare (shared by most ecologists of that time; Boucher, 1985a) is no longer generally accepted, the problem he defined remains with us. Indeed, it is the classical group-selection problem, as observed by D. S. Wilson (1983b).

Recent years have seen considerable advances in clarifying both the terminology and the reality of positive interactions, but there remains a considerable linguistic gap between technical and colloquial discourse. Cooperation is still used as a general term for all sorts of positive interactions; given the word's use in everyday speech, this is unlikely to change. The situation with mutualism is more hopeful, as biologists have come to distinguish it clearly from symbiosis in the technical literature. Many textbooks, however, particularly at the elementary and secondary school levels, perpetuate the confusion of earlier decades. The situation is even worse in general interest publications, where symbiosis is often used to denote little more than the juxtaposition of different elements. The battle for clarity is far from won.

NATURAL SELECTION: HISTORICAL PERSPECTIVES

M. J. S. Hodge

"NATURAL SELECTION" entered the vocabulary of scientific and other languages, as is well known, following Charles Darwin's publications in 1858 and 1859, most familiarly his book *On the Origin of Species by Means of Natural Selection, or the Preservation of Favoured Races in the Struggle for Life* (1859b). What is less well known is that, in introducing the new term, Darwin made no sustained attempt to provide it with an explicit definition that could be consistently invoked on all the occasions of its subsequent use. The title of the *Origin* may seem to offer an authoritative, inaugurating definition that equates natural selection with the "preservation of favoured races in the struggle for life"; but in the passage in the body of the book that comes closest to an overt definitional explication different phasing is deployed, so that, most conspicuously, the word "races" is not present. Here Darwin argues that because "variations useful in some way to each being in the great and complex battle for life" do occur sometimes in thousands of generations in the wild, and because "many more individuals are born than can possibly survive," then we cannot doubt that "individuals having any advantage, however slight, over others, would have the best chance of surviving and of procreating their kind," while "any variations in the least degree injurious would be rigidly destroyed." It is, then, this "preservation of favourable variations and the rejection of injurious variations" that "I call Natural Selection" (Darwin, 1859b, pp. 80–81).

To understand the history of the term "natural selection" both before and after this moment in the *Origin,* we have, therefore, to look not for a sequence of explicit definitional equations but, rather, for the reasons why people, starting with Darwin himself, have felt themselves able to grasp and wield the concept adequately in the absence of consistent, authoritative definitional analyses of the term. In Darwin's own case, the term itself was a secondary matter; what really counted was his argument for the analogy that the term was coined to signify, the analogy between man's selection and nature's. An account of how the term features in his work has to be grounded in an account of how he reached and developed the argument for this analogy, an analogy that was sanctioned, for Darwin, by very distinctive presuppositions about the relations between man and art as contrasted with God and nature. Conversely, an account of how the term has come to feature in scientific and other languages since Darwin has to explain how the term could often be detached from this analogy, and so from the presuppositions that provided the rationale for the first coining and introduction of the term.

Going back one or more generations before Darwin, one finds that by 1800 the use of the verb "to select" was a commonplace in writers on animal and plant breeding. A pamphlet by John Sebright—later to be a decisive source for Darwin's own knowledge of such subjects—was entitled *The Art of Improving the Breeds of Domestic Animals,* and in it Sebright declared: "Were I to define what is called the art of breeding, I should say, that it consisted in the selection of males and females, intended to breed together, in reference to each other's merits and defects" (Sebright, 1809, p. 5). As the definition indicates explicitly, this selection was an intentional action by men, acting on judgments of merit and defect. Nor is this surprising. "Selection," in general, as historical dictionaries confirm, was often synonymous with "picking" and "choosing," paradigmatically free and deliberate human actions.

We may say, then, that no one would easily or inadvertently slip into talking of nature as a realm where anything like selection was located; and, indeed, we find few authors before Darwin making that transition. There were some, however, who not only asserted that nature includes processes that resemble what the artful breeder effects, but who also deliberately expressed that assertion in appropriate phrasing. Patrick Matthew (1831, p. 308) wrote of a "natural process of selection"; and Allen Thompson (1839, p. 473), having explained that breeders ought to practice a "judicious selection" of males and females, went on to say that in "a state of nature" the "strongest males take precedence of the weaker, and naturally select the first females (as occurs in deer)."

But Darwin's own eventual decision to talk in such ways was not prompted by coming upon any such precedents in the literature. When he first came to that decision, late in 1838, in his Notebook E (Darwin, 1987), it was for reasons that arose from his own, newfound conviction that, thanks to providential laws given to nature by God, there exists a natural process analogous to the art of breeding (Hodge and Kohn, 1986). From July 1837, and the beginning of his concentrated notebook theorizing about species origins, Darwin had been comparing and contrasting the formation of varieties and species in the wild with the production and improvement of races in domesticated species. It was only after some months that he came to make an overt contrast between *artificial* and *natural* ways for such races to be formed in domesticated species; and his principal example of artifice was indeed "picking," the practice of selection. Moreover, he now thought of "picked" races as monstrous and so not as adaptations. By contrast, those races of domesticated species that owed their production and persistence to the influence, through the inheritance of acquired characters, of distinctive local conditions of soil or climate rather than to selective breeding, he thought of as natural and as adaptations. At this stage of his thinking, it was, therefore, with the formation of such natural and adapted races within domesticated species that Darwin compared the formation of new species from old in the wild.

Later, in constructing his theory of natural selection, Darwin was thus rethinking and replacing entirely these earlier contrasts and comparisons. This rethinking only becomes explicit in his extant notes in December 1838. A few months earlier, at the end of September, in his famous reflections on Malthus' writing on superfecundity and the struggle for existence among warring tribes of ancient men, Darwin had concluded that the purpose in nature, the "final cause," of population pressure was "to sort out proper structure and adapt it to change" (1987, Notebook D, p. 135). But although this sorting was conceived of as a discriminative rejection and retention, it was not seen by Darwin to have any special analogy with the breeder's art of picking or selection. What led Darwin subsequently to embrace such an analogy seems to have been contemplation of sporting dog breeds, as breeds that were improved by selective breeding, but in ways that made them fitted to adaptive predatory ends rather than making them monstrous in character (Hodge and Kohn, 1986). It is no coincidence that Darwin's earliest extant invocation of selection in nature comes in the first week of December, when he reflects that "if nature had had the picking she would make such a variety [of webbed-footed water dog] far more easily than man" (Notebook E, p. 63).

Here we have nature selecting, in that we have a deliberate *metaphor* that has nature doing what man familiarly does, but doing it much better. It would be tempting to conclude that everything Darwin is ever going to say about natural selection must be studied under the heading of metaphor. But to succumb to that temptation would be to overlook the potential metaphors often have to be developed into analogies (Soskice, 1985). Certainly Darwin always cultivated the figurative talk whereby nature was said to select, and he himself saw these expressions as, indeed, so many metaphors. But he did not ever see his theory as exclusively characterizable in such a way. The *analogy* developed to complement the *metaphor* became for Darwin another way; and the articulation of those *principles* that had natural selection as a corollary was another again.

According to this third understanding, Darwin saw his new theory as embodied in a small number of explanatory principles. Shortly before the water-dog reflections, and probably immediately after he had first embraced the selection comparison, Darwin declared (Notebook E, p. 58) triumphantly, "Three principles will account for all," and listed three generalizations about heredity, variation, and superfecundity. The text (plausibly dated to November 27) is a telling one, obviously, not least because it was to serve as an ancestor to the many subsequent attempts, by Darwin and others, to identify what general tendencies in nature causally entail, and so suffice to explain, the origin of new species from old in the wild by means that can be compared with the selective breeding practiced by man.

The articulation of principles and the deployment of metaphor were soon explicitly conjoined with the argument for the selection analogy. In March 1839 Darwin is already doing what he will do in the *Origin*: insisting how effective selective breeding has been in man's hands, and then asking "Has nature any process analogous—if so she can produce great ends" (Notebook E, p. 118). Before this, in mid-December, he had begun to develop the analogical argument that would dominate all future exposition of the theory. Even this early, the characteristic argument by proportion was in play: nature's selection, being so much more precise, prolonged, and comprehensive than man's, can suffice for proportionately greater effects—for, that is, the formation of species rather than mere varieties.

In his pursuit of his principles, metaphors, and analogy, Darwin only gradually came to make the particular term "natural selection" a decisive element in the exposition of his theory. "Selection," in the breeder's sense, Darwin seems to have been using routinely by early 1839, as seen by its inclusion in his "Questions for Mr. Wynne" (Gruber, 1974, p. 425), the addressee for these questions being just such a breeder. Darwin's earliest known use of the phrase "natural selection" does not appear until a manuscript text of 1841, where its use is not marked by any signs of self-conscious linguistic innovation (Darwin, 1841, p. 9; Cornell, 1984, p. 341). Likewise, the term does appear, contrary to legend, in his "Sketch" of 1842, the first dry run for the *Origin,* but only once, late on, and with no accompanying sense of a special linguistic moment (Darwin, 1842, p. 80). The opening sections of the "Sketch" set out almost all the argumentation later familiar from the *Origin,* the argumentation first adumbrated in Notebook E: that a natural means of selection—the struggle for existence—entails the selective preservation of advantageous variants in the wild, in a process analogous to man's art of selective breeding; that nature may, therefore, be said to be forever scrutinizing every animal or plant for its advantages and disadvantages in the struggle to survive and so on. But in none of the opening argumentation is any need felt, in 1842, for the term "natural selection" as such. By contrast, by the late 1850s, Darwin was planning to use for his *opus magnum* the title *Natural Selection* (Darwin, 1975). The term had now come into its own as a title under which he would write and as the title for his theory.

One way for people to make up their minds about the *Origin* was to make up their minds about the term "natural selection." By the time Darwin was preparing his third edition (which appeared early in 1861), he felt obliged to counter a telling array of adverse criticisms:

> Several writers have misapprehended or objected to the term Natural Selection. Some have even imagined that natural selection induces variability, whereas it implies only the preservation of such variations as occur and are beneficial to the being under its conditions of life. No one objects to agriculturalists speaking of the potent effects of man's selection; and in this case the individual differences given by nature, which man for some object selects, must of necessity first occur. Others have objected that the term selection implies conscious choice in the animals which become modified; and it has even been urged that as plants have no volition, natural selection is not applicable to them! In the literal sense of the word, no doubt, natural selection is a misnomer; but whoever objected to chemists speaking of the elective affinities of the various elements?—and yet an acid cannot strictly be said to elect the base with which it will in preference combine. It has been said that I speak of natural selection as an active power or Deity; but who objects to an author speaking of the attraction of gravity as ruling the movements of the planets? Every one knows what is meant and is implied by such metaphorical expressions; and they are almost necessary for brevity. So again it is difficult to avoid personifying the word nature; but I mean by Nature, only the aggregate action and product of many natural laws, and by laws the sequence of events as ascertained by us. With a little familiarity such superficial objections will be forgotten. (Darwin, 1959, pp. 164–165)

Not surprisingly, Darwin's response did not end the discussion (Beer, 1983; Young, 1985). His strategy—of appealing to precedent in the prestige science of physics and chemistry, and of declaring that his metaphorical phrases were likewise equivalent to expressions with plain, literal meanings—required more consensus than there ever had been in the understanding of theological, metaphysical, and scientific language concerning God, nature, laws, forces, and causes. One source of trouble was that Darwin liked the term "natural selection" because it could be "used as a substantive governing a verb" (F. Darwin, 1887, vol. 3, p. 46). But such uses appeared to reify, even to deify, natural selection as an agent, which Asa Gray for one condemned. Darwin explained that he used the term "much as a geologist does the word denudation—for an agent, expressing the result of several combined actions" (F. Darwin and Seward, 1903, vol. 1, p. 126).

Eventually Alfred Russel Wallace, an ally who had independently arrived at what both men regarded as the same theory of species origins, persuaded Darwin to use Herbert Spencer's phrase the "survival of the fittest" as a synonym for natural selection, but one free from any of the metaphorical elements that Wallace was convinced had led people to misunderstand Darwin's own term (see FITNESS, and Paul, 1988). Wallace was that sure many of Darwin's favorite figures of speech about, for instance, nature's constant scrutiny of living beings, were a liability. However, although Wallace had reached the theory without deploying the comparison between man's and nature's selection, he was content enough to embrace that analogy and to make it prominent in his own book, *Darwinism* (1889). The efforts made by Darwin and his allies to control the import and implications of the term "natural selection" extended into languages other than English. Darwin told H. G. Bronn, who translated the *Origin* into German, that some scientists like the term "because its meaning is *not* obvious, and each could not put on it his own interpretation" (F. Darwin, 1887, vol. 2, p. 278). He hoped there was a German equivalent that would have the same virtue.

Even in the 1859 edition of the *Origin*, Darwin insisted on distinguishing natural from sexual selection, emphasizing that in sexual selection individuals were differing in their ability to secure mates, a competition that did not entail death for the losers. The distinction between the two kinds of selection was required, Darwin urged, because traits that were advantageous in competing for mates (think of the peacock's tail) might well be detrimental to chances of survival. The theory of sexual selection was fully developed in the *Descent of Man* (1871), the most overtly ideological of Darwin's writings, manifestly permeated as it is by the sexism, racism, and imperialism, and more generally the liberal capitalism, characteristic of the age.

The metaphorical, analogical, and ideological elements in Darwin's use of the term "natural selection" have challenged scientists since to define and deploy the term conformably to newer ideals of professional specificity. The quest for scientific propriety for the term, however, has not always led to a consensus about its meaning, even for scientists themselves.

One activity prompting innovations in the explication of natural selection in our century has been the mathematical analysis of selective breeding, natural and otherwise. Well before 1900, applied mathematicians, most notably Karl Pearson, had constructed abstract, mathematical theorems to represent the changes in mean characteristics brought about in a specified population by differential death and reproduction (Norton, 1973). The novelty in the work of R. A. Fisher, J. B. S. Haldane, and Sewall Wright in the 1920s was that the mathematics was to represent selective breeding and its effects under Mendelian inheritance. Accompanying this work was, accordingly, a new conceptualization of natural selection and other factors making for change as influences that disturb the Hardy-Weinberg equilibrium frequency of genes. Here there was needed a completely general demarcation of selection from other factors such as drift, migration, and inbreeding. To supply this need, Wright drew up a classification that assumed that the direction and amount of gene frequency change is always determinate, in principle, in any selective process. The conclusion was that selection, as such, is a "wastebasket" category, including all causes of directed change in gene frequencies not involving mutation or introduction of hereditary material from outside. Wright stressed that selection will comprise biological components as diverse as differential viability, dispersal beyond the breeding range, fertility differences, and so on (1955).

With any such analysis, some of Darwin's and Wallace's concerns came to seem irrelevant. Thus Fisher emphasized that superfecundity, with its corollary of great mortality, was not necessary for nonrandom differential reproduction, and so should not be thought intrinsic to natural selection. That Malthus' writing on superfecundity had apparently first induced Darwin and Wallace to conceive of a process of selection going on in the wild was, therefore, Fisher urged, no longer relevant to an understanding of natural selection itself, although Fisher did take Malthus' name to label a parameter central to his own theorizing (Fisher, 1930b, p. 47).

Reflections on the demands of mathematical analysis have not, however, addressed all the ambiguities latent in the term "natural selection." A quite general issue has still received no canonical treatment: what kind of a thing is natural selection, anyway? A law, a principle, a force, a cause, an agent, or all or some of these? The view that natural selection is a law has been countered with the view that it is a principle, while that conclusion has been countered in turn by an insistence that it is neither (Reed, 1981; Byerly, 1983; Hodge, 1987b). Many textbooks talk of selection as a force, and the notion that evolutionary theory is a theory of forces has been defended explicitly (Sober, 1984a). However, if natural selection is in some serious sense a force, then one might expect a close analogue to mass in the theory of this force. But no plausible analogue for mass seems identifiable (Endler, 1986). A better tack may be to consider natural selection simply as a causal process and to define this process by specifying the conditions causally necessary and sufficient for its occurrence. A definition of natural selection as a cause is appropriate because evolutionary biologists today are still concerned, no less than Darwin was, with the definitional and evidential challenges set by the old (eighteenth-century) *vera causa* ideal for all physical science: namely, defining a cause and then marshaling independent, empirical, evidential cases for the existence of that cause and for its adequacy and its responsibility for the effects it is taken to explain.

It is, after all, in meeting these last three evidential challenges that upholders of the theory of natural selection can meet the old jibe that their theory is somehow a tautologous truism rather than an informative claim about the way the world works (Sober, 1984a; Hodge, 1987b, 1989).

The demand for a quite general explication of selection, natural or otherwise, also arises because in our day theorists of all sorts of subjects—for example, immunology, conceptual change in science, animal learning, and evolutionary epistemology—have sought to articulate and recommend selectional models of whatever it is that is being modeled, where selectional change is often contrasted, à la Popper, with instructional change (such as evolution by means of Lamarckian inheritance). For these purposes, then, it is appropriate to ask whether there are some defining features—perhaps nonfortuitous differential reproduction in a population of entities or properties that may vary fortuitously—that serve to distinguish any causal process that may be called selection (Darden and Cain, 1989).

Concentration on causation can resolve old misgivings about the term "natural selection" itself. A student today who objected that the term implicitly ascribes intentions to nature would likely be told that it is not the intentions alone of the human breeder that make her selective breeding causally efficacious in changing a herd or flock. If the farmer merely looks over the gate and intends then nothing happens; it is what she does physically, in separating certain animals or killing or castrating them or whatever, that makes the difference causally. The appropriateness of the term selection as applied to nature is, therefore, due not to any mimicking by nature of the farmer's intentions but to the occurrence in the wild of causal interactions that are equivalent in their consequences to the physical interventions the farmer makes in a physical course of events on the farm.

The newer twentieth-century explications of natural selection that have accompanied the rise of mathematical, experimental, and ecological population genetics have not displaced the older figurative and rhetorical life of the term so evident in Darwin's own writings. Rather, that life continues to be extended in novel invocations of the "selfishness" of DNA or the "tinkering" achieved in adaptive evolution (Dawkins, 1986; Jacob, 1989). The conclusion must be that scientific supporters of natural selection have not seen this semantic, and even ideological, promiscuity as a ground for abandoning the term altogether. It is an open question, perhaps, whether we see this habit as a sign that all language, including professedly scientific language, is irreducibly metaphorical; or whether we see it as a sign that such a term endures in science, despite leading a life full of metaphorical license, because scientists remain confident that it can always be explicated in ways that capture only the meaning it has to have for the purposes of mathematical theory or empirical instantiation or causal explanation.

NATURAL SELECTION: CURRENT USAGES

John A. Endler

"NATURAL SELECTION" can be defined as a *process* that occurs if and only if these three conditions are present: the population has (a) variation among individuals in some attribute or trait (phenotypic variation); (b) a consistent relationship between that trait and mating ability, fertilizing ability, fertility, fecundity, and/or survivorship (fitness variation); and (c) a consistent relationship, for that trait, between parents and their offspring, which is at least partially independent of common environment effects (inheritance). If these three conditions are met, then one or both outcomes will occur: (1) the trait frequency distribution will differ among age classes or life-history stages to an extent beyond that expected from ontogeny (growth and development); (2) if the population is not at equilibrium, then the trait distribution of all offspring in the population will be predictably different from that of all parents, beyond that expected from conditions (a) and (c) alone. Conditions a (phenotypic variation), b (fitness variation), and c (inheritance) contain all of the biology, and the process (outcomes 1 and 2) results purely from probability and statistics—the correlated effects of the biological conditions (Endler, 1986).

Natural selection shares some properties with genetic drift, a process meeting conditions (a) and (c), but not (b), and also requiring small effective population size (see Endler, 1986, and RANDOM DRIFT). Natural selection is often confused with evolution. The original definition of evolution was descent with modification (Darwin, 1859a; Lincoln et al., 1982; see also EVOLUTION). As Ernst Mayr (1962) so clearly pointed out, this includes both the *origin* as well as the *spread* of new variants or traits. Evolution may thus occur as a result of natural selection, genetic drift, or both, as long as there is a continual supply of new variation (as a result of gene flow and mutation in the broad sense; see Endler and McLellan, 1988). Natural selection does not necessarily give rise to evolution; a population may be at an equilibrium that arises from a balance of selective and other factors. Whether or not a population is at equilibrium depends upon its evolutionary history as well as on current conditions (a), (b), and (c). Unlike evolution, natural selection is a nonhistorical process that depends only upon current ecological and genetical conditions. Evolution depends not only on these current conditions but also upon their entire history. Natural selection deals with frequency changes brought about by differences in ecology among heritable phenotypes; evolution includes this as well as random effects and the origin of these variants (for discussion, see Endler, 1986; Endler and McLellan, 1988).

The term "natural selection" is often used in more restricted senses than that given above: mortality selection, nonsexual selection, and phenotypic selection. These restricted definitions emphasize certain specific aspects or components of the process, and they partially overlap in meaning. They can be placed in broader perspective by reference to the general definition.

By the nature of the general definition of natural selection, the process can be broken down into separate components in two rather different ways depending upon (i) alternative subprocesses or (ii) component or sequential subprocesses, analogous to parallel and serial wiring in electronics, respectively. The restricted meanings of natural selection concentrate on these components (see also SEXUAL SELECTION).

Mortality selection. Natural selection is sometimes used to refer only to the effects of consistent phenotypic-specific mortality (for discussion, see Fisher, 1930a; Ghiselin, 1969a; Endler, 1986). It is certainly true that natural selection can proceed exclusively with fitness variation (condition b) alone, but this is incomplete. Consider an expanding population consisting of two genotypes, one of which is increasing faster than the other. Some researchers (e.g., Hailman, 1982; Darlington, 1983) do not consider this a case of natural selection, because there is no differential mortality. This peculiar restriction to mortality selection is probably a historical artifact of the seductive attractiveness of Spencer's "survival of the fittest." Mortality is a single component of fitness variation, or condition (b) for natural selection, and thus is a rather incomplete if not misleading definition. Its use can lead us to ignore other fascinating biological phenomena that cause natural selection.

Sexual and nonsexual selection. Darwin (1859a, 1871) made a careful distinction between natural selection and sexual selection: sexual selection is a result of differential mating success, including fertilization and pairing; natural selection involves all other components of fitness. For simplicity, I will refer to the latter as nonsexual selection. The distinction was made because traits favored by sexual selection may sometimes be disadvantageous or may be opposed by other components of natural selection (Darwin, 1871; Ghiselin, 1974a; Wade and Arnold, 1980). Thus the outcome, as well as the dynamics, can be quite different from what Darwin and many biologists would regard as "natural selection" (Fisher, 1930a; Lande, 1981; Kirkpatrick, 1982; Arnold, 1983; Bradbury and Andersson, 1987). Explicit as well as implicit differences of opinion abound on whether or not sexual selection is a subset of natural selection. In addition, to add to the confusion, other aspects of differential reproductive success (such as fertility) have often also been included with sexual selection. Because the consequences of mating success are so distinct, it is best to restrict "sexual selection" to its original meaning and its application to differential mating success, rather than to include all aspects of reproductive success (Ghiselin, 1974a; Wade and Arnold, 1980; Arnold, 1983; Bradbury and Andersson, 1987).

By the general definition of natural selection given above, however, sexual selection is a logical subset of natural selection. This is true because (i) mating ability is one of several alternatives in condition (b) for natural selection (fitness variation), and (ii) the definition of the process of natural selection takes no account of the details of its outcome; it merely states that if conditions (a)–(c) are met then age classes will have different trait frequency distributions, and frequencies will change among generations if the population is not already at an equilibrium. In fact there is no difference between sexual and nonsexual selection in the methods of demonstration or measurement (Arnold and Wade, 1984a,b; Endler, 1986; Clutton-Brock, 1988). In addition, other components of natural selection can oppose one another in the same way that sexual and nonsexual selection can. In the very general sense, sexual and nonsexual selection are subsets or aspects of natural selection, but at a lower level (considering their dynamics and outcomes) they are very distinct. Note also that mortality selection is a special case of nonsexual selection, and that, like mortality selection, sexual and nonsexual selection are sufficient by themselves for natural selection to proceed.

Phenotypic selection and response. Quantitative geneticists, animal breeders, and plant breeders decompose the process of natural selection into phenotypic selection and genetic (or "evolutionary") response (Fisher, 1930a; Haldane, 1954; Falconer, 1981; Lande and Arnold, 1983). Phenotypic selection is the *within-generation* change in trait distribution among cohorts (or the difference between the actual number of mates and the effective number of mates in sexual selection) and is independent of any genetic system or genetic determination. In terms of the general definition of natural selection, phenotypic selection requires conditions (a) and (b) for natural selection. The response is the genetic change that occurs as a result of phenotypic selection in combination with the genetic system, which requires condition (c) for natural selection; it is the *between-generation* change in trait distributions after phenotypic selection. This distinction is a very important and useful one (Falconer, 1981).

If there is no inheritance (condition c), the process of natural selection cannot occur. In spite of this, phenotypic selection is sometimes called "natural selection" (e.g., Lande and Arnold, 1983), primarily because natural selection works on phenotypes and not on genotypes (Mayr, 1963). But natural selection involves the differential survival and perpetuation of phenotypes, and perpetuation requires inheritance. Phenotypic selection determines the distribution of traits during reproduction, but inheritance is required to transform the new distribution into the next generation. To say that natural selection is synonymous with phenotypic selection is to trivialize it—this is tantamount to saying that there are differences among different phenotypes (see Endler, 1986, chap. 2, for discussion).

The restriction of natural selection to phenotypic selection results at least in part from an inconsistent distinction between evolution, natural selection, and genetic drift, and the implicit acceptance of evolution as merely the change in frequencies. It also accounts for the occasional use of "evolutionary response" in place of "genetic response." "Evolutionary response" is an unfortunate usage because natural selection does not necessarily result in evolution—at equilibrium there can be a genetic response to phenotypic selection every generation, but no change in trait distributions, that is, no evolution. Random genetic drift can also yield differences among age classes, which will appear to be phenotypic selection if only a few generations are examined. This apparent phenotypic selection will be followed by a genetic response as the random within-generation change is transformed into the next generation through the hereditary process. But that is not natural selection. It is condition (b) and not (c) that distinguishes natural selection from genetic drift; merely splitting off (c) (inheritance) is insufficient. To be logically consistent, we must either include genetic response as part of the process of natural selection, as in the general definition (Endler, 1986), or distinguish *three* processes: (i) phenotypic ("natural") selection; (ii) genetic response; and (iii) cumulative genetic change (evolution). It is simpler to regard phenotypic selection and genetic response as sequential subprocesses of natural selection.

To put this into a broader perspective, those who restrict "natural selection" to phenotypic selection also call natural selection, as defined in the first section, "evolution," and those who are more careful, call it "evolution by natural selection." But evolution is more than merely a change in trait distributions or allele frequencies; it also includes the *origin* of the variation (Endler, 1986; Endler and McLellan, 1988). And once again this begs the question of what to do when the general process of natural selection results in a dynamic equilibrium, with no change in frequencies. For this reason it might be preferable to use the general definition of natural selection, and define evolution as a process including natural selection as defined here as well as genetic drift and other processes that generate new variants (see especially Endler and McLellan, 1988).

As with the distinction between sexual and nonsexual selection, the distinction between phenotypic selection and genetic response is an important and useful one. It is based upon a subdivision of the general process of natural selection into two parts. But this subdivision is logically quite different from the distinction between sexual and nonsexual selection. Both sexual and nonsexual selection can by themselves result in the process of natural selection, but phenotypic selection and genetic response do not separately lead to the process. The former distinction (sexual/nonsexual) breaks the process of natural selection into two alternative subprocesses that differ in the details of condition (b) for natural selection. But the latter distinction (phenotypic selection/genetic response) breaks the process of natural selection into two sequential steps, the first dependent upon conditions (a) and (b), and the second on condition (c) for natural selection—natural selection cannot occur unless both steps are present.

In summary, there are three common restricted meanings of the term "natural selection": mortality selection, nonsexual selection, and phenotypic selection. The first two focus on specific components of fitness variation (condition b for natural selection); the last focuses on the first stage of the process of natural selection. The specificity has proved very useful in exploring various aspects of natural selection and evolution. If one uses one or more of these restricted definitions, however, one must make absolutely clear which meaning is intended.

NEUTRALISM

Motoo Kimura

IN THE CONTEXT of modern evolutionary biology, "neutralism" refers to the neutral theory of molecular evolution. This theory (Kimura, 1968; for details, see Kimura, 1983) claims that the great majority of evolutionary changes at the molecular (DNA) level do not result from Darwinian natural selection acting on advantageous mutants but, rather, from random fixation of selectively neutral or very nearly neutral mutants through random genetic drift, which is caused by random sampling of gametes in finite populations.

Here "selectively neutral" means selectively equivalent, that is, neither advantageous nor disadvantageous. In other words, the mutant forms involved can do the job equally well in terms of the survival and reproduction of individuals. The theory does not deny the role of natural selection in determining the course of adaptive evolution, but it assumes that only a minute fraction of molecular (i.e., DNA, RNA, and protein) changes are adaptive. The essential aspect of the neutral theory is not that the mutant alleles involved are selectively neutral in the strict sense. Rather, the emphasis is on mutation pressure and random drift as explanatory factors in molecular evolution.

The neutral theory also asserts that most of the intraspecific variability at the molecular level (including protein and DNA polymorphism) is essentially neutral, so that the majority of polymorphic alleles are maintained in the species by the balance between mutational input and random extinction. In other words, the neutral theory regards protein and DNA polymorphisms as a transient phase of molecular evolution and rejects the notion that the majority of such polymorphisms are adaptive and maintained by some form of balancing selection. The underlying assumptions of this theory are sufficiently simple so that population genetical consequences can readily be worked out mathematically, enabling them to be tested by observations and experiments.

In so far as the neutral theory does not deny the possibility that some changes are adaptive, it is not antagonistic to the Darwinian theory of evolution by natural selection. However, because of its emphasis on mutation pressure and random drift, and because of its accent on negative selection rather than on positive Darwinian selection, the neutral theory clearly differs in its theoretical framework from the traditional neo-Darwinian or synthetic theory of evolution.

The proposal of the neutral theory (Kimura, 1968) led to a great deal of controversy, with strong opposition expressed by the neo-Darwinian establishment. This is often referred to as the "neutralist-selectionist controversy," which stimulated much research not only in molecular evolution but also in population genetics. The neutral theory also triggered reexaminations of the orthodox synthetic theory of evolution that had earlier appeared to be firmly established.

In order to derive some simple but basic equations representing the essential features of the neutral theory, let us consider an evolutionary process in which molecular mutants are substituted one after another at a gene locus (or a certain region of DNA) within the species. Each such substitution is made up of a sequence of events in which a rare mutant form that has appeared, usually singly represented, in the population finally spreads through the whole population. If such fixations are caused by random drift acting on selectively neutral mutants, we have the following formula for the rate of evolution per generation:

$$k_g = v_T f_0. \tag{1}$$

In this equation, v_T is the total mutation rate and f_0 is the fraction of neutral mutants. In other words, under the neutral theory the evolutionary rate is equal to the mutation rate of neutral alleles. Note that k_g represents the rate per generation at which molecular mutants are substituted in the course of evolution within the lineage. Each of these events takes a long time—four times the effective population size, on the average (Kimura and Ohta, 1969). Advantageous mutations may occur, but the neutral theory assumes that they are so rare that they may be neglected from our consideration. Thus, $(1 - f_0)$ represents the fraction of definitely deleterious mutants that are eliminated from the population without contributing either to evolution or to polymorphism. What is remarkable in the above formulation is that the rate of evolution is independent of population size and environmental conditions.

When we estimate the actual rates of evolution through comparative studies of protein or DNA sequences, we usually express the evolutionary rate by taking one year as the unit length of time, while the mutation rate is usually measured per generation. With this in mind, formula (1) may be modified to give the evolutionary rate per year so that

$$k_1 = (v_T/g)f_0, \tag{2}$$

where g is the generation span, and therefore v_T/g is the total mutation rate per year.

Let us now consider intraspecific variability. If we assume the infinite allele model of Kimura and Crow (1964), that is, if we assume that whenever mutation occurs at a locus, it leads to a new, non-preexisting allele, then at equilibrium in which mutational input and random extinction of neutral alleles balance each other, we have the following formula for the average heterozygosity per locus:

$$\overline{H}_e = \frac{4N_e v_0}{4N_e v_0 + 1}, \tag{3}$$

where N_e is the effective population size and v_0 is the mutation rate for selectively neutral alleles, so that $v_0 = v_T f_0$.

Ample evidence has now accumulated to show that molecular evolution is distinguished from phenotypic evolution by two remarkable features. They are: (i) constancy of the rate, that is, for each protein or gene region, the rate of amino acid or nucleotide substitutions is approximately constant per site per year (known by the term "molecular evolutionary clock"), and (ii) "conservative nature" of changes, that is, functionally less important molecules, or portions of molecules, evolve faster than more important ones.

The first feature, constancy of the rate, may be explained by the neutral theory by assuming that v_T/g remains the same (constant) among diverse lineages and over time for a given protein or gene, for which f_0 is assumed to be constant. In other words, the theory assumes that for a given gene, the rate of production of neutral mutations per year is nearly constant among diverse organisms whose generation spans are very different. Note that mutation here refers to changes that lead to DNA base replacements, but not to lethal and "visible" changes. These latter types of mutations, whose incidence has been known to be generation dependent, are now suspected to be largely caused or controlled by various movable genetic elements such as transposons and insertion sequences in the genome. In contrast, it is likely that errors in DNA replication and repair are the main causes of DNA changes that are responsible for molecular evolution. Thus the mutation rate for nucleotide substitutions may depend on the number of cell divisions in the germ lines, particularly in the male line, and this will make the molecular mutation rate roughly proportional to the rate per year. Experimental studies on this subject are much needed.

The second feature of molecular evolution, the conservative nature, has now been well established. Mutant substitutions that cause less drastic changes in the existing structure and function of a molecule occur more frequently than those which cause more drastic changes. This is easy to understand from the neutral theory, because the less drastic or more conservative the mutational change, the more likely it is to turn out to be selectively neutral. This means that for more conservative changes the values of f_0 are larger. A great deal of opposition was voiced by the neo-Darwinian establishment when such an explanation was first proposed (Kimura and Ohta, 1974), but it has now become common knowledge among molecular biologists. It is now a routine practice to search for various signals by comparing a relevant region of homologous DNA sequences of diverse organisms and to pick out a constant or "consensus" pattern, but to disregard variable parts as unimportant.

Regarding the constancy of the rate, it should be pointed out, from the standpoint of the neutral theory, that a universally valid and exact molecular evolutionary clock would exist only if, for a given molecule, the mutation rate for neutral alleles *per year* (v_0/g) were exactly equal among all organisms at all times. Any deviation from the equality of neutral mutation rate per year makes the molecular clock less exact. This means that departure from exact clockwise progression of molecular evolution by no means invalidates the neutral theory.

The outpouring of DNA sequence data during the last decade has added some very strong pieces of evidence for the neutral theory. It has now been well established that synonymous base substitutions within codons, which do not cause amino acid changes, occur at a much higher rate in evolution than amino acid–altering substitutions. It has also been found that evolutionary base substitutions in "introns" occur at a rate comparable to that of the synonymous ones or even higher. Considering that natural selection acts on the phenotype of the organism, these observations showing a preponderance of synonymous and other silent substitutions suggest that *molecular changes that are less likely to be subject to natural selection occur more rapidly in evolution.* This can readily be explained by the neutral theory, because such molecular changes must have a higher chance of being selectively neutral (i.e., larger f_0) and therefore neutral evolution occurs at higher rate.

It was predicted (Kimura, 1977), based on the neutral theory, that the maximum evolutionary rate is set by the mutation rate ($k \leqslant v_T$), and that the maximum rate is attained when all the mutations are selectively neutral ($f_0 = 1$). A dramatic example vindicating this prediction was the discovery of very high evolutionary rates of pseudogenes (or "dead" genes) that have lost their function. The best-known example is the pseudo α-globin genes in the mouse. The estimated rate in globin pseudogenes is about $k = 5 \times 10^{-9}$ substitutions per nucleotide site per year in mammals. What is really interesting is that the rates of substitution are equally high in all three codon positions. Moreover, pseudogenes accumulate deletions and additions at very high rates, suggesting that they have been liberated from the constraint of negative selection, and that they are on the way to disintegration by accumulating various mutational changes.

Regarding intraspecific variability at the molecular level, it is expected, from the neutral theory (cf. Equations 2 and 3) that genes or portions of DNA that evolve more rapidly must show higher intraspecific variability. Recent observations on DNA polymorphism in natural populations of fruit flies are consistent with this expectation: on the whole synonymous and other silent sites are much more polymorphic than amino acid–altering sites.

The predominant role played by mutation pressure in molecular evolution, quite in line with the neutral theory, has become increasingly evident from recent studies. One of the most remarkable examples demonstrating this is a very rapid evolutionary change observed in genes of RNA viruses (such as influenza A virus genes and retroviral oncogenes), which are known to have very high mutation rates: RNA viruses show evolutionary rates roughly a million times as high per year as those of DNA organisms. This can readily be explained by noting that in Equation (2) the value of v_T is about a million times higher in RNA genomes than in DNA genomes.

The concept of mutation-driven neutral evolution is also useful in understanding the remarkable evolution of the deviant coding system recently discovered in the A-T rich bacterial species *Mycoplasma capricolum* (see Kimura, 1989).

Meanwhile, the main idea of the neutral theory has been used by Freeman Dyson (1985) to develop a new theory of the origin of life. According to the Dyson theory, an active protein evolved first in an Oparin-type primitive cell through a process similar to random frequency drift in a finite population. The theory assumes also that the RNA gene emerged later in the cell as a parasite. Irrespective of whether the Dyson theory is valid or not, it is possible that chance in the form of random genetic drift has played a very important role in the origin of life. This means that the most prevalent evolutionary changes that have occurred at the molecular level since the origin of life on earth are those that were caused by random genetic drift rather than by positive Darwinian selection.

In the past, our scientific worldview was largely based on knowledge in the physical sciences. With the advent of molecular biology, the relative influence of the biological sciences has been much enhanced. As a part of this trend, studies of molecular evolution, including molecular population genetics in general and the neutral theory in particular, have generated important new resources for the formation of our worldview. The perspective of the neutral theory suggests an alternative to the Darwinian term "survival of the fittest"—namely, "survival of the luckiest." This alternative (proposed by Kimura, 1989) serves to emphasize the importance of good fortune for success in evolution.

NICHE: HISTORICAL PERSPECTIVES

James R. Griesemer

ALTHOUGH Graham Bell has suggested that "niche" is "a term perhaps best left undefined" (Bell, 1982, p. 510), the niche concept has been an important organizing idea in ecology ever since its introduction by Joseph Grinnell and Charles Elton early in this century. Some have heralded the niche as a foundational concept and one of the most important theoretical developments of modern ecology (Hutchinson and Deevey, 1949, quoted in Krebs, 1978, p. 231). Others have criticized it as tautological (Peters, 1976), or speculated that the concept would "probably turn out to be unnecessary, allowing an always welcome simplification of ecological jargon" (Margalef, 1968, p. 7). Differing insights rest, in part, on the heterogeneity of uses to which the niche concept has been put during its "curious" history (see Cox, 1980).

The common meaning of the term "niche," according to the Concise Oxford English Dictionary, is a "vertical recess in wall to contain statue, vase, etc.; (fig.) place destined for person's occupation"; it is probable that early ecological meanings derive from the sense of a special place suited to its occupant. The history of the meanings and uses of the niche concept trace to nineteenth-century ideas of the balance of nature, design, and the superorganism.[1]

1. See Egerton (1977, 1983), McIntosh (1985), and Worster (1977) for general histories of ecology and Kingsland (1985) on population ecology; see also the symposium edited by Maienschein et al. (1986). For some related concepts, see COMMUNITY and EXTINCTION. Historical reviews of the niche concept can be found in Colwell and Fuentes (1975), Cox (1980), Diamond (1978), Egerton (1977), Gaffney (1975), Giller (1984), Hutchinson (1957, 1965, 1975, 1978), Kingsland (1985), MacArthur (1968, 1972), McIntosh (1985), Pianka (1974, 1976), Schoener (1989), Udvardy (1959), Vandermeer (1972), and Whittaker et al. (1973). Whittaker and Levin (1975) contains reprints of many seminal papers on the niche concept and Hazen (1964) reprints a number of additional important papers. Austin (1983) and McIntosh (1985) review the plant ecologists'

Over the last century there have been three major shifts in the meaning of "niche": (1) from the nineteenth-century focus on a species' place in the balance of nature to concepts of the specific place or role of a population or species in an environment or community, (2) from the concept of place or role to a formal, geometric definition of the occupant of a place in terms of idealized environmental dimensions, and (3) from the formal definition to an operationalized description in terms of actual resource utilization (Schoener, 1989). In addition, uses of the niche concept have shifted as interests of animal ecologists changed from the geographical basis of speciation and evolution to problems in community structure, to the ecological dynamics of laboratory populations, and to the evolutionary dynamics of communities. More recently, some biologists have argued for a return to some of the earliest conceptions of the niche (James et al., 1984). Others have questioned some of the perceived consequences of early conceptions of the niche, particularly the heavy emphasis on the role of competition in structuring niches (Simberloff, 1982, 1984) as well as the concept of a vacant niche (Herbold and Moyle, 1986; Colwell on NICHE).

Roswell Johnson (1910) is probably the first biologist to use the word "niche" to mean the place of an organism in nature (Gaffney, 1975; cf. Cox, 1980, and Schoener, 1989). Joseph Grinnell, however, was the first to develop a full account of the niche concept and to incorporate it into a broad program of biological research. Grinnell and Charles Elton introduced the niche concept independently to designate a place or role in an environment or community (Grinnell and Swarth, 1913; Grinnell, 1914, 1917a, 1924, 1928, excerpts reprinted in 1943; Elton, 1924, 1927, 1946). Grinnell's focus on environment and Elton's on community reflect their differing theoretical programs.

Their ideas about niches extended Darwin's metaphor of a world so filled with species that the improvement of one by natural selection often implied the "wedging" out of another (see Egerton, 1977). Grinnell's focus on problems of distribution and evolution led him to think of the niche as a unit whose limits are set by factors of food supply, shelter (including refugia from enemies and safe breeding places), competition, parasitism, temperature, humidity, rainfall, insolation, nature of the soil, and others (Grinnell, 1917b, 1928). Elton's focus on food relations as the fundamental structure underlying all animal communities led him to think of niches

---

related concept of continuum. This list is far from comprehensive and the literature cited is far from agreement on many issues. Claims made for the first ecological use of the term "niche," for example, range from 1910 (Cox, 1980; Gaffney, 1975) to 1924 (Vandermeer, 1972). Schoener (1989) compares a variety of historical meanings of the term in relation to its role in understanding competition and an assessment of the modern niche theory and its problems. His paper complements the focus on Grinnell and Elton presented here.

in terms of roles in food chains, such as herbivore and carnivore, although he was aware of many factors influencing these roles.

Grinnell: Geography, evolution, and the niche. Joseph Grinnell was interested in how the evolution of the environment brings populations of animals into changing association with other plant and animal species and thus drives organic evolution and speciation (Grinnell, 1924). He used "ecologic niche" and "environmental niche" as equivalent terms (Grinnell, 1924, p. 227).

Grinnell's approach to natural history was based on late nineteenth-century traditions of biogeography, systematics, and Darwinian evolution (Mayr, 1973; Star and Griesemer, 1989; Griesemer, 1990). In considering the physical and biotic environment of a species, he developed an ecologic hierarchy of world realms, regions, life zones, faunal areas, (plant) associations, and niches to parallel the systematic hierarchy from kingdoms to subspecies (see Grinnell, 1928). Each level of the hierarchy picked out relevant environmental factors that might serve as causes of speciation through effects on the distribution and abundance of species and subspecies. Abiotic factors were mostly associated with higher levels, for example, temperature with life zones and humidity with faunal areas. At the lowest level niches were associated with a complex of biotic and abiotic factors and were unique; each species occupied one niche, that is, one set of hierarchically characterized abiotic factors plus plant and animal species that served a variety of roles as elements of each species' environment. It is Grinnell's *hierarchical* concept of environmental determinants of distribution that led him to characterize the niche as the *ultimate* distributional unit (Grinnell, 1928).

Grinnell's most cited paper on the concept of niche is "The Niche-Relationships of the California Thrasher" (Grinnell, 1917a), although he first used the term "niche" in his 1913 Ph.D. dissertation (see Grinnell, 1914). (The 1917 paper is the first of Grinnell's in which "niche" appears in the title, probably explaining why some authors claim 1917 as the year of Grinnell's introduction of the term—e.g., Udvardy, 1959.) It seems probable that the term emerged from discussions among the active group of faculty and students at Stanford University attracted by the evolutionist David Starr Jordan and inspired by the western biogeographic work of C. Hart Merriam, whose life-zone concept provided a starting point for much of Grinnell's work.

The term "niche" first appears in Grinnell's published work in a paper coauthored with Harry Swarth, whose later work at the California Academy of Science on speciation in Galapagos finches predates Lack's (Grinnell and Swarth, 1913; see Cox, 1980, and Kingsland, 1985). Grinnell and Swarth characterized the niche as a subdivision of association: "As with zones and faunas, associations are often capable of subdivision; in fact such splitting may be carried logically to the point where but one species occupies each its own niche" (1913, p. 218).

Grinnell's idea of the niche as an ultimate distributional unit flows from his conception of "competitive exclusion," usually attributed to G. F. Gause (e.g., Hutchinson, 1957, calls it the Volterra-Gause principle; Udvardy, 1959, and Hutchinson, 1965, acknowledge Grinnell). Darwin had articulated the core idea in *On the Origin of Species* and Grinnell presented a similar view: "It is only by adaptations to different sorts of food, or modes of food getting, that more than one species can occupy the same locality. Two species of approximately the same food habits are not likely to remain long evenly balanced in numbers in the same region. One will crowd out the other; the one longest exposed to local conditions, and hence best fitted, though ever so slightly, will survive, to the exclusion of any less favored would-be invader" (1904, pp. 375–377).

Elton: Food chains, animal communities, and the niche. Charles Elton's widely read book, *Animal Ecology,* introduced a generation of scientists to a modern conception of animal community (Kingsland, 1985). The core of Elton's theoretical outlook was that ecology is the study of what organisms are *doing* in their environments. By studying the circumstances and limiting factors under which animals do what they do, Elton thought it would be possible to discern "the reasons for the distribution and numbers of different animals in nature" (Elton, 1927, p. 34).

Elton's focus on the overriding importance of food relations led him to characterize the niche primarily in terms of populations' functional roles in a community as eaters and eaten: "Animals have all manner of external factors acting upon them—chemical, physical, and biotic—and the 'niche' of an animal means its place in the biotic environment, *its relations to food and enemies.* The ecologist should cultivate the habit of looking at animals from this point of view as well as from the ordinary standpoints of appearance, names, affinities, and past history. When an ecologist says 'there goes a badger' he should include in his thoughts some definite idea of the animal's place in the community to which it belongs, just as if he had said 'there goes the vicar'" (1927, pp. 63–64).

Elton's interest in functional roles stemmed from his concern to develop a coherent account of food chains and cycles and to explain his new concept of the "pyramid of numbers," which relates the relative sizes and numbers of organisms to their position in a food chain (1927, chap. 5). As such, he was less interested in species per se than in comparisons among local community structures that would support generalizations about food chains, size, and relative abundance. Nevertheless, Elton was aware of a wide variety of factors influencing niches, and he often mentioned microhabitat factors such as suitable soil types for nesting birds and places in which different species feed.

Despite the fact that Elton and Grinnell both considered biotic and abiotic factors, many textbooks distinguish the two by calling Grinnell's concept the "habitat" niche and Elton's the "functional" niche, as if to suggest that Grinnell's concept was primarily abiotic and Elton's biotic (Krebs, 1978; Ricklefs, 1979; cf. Schoener, 1989, for a counterargument). Whittaker et al. (1973, p. 322), following Dice (1952) and Clarke (1954), attempt to distinguish "place" and "functional" concepts of niche, arguing that the functional concept is the more fundamental and that part of the confusion of the two stems from the mixture of habitat and niche dimensions of Grinnell's (and Hutchinson's) views (see also Pianka, 1974, pp. 185–187). Grinnell's and Elton's niche concepts are better distinguished against the backdrop of their differing theoretical aims of explaining speciation versus community structure, however, rather than focusing on a few aspects of their explicit definitions of "niche" (see also COMMUNITY).

Grinnell and Elton both identified the niche as the place/role a species happens to occupy in an environment, not a property of the occupying species as in the Hutchinsonian and modern resource utilization conceptions of the niche. Grinnell wrote, "if a new ecologic niche arises, or if a niche is vacated, nature hastens to supply an occupant, from whatever material may be available. Nature abhors a vacuum in the animate world as well as in the inanimate world" (1924, p. 227).

The principal point on which Grinnell and Elton appear to differ is whether more than one species can occupy a single niche, but this stems from Elton's less precise characterization of the niche; a comparison of their examples shows the two concepts to be rather similar (Schoener, 1989). Grinnell concluded that niches are unique because his taxonomical approach took all important environmental factors into account. Elton identified functional food relations in a community loosely enough to suggest that niches "are only smaller subdivisions of the old conceptions of carnivore, herbivore, insectivore, etc., and that we are only attempting to give more accurate and detailed definitions of the food habits of animals" (1927, p. 64). As functional roles, niches could be filled by distinct (but often closely related) species in different communities. Elton suggested a concept of ecological "equivalents," arguing, for example, that an arctic niche filled by foxes that eat guillemot eggs and remains of seals left by polar bears is filled in tropical Africa by hyenas that eat ostrich eggs and remains of zebras killed by lions (1927, p. 65).

However, just as Elton recognized that factors other than food were often important in determining the species in a community, Grinnell fully recognized the functional nature of niches in his version of ecological equivalents, calling attention to "the great number of *similar* ecologic types of animals which are developed in widely separated regions and which are derived from *unrelated* stocks. The Kangaroo Rat of our deserts corresponds ecologically to the Jerboa of the Sahara; but it is derived from squirrel ancestry, while the Jerboa is more nearly related to house mice" (Grinnell, 1924, p. 227). Grinnell's conclusion that the niches of distinct species could be ecologically similar, but not identical, probably followed from his greater attention to the vast number of determining environmental factors.

Grinnell tended to focus on the distribution of *species* into niches because he was an evolutionary biologist whose first concerns were speciation and the relationship between the evolution of the environment and organic evolution (Grinnell, 1924; cf. Griesemer, 1989). Populations of the same species may be distributed among a number of local communities of the sort that interested Elton. Grinnell placed great emphasis on subspecies because these are of greatest interest to evolutionists. Elton's interest in evolution appears to have had more to do with population numbers and the possible role of ecologists in solving the problem of how nonadaptive traits might spread. Elton suggested that release from competition after a population crash might allow nonadaptive traits to spread before they are scrutinized by selection (1927, chap. 12).

Early empirical studies of the niche. Empirical studies of competition in the laboratory and the field were essential to development of the niche concept (Hutchinson, 1978; Kingsland, 1985; McIntosh, 1985). G. F. Gause (1934) brought together Elton's concept of the niche and V. Volterra's mathematical descriptions of two-species interactions in predator-prey systems to found an important branch of the experimental investigation of competition. Gause's approach was to construct simple laboratory systems based on Volterra's equations in order to see whether the predicted oscillations of predators and prey could be realized. A related but different approach was pioneered by Thomas Park in experimental studies of competition among species of the flour beetle genus *Tribolium* (e.g., Park, 1948, 1954b; for reviews, see Neyman et al., 1956, and Park, 1962; see Park, 1939, for a contrast between Gause's and Park's approach).

Experimental studies of competition had two important consequences for the concept of the niche. First, they focused attention squarely on central theoretical problems such as the dynamics of competition and held out hope for an ecological theory based on general principles. Second, they reinvigorated the competitive exclusion principle, which stated that "complete competitors cannot coexist" (Hardin, 1960, p. 1292). Complete competitors are usually understood to be species that have identical ecological niches. Prior to the experimental studies, competitive exclusion was perceived to be a qualitative principle so obvious as to seem uninteresting (Hutchinson, 1975; Kingsland, 1985). Also, Grinnell's earlier use of competitive exclusion in an integrated evolutionary view had virtually disappeared as geneticists took over the development of evolutionary theory and systematists developed the new speciation theory. Gause's and Park's experiments showed that the concept of niche, in the guise of determinants of relations of competitive exclusion, was central to an understanding of population dynamics and the evolutionary structuring of communities.

Some disagreement over the utility of the niche concept stems from apparent circularity of the competitive exclusion principle. If species are observed to coexist, then by the competitive exclusion principle they must have different niches to avoid competition, whether these differences are discernible or not. If species do not coexist, then they must overlap in their niches and competition prevents, or would prevent, coexistence. Such explanations appear to be "competitionist story-telling" after the fashion of adaptationist story-telling: assume that competition is ubiquitous and you will find that you can explain all community relations in terms of competition. G. E. Hutchinson (1957, p. 418) tried to rescue the competitive exclusion principle by reformulating it as an empirical claim: realized niches do not intersect.

There are still dangers of circular reasoning about niches based on studies of resource utilization, however: if species coexist and one resource dimension indicates too much overlap to permit coexistence, perhaps two resource dimensions will separate them; if two will not do, try three (see also RESOURCE). The competitive exclusion principle, and with it the niche concept, is thus reduced to trivality on the presumption that no two species will be identical in all respects, and therefore some resource dimension will explain coexistence (Hardin's "axiom of inequality"). In addition to conceptual problems, many naturalists were skeptical of the relevance of "bottle experiments" to natural populations (see, e.g., discussion by Mayr in Park, 1939, p. 254), which could only be established by attention to the problem of explaining coexistence among species in natural communities.

David Lack (1947a; see also Kingsland, 1985) studied Galapagos finch species that seemed to share niches, prompting him to give a nonadaptive explanation but then to change his interpretation in light of fine discriminations in resource utilization based on bill size. Stephen Jay Gould (1983) identifies this shift as part of a broader "hardening" of the evolutionary synthesis between the 1940s and 1960s against nonadaptive explanations.

Hutchinson and his student Robert MacArthur also raised questions about how the evolution of niches permits coexistence. Hutchinson's famous "paradox of the plankton" concerned the partitioning of an apparently homogeneous environment so as to permit coexistence of similar planktonic species (Hutchinson, 1961). His concept of limiting similarity expressed the limits of niche overlap that permit coexistence in measurable terms (Hutchinson, 1959; Abrams, 1983). MacArthur made an equally famous study of warblers that showed that careful field observation may reveal how species avoid competition on the slenderest of niche separations (MacArthur, 1958). Equally important were the efforts by Robert MacArthur, Richard Levins, Jonathan Roughgarden, Thomas Schoener, and others to develop operational measures of niche metrics that could be applied to field studies (reviewed in Colwell and Fuentes, 1975; Schoener, 1989).

Hutchinson's geometrical abstraction of the niche. G. E. Hutchinson (1957, 1959, 1965, 1975, 1978) formalized the niche concept in terms of the occupation of a hypervolume of a phase space whose dimensions represent all the environmental factors acting on organisms. This represented a radical shift in the meaning of the niche concept from the place or role of a species or population in a community to the environment utilization properties of the occupying species (see Schoener, 1989). Although the credit for this shift is usually accorded to Hutchinson for his famous "Concluding Remarks" of 1957, Hutchinson himself credits MacArthur for the concept of fundamental niche.

The *fundamental* niche represented the range of environmental factors that would permit the occupying species to persist indefinitely. Points in the physical environment (biotope) map onto points in this abstract niche space. The *realized* niche represented that fraction of the fundamental niche in which the species actually persists, that is, the actual part of the fundamental niche that does not overlap that of other species plus that overlapping part in which the species can persist by excluding competitors. The realized niche is, therefore, defined in relation to a set of other species.

Hutchinson's conception shifted the focus away from the (possibly empty) place or role and toward the occupant of that place in a community, defined by the total multivariate range of permissive environmental conditions, including the presence of other species that may compete with the occupant for resources. Nevertheless, whether niches are environmental or populational attributations (see Colwell on NICHE) depends on how "resources" are treated. Because Grinnell and Elton emphasized other species as well as physical resources such as light and heat, "empty niches" can clearly exist in the sense that the species serving as resources might occur without the presence of a given species. Competition became of central importance in understanding the realized niche, because competitors interacted to restrict each other's occupancy of regions of overlapping niche space.

Hutchinson's definition is of the niche *of a species*. The region that constitutes the fundamental niche is characterized either set-theoretically or geometrically, usually in terms of the set of points (combinations of environmental factor values) at which a species persists. Hutchinson's abstraction is static, atemporal, and does not immediately suggest how to represent variability of organisms in their utilization of the environment over time or variability of populations or species in space and time. Thus the nonoverlap of realized niches appears to be a logical consequence of the definition of fundamental niches such that species persist at points in niche space. If organisms range over an array of points in utilization space, it is less clear whether overlap is logically excluded in all competitive situations.

Others, including MacArthur (1968), sometimes mistakenly characterized Hutchinson's niche as a property of organisms. Because Hutchinson's followers were interested in the evolution of niche properties, allowance had to be made for variability. The dimensions of niche space came to be interpreted as actual resource utilization, with species represented in terms of clouds of points or probability densities that could be partitioned into "within" and "between" organism components (see, e.g., Roughgarden, 1972). Their work represents a final shift in the meaning of niche.

Modern "niche theory," as founded by MacArthur and Levins and others, shifted the focus from the permissive range of environmental conditions to the actual resource utilization distribution of species (Levins, 1966, 1968; MacArthur and Levins, 1967; MacArthur, 1968, 1972b; see also Hutchinson, 1978, and Schoener, 1989, for further review and discussion). This shift not only resolved some aspects of the question of how to operationalize the Hutchinsonian niche but addressed the conceptual problem of how to integrate the niche concept with ideas about the evolution of niche properties such as niche breadth, niche overlap, and limiting similarity as these relate to distributions of resource use within and among individuals of populations of possibly competing species.

The mathematical characterization of niche properties or metrics rests on both the older and the newer conceptions of the niche. The old ideas about place and role give the logical conditions under which competitive exclusion and other consequences of interaction among organisms within and among species can be drawn. These in turn are given operational, mathematical expression in terms of the Hutchinsonian concept of a multidimensional niche space and its operationalization in the resource utilization distribution. Much of competition theory, for example, is developed in terms of competition coefficients expressing the relative effects of members of one species on another over a range of environments. These coefficients include terms that can be interpreted as measures of niche breadth and overlap (Colwell and Futuyma, 1971).

The logical connection between old and new concepts of the niche should caution students of ecology and evolution against assuming that the mathematization of niche theory has somehow made the older qualitative views superfluous or outmoded. Evolutionists have long recognized that study of their historical predecessors can provide considerable insight into modern concepts. Ecologists have not devoted the same attention to their own past, and one is tempted to speculate that some current controversies in theoretical ecology reflect a lack of attention to the historical roots of ecology's central concepts.

NICHE: A BIFURCATION IN THE CONCEPTUAL LINEAGE OF THE TERM

Robert K. Colwell

WHATEVER THE subtle differences between Joseph Grinnell's and Charles Elton's concepts of the niche (see Griesemer on NICHE; Schoener, 1989), both clearly identified the niche as *an attribute of the environment*. For example, Grinnell (1917a) wrote that the niche of the California Thrasher was "one of the minor niches which with their occupants all together make up the chaparral association." Elton (1927), noting that the arctic fox and the spotted hyena eat both bird eggs and carrion, concluded that each of these two species therefore occupied "the same two niches" (the egg-eater niche and the carrion-feeder niche).

Although they often focused on particular species as examples, it is clear that neither Grinnell nor Elton saw these accounts principally as descriptions of niches themselves, but as examples of how animals filled them. Each niche was a role in the ecological drama that different actors might fill at different times or in different places. In ecological time, a "vacant niche" was an opportunity for survival and reproduction. In evolutionary time, it was an opportunity for adaptation or for speciation (Levins and Lewontin, 1985, pp. 67–71). I will refer to the Grinnell/Elton niche as the *environmental niche* concept, a term actually introduced by Grinnell (1924).

G. E. Hutchinson (1957), in a move that Schoener (1989) aptly calls "revolutionary," explicitly redefined the niche as *an attribute of the population* (or species) *in relation to its environment*. His hyperspatial niche, whatever its complexities, was at base simply an ecological description of the phenotype of some particular population or species. Just as genotypes and their frequencies within a population can be conceptualized as a "gene pool," the Hutchinsonian niche of a population describes ecological aspects of the phenotypes of a population, including physiological characteristics and relations with other species (Hutchinson, 1957; Colwell and Fuentes, 1975). In this view, the ecological drama has become improvisational theater—each species writes its own script, scene by scene, as the drama unfolds. Ecological opportunities take the form of available resources (in the broadest sense), not vacant niches. The exploitation of evolutionary opportunities are described as "niche shifts" or adjustments (e.g., Roughgarden, 1976), not as a process of adaptation *to* a niche. I will refer to this more recent view as the *population niche* concept.

Ecological equivalents. The concept of the niche as an attribute of the environment arose in part as a means of accounting for "ecological equivalents"—different species that function in similar ways in different regions or in different habitats, as in Elton's example of the arctic fox and the spotted hyena (Grinnell, 1924; Elton, 1927). For example, many striking parallels in morphology, behavior, and interspecific relations link four unrelated lineages of nectar-feeding birds on different continents—hummingbirds (Trochilidae) in the New World, sunbirds (Nectarinidae) in Africa and Asia, honeyeaters (Meliphagidae) in Australia and New Guinea, and honeycreepers (Drepanididae) in Hawaii (see *American Zoologist* 18: 681–819). According to the environmental niche concept, these groups of birds (or particular members of them) "fill the same niche" in different places. In the terminology of the population niche concept, the niches *of* these birds, which live in different places, are remarkably similar.

But the distinction between these two concepts of niche is merely a trivial question of words. If the niche is an attribute of the population then the statement that two species "have similar niches" represents nothing more than a shorthand *description* of ecological, morphological, and behavioral similarity. The similarities themselves are phenomena requiring explanation in terms of ecological and evolutionary processes. In contrast, if niches are attributes of the environment, the statement that two species "fill the same niche" in different places or times may be taken not only as a description of ecological similarity but also, in itself, as an *explanation* for similarity and a statement about the repeatability of environmental structure (e.g., Price, 1984; Giller, 1984, p. 19).

The problem of vacant niches. If niches are attributes of the environment, then they are equally meaningful whether filled or vacant. If the niche is an attribute of the species or population, then the idea of a *particular* vacant niche is either meaningless (in prospect) or at most (in retrospect) implies recent extinction or local disappearance. In the environmental view, a niche may be altered as changes in the composition of communities or in the abiotic environment affect the kinds of resources available and the relations between species (a point that Grinnell made especially clear)—but the niche remains a role largely independent of its occupancy (Grinnell, 1914; Taylor, 1916; Price, 1984; Lawton, 1982, 1984). Like an actor playing Hamlet in repertory, another actor may replace him if he is run over by a truck. Different actors may play Hamlet in somewhat different ways, and approaches to the role may change over time as the social context of the theater changes, but the role itself exists apart from the performance.

In contrast, if the niche is viewed as an attribute of the population, temporal changes in available resources or relation between species are treated as pressures that directly alter the niche of a focal species, through phenotypic plasticity or genetic adaptation. If the species approaches extinction as a result of such environmental changes, it does so because the characteristics of its niche fail to adjust or adapt sufficiently, not because the niche it once occupied has changed and left the species behind.

The environmental niche concept, including the corollary notion of identifiable vacant niches, has persisted up to the present, in sometimes uneasy coexistence with the population niche concept (e.g., Herbold and Moyle, 1986). Even Hutchinson, in his seminal paper formalizing the population niche concept, speculates whether species-poor faunas of waterbugs imply the existence of empty niches; he also states that the "rapid spread of introduced species often gives evidence of empty niches" (1957). In his most recent and by far most extensive treatment of the niche, however, this usage has disappeared entirely, except in historical discussion of the work of Grinnell and others (Hutchinson, 1978).

The popular media, especially natural history documentaries made for television, consistently rely on the concept of the environmental niche. Moreover, the idea of vacant niches is often used as an appealing explanation for the familiar phenomenon of introduced pests. For example, D. Quamman (1987, p. 30) writes: "Every species of organism requires its own ecological niche . . . whether it happens to be a native creature in an ancestral habitat or an invader in a new one. If no niche is available, the invader species will meet ecological resistance . . . [but] each vacant niche is an invitation to some nasty biological surprise."

Although many academic ecologists use the niche concept in its older, environmental, sense as well and its population sense in introductory teaching and in casual discussion, some ecologists (e.g., Price, 1980, 1984; Lawton, 1982, 1984; Walker and Valentine, 1984) continue to use the environmental niche concept and the related idea of vacant niches in serious treatments of the empirical issue of whether or not communities are "saturated." Two kinds of data are cited as evidence for the existence of *particular* vacant niches (as opposed to simply the existence of unused resources or "empty niche *space*"), and by extension, for the prior existence of environmental roles.

Some studies point to the absence of certain quite specialized organisms where they might be expected on the basis of their presence in similar environments elsewhere. J. H. Lawton (1982, 1984), for example, found that certain classes of herbivorous insects (e.g., miners of the "leaves" or pinnae) present on bracken (*Pteridium aquilinum*) in Britain are absent on plants of the same species in New Guinea and New Mexico. His hypothesis is that the "pinna-miner niche" (among others) on bracken is vacant in the latter two regions, probably because no appropriate insects have appeared there. Numerous similar proposals have been made for other cases of specialized organisms by P. W. Price (1980, 1984).

Proposing that a particular niche is vacant, in this sense, means that all the resources required for survival of an appropriate but absent species are currently available but not being utilized and that the introduction and establishment of an appropriate species would have no significant effect on the rest of the community, other than direct effects on the resources themselves. Testing the hypothesis by making experimental introductions, aside from its ethical and legal problems, may confirm the hypothesis, but cannot disprove it. If an introduction fails to become established, or becomes established but displaces or replaces one or more established species, the argument can always be made that the wrong species was chosen, or even that the appropriate species has not yet evolved.

Testing the hypothesis that one community has a vacant niche that is filled in another without experimental introductions is even more onerous, carrying a heavy burden of proof of *ceteris paribus*—many competing hypotheses about other differences in the two cases must be disproved. In the case of Lawton's bracken, these include climate, natural enemies, plant genotype and phenotype, and indirect effects from other bracken herbivores mediated through plant physiology, including both nutrients and defense substances. Nonetheless, in principle the hypothesis is testable, and other evidence does suggest that vacant niches, in this well-defined sense, may well be common among specialized herbivores (see Strong et al., 1984). The rigorous disproof of their nonexistence, however, is most daunting.

Lawton's intrinsically interesting study, if conducted by an adherent to the population niche concept, would probably not be framed in terms of niche language at all. Instead, the question would be phrased in terms of conditions and resources: "Are the environmental conditions (plant genotypes and phenotypes, microclimate, enemies) similar for bracken insects in Britain and New Guinea? If so, are the resources (photosynthate, other nutrients, anatomical access) used by stem borers on British bracken left unused in New Guinea bracken?" If so, then a British stem borer would be expected to become established on bracken in New Guinea, if introduced, without significantly affecting any other member of the bracken insect community.

Vacant niches and introduced species. The second line of evidence cited by academic proponents of the vacant niche idea comes from actual introductions of exotic species, precisely along the lines suggested in

Quamman's (1987) popular account. Walker and Valentine (1984), for example, cite Simberloff's (1981) analysis of recorded introductions to make the claim that "there are indeed empty niches in which introduced species can take up occupancy without disturbance of the resident community, and that successful introductions often occur in unoccupied niches" (p. 887). Herbold and Moyle (1986), however, after reviewing Simberloff's (1981) sources in detail, offer persuasive arguments that his study "offered poor support for a model of vacant niches." An introduced species, they conclude, nearly always "rearranges the community, rather than slipping into an empty slot" (p. 757).

With Simberloff's study as support, however, Walker and Valentine (1984) use rates of diversification of fossil lineages to predict maximum potential species numbers based on the assumption of logistic increase in species number. They estimate the proportion of "empty niches" as the difference between these numbers and estimates of extinctions over the same period for these lineages. The entire procedure rests on two unstated and unsupported assumptions: (1) that the "missing species" somehow differ in ecological role from those actually present and (2) that speciation, *without* extinctions, would nonetheless come to a halt at some maximum diversity for each lineage.

In summary, although often criticized by proponents of the population niche concept (e.g., Colwell and Futuyma, 1971; Whittaker et al., 1973; Levins and Lewontin, 1985; Herbold and Moyle, 1986), the environmental niche concept (and the notion of vacant niches in particular, in the sense of Price and Lawton) cannot be dismissed simply by asserting that the population niche concept is "correct," and thus, by definition, no niche exists without a species in it. *An empirical claim* that particular niches exist apart from the species that fill them is extremely difficult to test, however, because it is virtually unfalsifiable, even in the best of cases. In itself, the fact that any imaginative naturalist can describe an unlimited number of unfilled niches for which plausible organisms might exist casts serious doubt on the operational utility of the environmental niche concept in its broadest sense.

The niche and the public. The environmental niche concept has recently figured in public policy debates in regard to the environmental use of engineered organisms (e.g., Davis, 1984; Simberloff and Colwell, 1984). Influential advocates of little or no government regulatory oversight of such uses (e.g., Davis, 1984) have made the claim that organisms (particularly microorganisms) in nature have had eons to become perfectly "adapted to some ecological niche," so that any genetic alteration made to them through genetic engineering "is infinitely more likely to impair rather than to improve the adaptation," producing poor competitors that will be lucky to survive at all. Thus, they argue, there is little risk involved in the environmental introduction of the genetic novelties produced by biotechnology—even those intended for long-term survival in the open environment.

In fact, the view of microbial biologists generally seems to be that niches are fixed and microorganisms evolve to fill them (e.g., Brock, 1985). In support of this view, the resources used by microbes are often chemically simple compounds that indeed can be metabolized in only a very limited number of ways. Non-nutritional aspects of microbial ecology (such as interspecific interactions), however, which may well be less determined by the environment, are too often ignored (Tiedje et al., 1989), in spite of their relevance to the niche.

In addition to using the environmental niche concept as a way of conveying an image of the "wisdom of nature," the popular media frequently imply that different organisms perform their niche-defined duties for the good of the community or ecosystem as a whole. Evolutionary ecologists reject this point of view as a generality for lack of either empirical evidence or plausible evolutionary mechanism, although under special circumstances species may evolve to one another's mutual benefit (D. S. Wilson, 1980). In the case of microbial communities, for example, species are sometimes organized functionally into "consortia" in which each member of a set of species performs a different, essential step of a chain of chemical transformations. To the degree that such species are interdependent and alternative pathways are not possible, microbial consortia do seem to approach the "beneficent" ideal of the natural history television scriptwriter.

The use of the term "niche" in business jargon adds further complications to the popular understanding of the ecological niche (e.g., Anonymous, 1989; Bradburd et al., 1989). A business management journal once republished (Colwell, 1985a) a long excerpt from an article of mine that had first appeared in the popular magazine *Natural History* (Colwell, 1985b). The original article, about the natural history of hummingbird flower mites, had no business relevance whatsoever, and never even used the word "niche"—but the excerpt in the business publication was entitled "Life in the Niches." The editor's intent was to show his readers the arcane ways in which these obscure animals make a living, to suggest how imaginative one must be to "find an empty niche" in business. It is worth noting that the idea of filling or "carving out" a vacant market niche, unlike the strict criteria for filling a vacant ecological niche, in no way precludes the "disturbance" or even the "extinction" of the resident occupants of other market niches. For example, the rapid ascendancy of facsimile machines suggests that they have filled a vacant market niche, but the effect on the telex and overnight mail delivery industries has been significant.

The notion of the vacant niche has its folk counterpart in the long-standing and misguided tradition of introducing exotic organisms that are "missing" from the colonial homes of immigrant settlers, too often with disastrous results. In mainland America and in Hawaii, for example, societies of bird fanciers used to exist for the sole purpose of introducing European birds to the New World (Moulton and Pimm, 1983). Hunting and fishing interests have been responsible for an endless series of destructive and ill-considered introductions of game animals around the world, often with the justification that no sufficiently similar organisms were already present (Elton, 1958; Moyle, 1986).

The niche and competition. Historically, the concept of the niche as an attribute of a population has been inextricably connected with empirical investigations and modeling of the role of competition among species in community structure. The development of the body of work known as "niche theory" (reviewed by Schoener, 1986, 1989) began with the work of Robert MacArthur and Richard Levins (1967), which focused on a noncircular restatement of the "competitive exclusion principle." (See Griesemer on NICHE, Kingsland, 1985, and Schoener, 1989, for reviews of the history of competitive exclusion.) Qualitatively stated, their idea of "limiting similarity" was simply this: the more ecologically similar two competing species are, the less likely they are to coexist, so there must be some limit to the ecological similarity of coexisting species. The degree of similarity, and thus the level of competition, was measured as niche overlap in relation to niche breadth, for which many theoretical formulations and empirical measures have been devised and explored in the intervening years.

Thus "niche theory" has become synonymous with competition theory, but for reasons of history rather than logic. In contrast, the environmental niche concept, ever since its origin in both Elton's and Grinnell's writings, has explicitly included not only competition but predation as key interactions defining the ecological roles of species—both "food and enemies." Eduardo Fuentes and I attempted to broaden the population niche concept to include predation, parasitism, and mutualistic interactions as well as competition in our review of experimental work that could be viewed as pertinent to the niche (Colwell and Fuentes, 1975). Judging from subsequent history of theoretical and empirical use of the population niche concept (e.g., Schoener, 1989), however, our success has been minimal.

Because of its close ties to competition theory, the fate of the population niche concept has been inevitably tied to the shifting fortunes of interspecific competition as an explanation of community structure. But the word "niche" need not imply any particular view on the importance or irrelevance of interspecific competition in communities. The identification of niche theory with competition theory, however, together with the desire of most ecologists to avoid any a priori presumption that competition is important in communities, appears to have produced a strong decline in the use of niche terminology in community ecology. Whether this decline will lead to the extinction of this conceptual branch of the niche remains to be seen.

PARSIMONY

Elliott Sober

THE PRINCIPLE of parsimony ("Ockham's razor") is a methodological injunction that has figured in many scientific debates concerning how rival hypotheses should be evaluated. The principle says that the greater parsimoniousness of one hypothesis over another counts in favor of the former, when the question is which of the rivals should be regarded as true. Sometimes the principle is stated by saying that if two hypotheses are both compatible with the observations (or both "fit" the observations equally well), then the more parsimonious one has the better claim to be regarded as true.

What makes a hypothesis parsimonious? More parsimonious hypotheses postulate fewer processes, entities, or events of some specified type than do their rivals. What those processes, entities, and events are, of course, varies from one scientific problem to another. More parsimonious hypotheses are standardly said to be simpler.

One prominent recent example of an appeal to parsimony in evolutionary biology is G. C. Williams' (1966) landmark argument against group selection hypotheses. Williams claims that hypotheses of adaptation are "onerous" and should be invoked sparingly. To explain why airborne flying fish return to the water, for example, one does not need to invoke a specific adaptive mechanism; rather, the mere physical fact of gravity suffices. Williams claims that this methodological idea entails that lower-level selection hypotheses are to be preferred over higher-level ones. A trait is better explained by viewing it as an individual adaptation than as a group adaptation. The reason, Williams says, is that the former is more parsimonious.

An example that Williams uses is the behavior of musk oxen when attacked by predators. The males in the herd stand shoulder to shoulder to form a circle and face the attack; the females and young are protected at the center. Males thereby help protect females and young to whom they are not related. How can the idea of group adaptation explain this apparently altruistic act? Perhaps groups in which males behaved altruistically went extinct less often and founded more colonies than did groups in which the behavior was absent. The behavioral trait evolved because it was advantageous to the group, even though it was disadvantageous to the individuals possessing it.

Williams views this explanation in terms of group selection as unparsimonious. He prefers the following individual selection explanation. Each individual deploys a policy that determines in a given circumstance whether it engages in fight or flight. Such policies evolved because they were good for the organisms possessing them. It is individually advantageous for an individual to flee from a decisively stronger attacker, but to stand and fight an attacker who is less intimidating. A prediction of this idea is that there are attackers that lead adult males to behave one way and females and young to behave the other. Williams advances the conjecture that the wolves that attack musk oxen fall in this category. The wagon-training behavior is not a group adaptation but the statistical summation of individual adaptations.

Williams sees this appeal to parsimony as providing a reason to think that group selection hypotheses are false and lower-level selection hypotheses are true. He does not reach an agnostic conclusion about musk oxen; after describing the group selection and the individual selection hypotheses, he concludes that we should prefer the latter.

The principle of parsimony, both in and out of evolutionary biology, has been used in this way. The claim has been advanced that more parsimonious hypotheses are not simply good candidates for further investigation; nor are they merely easier to grasp and think about. They are also more likely to be true.

Williams' parsimony argument is no longer the keystone of the case against group selection. In the years following his book, a number of biologists investigated mathematical models to see under what circumstances an altruistic character could evolve and be maintained by group selection. The consensus reached was that this is possible only within a narrow range of parameter values. Although some of the assumptions in the models seem a priori to bias the question posed (Wade, 1978), this conclusion is still taken seriously by the majority of evolutionists.

The most influential argument against group adaptation at the present time is thus a biological one, not a purely methodological one. It isn't that the hypothesis of group selection is less parsimonious than the hypothesis of individual selection. The idea, rather, is that group selection requires an improbable coincidence of biological circumstances to occur in nature. Parsimony helped dislodge the idea of group selection, but the hegemony of this attitude is presently maintained by other means.

Even though parsimony no longer occupies center stage in the debate about units of selection, important questions can be raised about the use of parsimony in Williams' argument (Sober, 1984a). Why should the fact that one hypothesis is more parsimonious than another be a reason to think that the first is true and the second is false? Does the use of parsimony in evolutionary biology presuppose that evolution proceeds parsimoniously? If so, does our knowledge of the evolutionary process support or confirm this supposed prerequisite?

Even if the units of selection controversy is now fairly independent of these questions about parsimony, the same cannot be said of the controversy about phylogenetic inference. During the 1960s three different groups of biologists formulated the idea that the most plausible hypothesis of the phylogenetic relationships that obtain among a set of taxa is the one that most parsimoniously accounts for the taxa's observed characteristics. R. A. Fisher's students A. Edwards and L. Cavalli-Sforza (1963), who approached the problem of phylogenetic inference from the standpoint of the idea of likelihood, stated the idea but said that its presuppositions are none too clear. J. Camin and R. Sokal (1965), the latter a founding father of phenetics, described the idea and asserted that the parsimony idea implausibly presupposes that evolution proceeds parsimoniously. And W. Hennig (1965), laying down the fundamental ideas of cladistics, espoused an "auxiliary hypothesis," according to which shared derived characters (apomorphies) must be assumed to be homologous, unless there is specific evidence to the contrary. Although Hennig never used the term "parsimony," his view that apomorphies constitute *the* source of evidence leads directly to the idea of phylogenetic parsimony, which now is the central cladistic approach to phylogenetic inference (cf., e.g., Eldredge and Cracraft, 1980; Nelson and Platnick, 1981; and Wiley, 1981).

To understand how the concept of parsimony is used to make phylogenetic inferences, consider three species, *A*, *B*, and *C*. Suppose that none of these is ancestral to any of the others. We wish to infer which two of these form a monophyletic group apart from the third. The two hypotheses we will consider, illustrated in the accompanying figure, are (AB)C and A(BC). The (AB)C hypothesis says that *A* and *B* have an ancestor (represented by a node in the branching diagram) that was not an ancestor of *C*; it says that *A* and *B* are members of a monophyletic group to which *C* does not belong. The A(BC) hypothesis says that *B* and *C* form a group apart from *A*.

Given that direct observation of the evolutionary branching process is not an option, we might hope to choose between these two hypotheses by observing the characteristics exhibited by the three species. Consider a character that comes in two states, denoted by 0 and 1. For example, if *A* and *B* were two species of bird and *C* were some species of crocodile, the

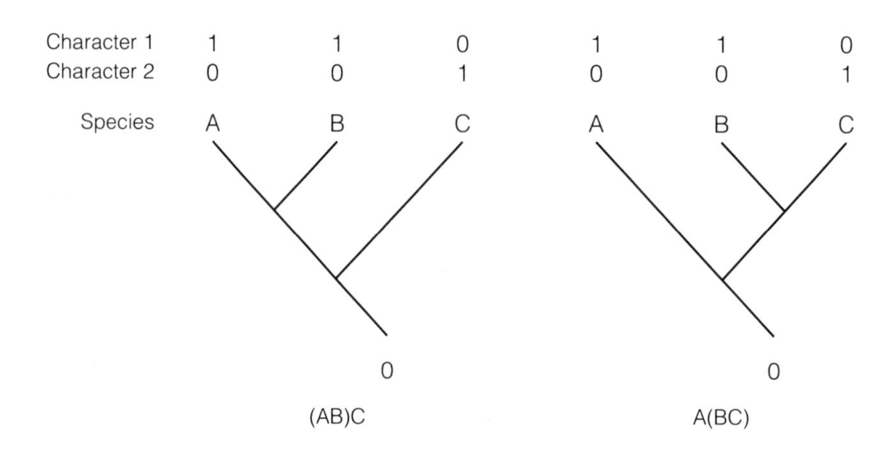

character might be "mode of locomotion," whose two states might be "not flying" and "flying." We observe that *A* and *B* fly, but *C* does not. How does this observation help us choose between the (AB)C and the A(BC) hypotheses?

Parsimony addresses this question as follows. Suppose we know the character's *polarity.* That is, suppose we know which character state was present in the species ancestral to all three taxa. An example of this sort of knowledge would be the proposition that the ancestor common to crocodiles and birds did not fly. If so, then the presence of flying in *A* and *B* counts as a derived (apomorphic) similarity, not an ancestral (plesiomorphic) one.

Character 1 in the figure provides an example of a derived character state present in *A* and *B* but absent in *C.* Consider how the (AB)C hypothesis could explain the character states found at the ends of the branches of the diagrams. For those character states to be present at the ends of the branches, given that the base of the diagram is in state 0, only a single change in character state (from 0 to 1) is needed. For A(BC) to explain the observations, however, at least two changes are needed. Each hypothesis is consistent with the number of changes being greater than the *minima* just stated. Because (AB)C requires only one evolutionary change in state, while A(BC) requires two, (AB)C is judged to be the more parsimonious explanation of Character 1.

This discussion of Character 1 illustrates why considerations of parsimony imply that derived similarities should be interpreted as evidence of phylogenetic relatedness. If we turn now to Character 2, we can see why parsimony implies that ancestral similarities are not evidence of relatedness. If the ancestor is assumed to be in state 0, then Character 2 has *A* and *B* exhibiting an ancestral character state that is not present in *C.* The

(AB)C hypothesis can explain this sort of observation by postulating a single evolutionary change in state, *and the same is true of the A(BC) hypothesis*. The two hypotheses provide equally parsimonious explanations of Character 2.

Parsimony in the phylogenetic inference problem is measured by counting the number of changes in character state that a tree topology requires to generate the observations at the tips. The principle of phylogenetic parsimony is simply the idea that derived similarities are evidence of relationship and ancestral similarities are not. Because of this dismissal of ancestral similarity as evidentially meaningless, defenders of parsimony disagree with those who use overall similarity to infer relationships.

A heated biological controversy has ensued since the 1960s. Much of the criticism elaborates the claim advanced by Camin and Sokal (1965)—that the use of parsimony presupposes that evolution proceeds parsimoniously. The allegation is that parsimony assumes that change is very improbable. J. Felsenstein (1983) has attempted to show that parsimony makes this assumption by subjecting phylogenetic inference to a likelihood analysis; he claims that the most parsimonious tree can be assumed to be the tree of maximum likelihood only when rates of change are assumed to be very low.

It is generally agreed that if parsimony were to have this presupposition, the method would be quite suspect. The reason is that convergent evolution is a familiar phenomenon. And as Felsenstein has stressed, systematists who use the parsimony criterion often find that the most parsimonious tree requires multiple originations on many of the characters it analyzes.

Defenders of parsimony have denied that their favored method makes this presupposition. E. O. Wiley (1981), for example, says that nothing is assumed beyond the idea that the species considered are the product of descent with modification.

One prominent defense of parsimony has been Popperian. N. Eldredge and J. Cracraft (1980), G. Nelson and N. Platnick (1981), and Wiley (1981) all argue that the most parsimonious hypothesis is the one that is most falsifiable; it is the one that requires the fewest ad hoc hypotheses. They see parsimony as a basic ingredient in the scientific method—a principle that does not require substantive assumptions about the process under investigation.

In the same vein, J. Farris (1983) has argued that phylogenetic parsimony favors hypotheses of greater explanatory power. He also emphasizes the difference between minimizing assumptions and assuming minimality; the use of parsimony, he says, involves the former, not the latter.

Sober (1988) reviews these and other arguments concerning phylogenetic parsimony. He proves within a likelihood framework (in an idealized model in which evolutionary rates are assumed to be uniform) that apomorphic resemblances are evidence of relatedness, regardless of whether change is expected to be rare or common. He also shows, within the same model, that plesiomorphic resemblance is not evidentially meaningless. Sober further establishes that derived similarity is better evidence of relatedness than is ancestral similarity, within a broad, though not universal, range of parameter values.

Parsimony involves different considerations in different problem contexts. Williams made a judgment about the complexity of selection processes in arguing that individual selection hypotheses are more parsimonious than hypotheses of group selection. Defenders of phylogenetic parsimony count the number of evolutionary changes in character state that a tree topology requires to explain observed character distributions. The possibility remains, however, that underlying this diversity of subject matters, there is a single methodological principle whose validity does not depend on special assumptions about the subject matter considered. This possibility is sometimes expressed by saying that the principle of parsimony is "purely methodological"—that it requires no substantive assumptions about the special subject matters to which it is applied.

To discuss whether there is a single "principle of parsimony," invariant over subject matter, would require investigation into the theory of scientific inference. Such an inquiry is more often pursued by philosophers than by biologists, yet it is of central importance to the methodological issues canvassed here. For disputes about the reasonableness of parsimony considerations in a specific problem setting inevitably lead to broader questions about the relevance of parsimony to scientific inference as a whole. In this respect as well as in others, it is natural to see scientific issues flowing into philosophical ones.

PHENOTYPIC PLASTICITY

Deborah M. Gordon

CHARLES DARWIN wrote in 1881, "I speculated whether a species very liable to repeated and great changes of conditions might not assume a fluctuating condition ready to be adapted to either condition" (1881, p. 6). This is a speculation about phenotypic plasticity; its terms are not defined. What are the time scale and consequences of a species' "fluctuating condition"? Since Darwin's time, the notion of phenotypic plasticity has taken on several different meanings. Here I will use "plasticity" and "flexibility" interchangeably, to mean simply a capacity for change or transformation.

Usage: What "phenotypic plasticity" can mean. In the literature on phenotypic plasticity and its evolutionary consequences, the meaning of terms such as "plasticity," "flexibility," and "change" can vary along three axes:

1. *Time scale: across or within generations.* "Phenotypic plasticity" can be considered the source of changes occurring in evolutionary time, that is, on the time scale of many generations, or of changes occurring within the lifetime of an individual, within a single generation.
2. *Variation: among or within populations.* Phenotypic plasticity can be seen as a source of variation among populations. Examples of this would be variant generations or variant populations in distinct environments. Plasticity can also be seen as a source of individual variation within a population and/or within a generation.
3. *Relation to environmental change: continuous versus discrete, and reversible versus irreversible.* Both environments and organisms can change in ways that are continuous or discrete. Temperature and height are gradients along which environments and organisms might undergo continuous change. Examples of discrete transitions

> would be those between a marine and terrestrial environment, or between the larval and pupal life history stages of an insect. Change can also be reversible, such as changes of leaf position in response to light, or irreversible, as in transitions between life history stages. Many combinations are possible along the continuous/discrete, reversible/irreversible dimensions, but two are most common in the literature. Plasticity is generally viewed either as a progressive response to continuous environmental change, or as an irreversible response triggered by a particular environmental event.

When phenotypic plasticity is viewed as a source of variation within a generation, it draws strongly on the idea of a norm of reaction, that is, heritable phenotypic variation among individuals of a single genotype, elicited by variation in environmental conditions (Gause, 1947; Gupta and Lewontin, 1982). For example, for Meyer (1987), morphological variation within populations of cichlid fishes is an example of phenotypic plasticity; so, too, for Silvertown (1984), is variation within clutches in seed germination times; and, for Palumbi (1984), variation among sponge populations in tissue structure. Phenotypic plasticity, as these authors use the term, is almost equivalent to a norm of reaction; individual variation in phenotype is caused by environmental variation. The difference from a norm of reaction is that here the observed variation, seen to be an expression of phenotypic plasticity, is not necessarily among individuals that are genetically identical with respect to the character being considered. In most cases in which the notion of phenotypic plasticity is used in this way, the genetics is not well enough known to justify the assertion that all individuals are genetically identical.

A range of environments might permit individuals to exhibit a range of phenotypes in two ways. First, an individual might respond to environmental change by all-or-nothing, irreversible changes in its phenotype. Alternatively, an individual might be flexible, capable of various, reversible transformations in response to a range of fluctuating conditions. Bradshaw (1965) and Smith-Gill (1983) consider this distinction to be crucial in any discussion of the evolution of phenotypic plasticity. They use "plasticity" to refer to irreversible change, and "flexibility" to mean a reversible response to fluctuating conditions, but this usage is by no means universal.

As it is used in quantitative genetics models (Via and Lande, 1985), phenotypic plasticity refers to variation from one generation to the next. Measures of the variation are used to make inferences about evolutionary changes of gene frequencies in a population. The value of a phenotypic character is averaged over all individuals in a population in a given environment. Any particular individual's value of this character will deviate from the population average. In the absence of interactions, an individual's value can be specified as the average value, plus deviation from the average due to genetic differences from the average genotype, plus deviation from the average due to differences of the given environment from the average environment.

Consider two populations in two environments. Average values of a phenotypic character differ in these two populations. This may be because they are genetically identical but responding to two different environments; phenotypic differences would then be due to the reaction norm of that genotype. A second possibility is that the two populations are genetically different, and the phenotypic differences are due only to these genetic differences; environmental differences are irrelevant. A third possibility is that each population contains only one genotype, the genotypes of the two populations differ, and the two environments are the same. Then phenotypic differences would be due to differences in the way that each genotype responds to the same environment. Such differences would be the result of a gene-environment interaction. Via and Lande use "phenotypic plasticity" to mean differences between populations in gene-environment interactions. Phenotypic plasticity has both a genetic and an environmental component because it is based on genetic differences in response to environmental conditions. In the quantitative genetics view of plasticity, each individual experiences only one environment, and the value of the phenotypic character of that individual is assumed to persist throughout its life. Stearns (1982) uses "phenotypic plasticity" in this sense; in his formulation, differences between populations in different environments are due either to phenotypic plasticity (related to gene-environment interactions) or to selection (related to changes in gene frequency alone). Similarly, Lynch (1984) uses the term to describe morphological changes in successive generations of *Daphnia* that result from changes in environment.

Plasticity in the quantitative genetics sense is sometimes said to be a change in the mean reaction norm, from one population to another. This can be confusing because the notion of reaction norms is used elsewhere to draw attention to variation within a population. To specify a reaction norm is to specify not only the range of variation of some character within a population but also its mean value. Thus when a reaction norm of a character differs from one population to another, the mean value of that character may also differ. In quantitative genetics models, what is of interest is a difference in the average value of a phenotypic character from one population to another. This difference in average value is called phenotypic plasticity when it is caused by different reaction norms.

The important point here is that phenotypic plasticity, in the quantitative genetics sense of the term (e.g., Via and Lande, 1985), may not contribute to variation among individuals within a population or generation.

It causes one generation as a whole, on average, to look different from the preceding one. Thus studies that use phenotypic plasticity to account for variation among individuals within a population (e.g., Meyer, 1987) are only tenuously related to those (e.g., Lynch, 1984) that use plasticity to account for differences between successive generations. Authors such as Meyer (1987) use variation among individuals within a population to demonstrate phenotypic plasticity. But for authors such as Lynch (1984), this can be irrelevant: plasticity permits variation across different populations in different environments. These different populations could be successive generations.

Measuring phenotypic plasticity. It is important to distinguish measurements of change from measures of variation. Change is a process that occurs over time; measures of variation are made at one slice in time. Change may produce variation. Individuals responding to a variety of environmental conditions may come to differ in phenotype, and this may produce variation within a population. A measure of variation may not specify the process that caused individuals to vary. To measure variation, one examines how individuals vary in static attributes, measured at a particular time. This approach was employed by Schlichting and Levin (1986), who measured certain morphological characters (i.e., height, leaf width, etc.) in several plant species. They measured these characters in several groups of individuals of each species. Each group had been raised in a different environment. Genetic variation was not considered. Within a species, variation across groups was considered to be a measure of response to different environments and thus a measure of that species' phenotypic plasticity.

A second way to measure change is to follow individuals through time, in different environmental conditions, and to measure the extent to which the phenotype of each individual responds to these differences. This is the approach I used (Gordon, 1991) in studies of flexibility in the foraging behavior of ant colonies. Harvester ants engage in a characteristic sequence of tasks each day. I measured changes in this temporal pattern in response to changes in environmental conditions such as food availability or the presence of other ant species. A colony's behavioral flexibility provides a repertoire of potential responses to certain types of environmental change. Degree of flexibility is measured as the extent to which the temporal pattern of each individual colony's behavior actually changes over a range of conditions. Colonies vary in flexibility.

The relation between variation and change is not one to one. Variation within a population can be a consequence of individual responses to differences of environment. Variation that appears in this way has been called "fabricational noise" (Seilacher, 1973). For example, Roth (1989) showed that differences in tooth morphology in elephants were not species-specific characters, but instead caused by differences in wear on the teeth as they developed, depending on what the elephants ate. For another example, consider two conspecific populations of insects. Each lives on a different houseplant. One plant, and thus one population, is moved toward the sun; the second is kept out of the sun in uniform temperature conditions. In this insect species, warmth enables the larvae to grow faster, resulting in adults of a larger size. In the first population, larvae on the sunny side of the plant grow larger than those on the shady side. If one measures the head width of all adults on the plant in the sun, there may be greater variance than in the second population. One could use this measure to assess the plasticity of the insect growth in response to temperature. Greater variance would mean a greater capacity for change in response to changed conditions.

But variation cannot always be attributed to phenotypic response to differences in environment. A population of insects all raised at exactly the same temperature may vary in size. Moreover a change in the phenotypes of individuals within a population does not necessarily lead to a larger variance. Warmth may cause the larvae of the insect species to grow faster, but a fertilizer used on the houseplant may impede their growth. Both temperature and fertilizer affect the size of adults, but their combined effects may not produce a population more varied than one living on a cold, unfertilized plant.

"Variation" is often meant to refer to differences among individuals in some static attribute. For example, the tulips in my garden vary in the sizes of their flowers. Tulips open and close their petals in response to changes in the amount of available sunlight. Ambiguity arises because it is possible to make static measurements of dynamic events. Suppose I go out and measure the flower diameters of all the tulips in my garden. These measurements will vary. Is the variation due to phenotypic plasticity, in the sense that some tulips open more in the sunlight than others, or do adult plants differ in size, or both? To answer these questions, it would be necessary to measure individuals at different times. To measure variation in phenotypic plasticity, it would be necessary both to measure individuals at different times and to compare the trajectories of different individuals. Two sets of instantaneous measures, made at different times, will give different measures of variation among individuals. For example, if I measure the tulips at night, when all are closed, I may see less variation in flower size than if I measure the tulips during the day. Unless the relation between temporal change and synchronic variation is considered, measures of phenotypic plasticity may be measures of variation among individuals in fixed attributes.

Evolution and phenotypic plasticity. In general, two kinds of evolutionary arguments have been made about phenotypic plasticity. One is concerned with selection of plasticity itself, and its evolutionary advantages, or disadvantages. In this case, plasticity is usually viewed as a characteristic of individuals, within generations, leading to variation within populations. Such views usually consider the relation between the time scale of phenotypic change and the time scale of environmental change. The second line of argument is concerned with the ways that natural selection could fix particular characters in a population, and how such selection would be affected by plasticity (Stearns, 1989). In this view, plasticity occurs on the scale of many generations, leading to variation across, not within, populations, and environmental change takes place on a time scale similar to that of generations.

In the first case, when plasticity is seen as a source of change in an individual's phenotype, during that individual's lifetime, discussion of the evolution of plasticity is often about the advantages of flexibility. Here discussion tends to follow the framework set out by Levins (1968): what is the optimal relationship between the time scale, or spatial scale, of environmental change, and the scale of phenotypic change? The relation between the time scale of environmental change and generation time is crucial. When the time scale of environmental change is much longer than generation time, selection may allow organisms to evolve faster to a fixed, optimal phenotype than would flexibility within the lifetime of individuals. When environmental change is much more rapid than generation time, the ability to change may be optimal. For example, if it is alternately cold, then hot, each year, and a generation lasts for many years, hibernation may be preferable to a thick coat that would be too hot in the summer. But if the environment becomes colder on a scale of thousands of generations, a species may evolve thick fur rather than the capacity to hibernate. In general, such arguments (reviewed in Sultan, 1987) conclude that the possibility of changing phenotype within an individual's lifetime permits organisms to respond to changing environments in ways that enhance their reproductive success. The capacity for change is itself seen as a good thing when environments are not stable.

The capacity for change may also be disadvantageous. One such argument comes out of earlier discussions of homeostasis, in which an organism's capacity to stay the same in a changing environment was seen as advantageous (Waddington, 1957b; Caswell, 1983). Here again the notion of plasticity is used in the norm-of-reaction sense, as a source of variance within generations. An organism that must change its phenotype when the environment changes is seen as one unable to maintain a stable internal milieu.

Marshall and Jain (1968) suggest a trade-off between genetic polymorphism and individual homeostasis. Polymorphism provides for the possibility that there will exist, somewhere in the population, the right or best genotype for a particular environmental condition. Plasticity either (within generations) allows individuals to change to meet changing conditions or (from one generation to the next) allows a whole generation to change to meet new conditions. For example, Bradshaw (1965) suggests that recurrent variation in selection pressures leads either to the evolution of the capacity for reversible change or to genetic polymorphism. Smith-Gill (1983), however, views reversible change as nonadaptive, but considers characteristics that provide for environmentally induced irreversible change, or "genetically controlled switches," to have evolved because they have provided an evolutionary advantage.

It has also been argued that phenotypic plasticity not only permits but actually speeds up evolutionary change. An example is Waddington's (1942) notion of genetic assimilation in response to environmental change. By increasing variance (either within populations or between generations) and providing phenotypic options for selection to choose from, heritable phenotypic plasticity provides opportunities for selection—analogous to the way that genetic variation does. In this sense, discussion of the advantages of phenotypic plasticity has clear analogies to earlier discussions of the advantages of genetic variation, of heterozygosity, and of the advantages of generalist over specialist strategies (see, e.g., Lewontin, 1974a).

The evolution of plasticity requires variation in plasticity across individuals in a population. (One could also consider variation across groups.) One way to talk about the evolution of plasticity is to ask how well a genotype capable of change can compete, over evolutionary time, with a more fixed one. For example, Maynard Smith (1988) has argued that in certain environmental conditions, a genome producing a mixed strategy or flexibility can replace a more optimal, but more fixed, genome. West-Eberhard (1989) reviews arguments for the role of plasticity in promoting speciation.

Evolutionary arguments about phenotypic plasticity in the quantitative genetics sense, that is, intergenerational change in the average values of populations, are usually concerned with the ways in which phenotypes that originally arose through plasticity could eventually become genetically fixed in populations. Here the question is how selection will affect the way that a population responds to environmental change, under the assumption that all individuals respond in the same way. This line of argument tends to consider the adaptiveness of particular characters, and how such characters may evolve if they first appear as part of the repertoire of a phenotypically plastic species. For example, Harvell (1986) discusses the evolution of inducible defenses in this way. She asks, why have bryozoans not evolved a consistent response to predation by nudibranchs? Why does the defense have to be induced? Here flexibility is viewed not as an advantage but as an intermediate step on the way to the evolution of a more optimal fixed phenotype.

In this view, plasticity is sometimes considered to be an obstacle to evolutionary change. Plasticity can be seen as a bad thing, not because plasticity itself is undesirable, but because it may slow the evolution of a particular, desirable character (Wright, 1931; Anstey, 1987). Of a range of phenotypes, only some are optimal. If the production of phenotypes were irreversible, not subject to change, then selection could choose the best ones and eliminate the others. But plasticity allows less than optimal organisms to slip out from under selection by taking a more optimal form temporarily. Via and Lande (1985) model the effects of environmentally induced changes in phenotypes from one generation to the next. Many evolutionary outcomes are possible, depending on the correlations between phenotypic and genotypic variation. In some conditions, plasticity acts to prevent a particular character from becoming more frequent (in evolutionary time).

The diversity of opinions about the evolutionary advantages and disadvantages of phenotypic plasticity is not merely a consequence of the confusion about what phenotypic plasticity is. Opinions may be diverse even when meanings are unambiguous. But the multiplicity of meanings of phenotypic plasticity does make the evolutionary discussion more obscure. Confusion arises from several sources. One is ambiguity about the time scale, level of variation, and relation to environmental change implied by a particular use of "phenotypic plasticity." The second is empirical measures of variation, used to demonstrate the existence of phenotypic plasticity, that do not specify what processes caused that variation. Finally, those who see phenotypic plasticity as a source of changes from one generation to the next in the average value of certain characters tend to view the evolution of plasticity as a step toward the eventual fixation of those characters. By contrast, those who see phenotypic plasticity in terms of individual transformations within a generation are more concerned with the relation between the time scale of plasticity and that of environmental change.

Progress

Richard Dawkins

THE GREAT CHAIN of being maintains its stranglehold. One might have hoped that Darwin would have shaken us free of it. All lineages have had an exactly equal time to evolve since the dawn of life. So, even if there is some sense in which recent animals are more advanced than their dead ancestors—a generalization that is itself disputed (Nitecki, 1988)—evolution would seem positively antithetical to the idea that *contemporary* animals should be ranked on a "higher or lower" scale. Yet, far from being displaced by evolutionary thinking, as should have happened, it is almost as though evolution gave the ladder a new extension; the phrase "evolutionary scale," as in "moving up the evolutionary scale," is bandied around as though it were as meaningful as a scale of weights or lengths. Textbooks of zoology treat the phyla in a standard order, which is implicitly, sometimes even explicitly, taken to correspond to some sort of "phylogenetic scale." For example, Volume 1 of the leading English-language treatise on invertebrate zoology is titled *Protozoa through Ctenophora* (Hyman, 1940). The preposition "through" (the American equivalent of the British "up to and including") implies that the phyla form up in a natural order, and every zoologist knows what that order is. Yet in truth it would have been as sensible for Hyman to have treated the phyla alphabetically. Here I shall analyze the many confused meanings of "higher and lower animals." I shall then go on to show that there is confusion not only in the terms themselves but in the inferences drawn from them.

If there are two animals *A* and *B*, what can it mean to say that *A* is lower than *B*? It seems to be a phantasmagoric muddle-up of the following relations:

- *A* is ancestral to *B*.
- *A* resembles their common ancestor more than *B* does.
- *A* is simpler than *B*.

- *A* is stupider than *B*.
- *A* is less well adapted than *B*.
- *A* is less adaptable (or versatile) than *B*.
- *A* resembles "man" less than *B* does.

Separately these relations might, at least in principle, be made meaningful. I shall discuss the first three as examples of how. But even if they can be made meaningful, they do not all mean the same thing and are not necessarily consistent with one another.

A is ancestral to B. My grandfather is ancestral to me: my great uncle is not. If *A* is a particular individual Carboniferous fossil, it is conceivable that *A* is ancestral to me but it is statistically improbable. (And in the unlikely event that *A* is ancestral to me it is then virtually bound to be ancestral to you too, and indeed to all mammals, because the Carboniferous is a long way back and we mammals are relatively close cousins.) It is in general safe to say that, for any two individual animals about which we might feel tempted to use the relation "is lower than," the meaning "is ancestral to" is going to be false.

Strictly, the ancestor/descendant relation can be applied only to individual animals. But there is a temptation to apply it to taxa. For instance people might ask whether the species *Australopithecus afarensis* is ancestral to the species *Homo sapiens*. Presumably they would answer yes if any individual *A. afarensis* is ancestral to any individual (and presumably all) modern *Homo sapiens*. One can have a sensible discussion about this when one is talking about fossil species, such as *A. afarensis*. But what if there were surviving individuals whom we would wish to classify in the species *A. afarensis*? Obviously we cannot say that any of these living individuals is our ancestor, only that they share a common ancestor with us, and that ancestor would have been classified in the same species as they are but in a different species from us. In other words, they have changed less since the time of the common ancestor than we have. So this turns out to be a special case of the second of our seven relations.

A resembles their last common ancestor more than B does. I shall use "is more primitive than" as a synonym for this. I should point out, however, that the word "primitive" itself is subject to some of the confusion raised by the word "lower." For example, some people will automatically construe "primitive" to mean "simple" or even "stupid."

A is more primitive than individual *B* if their last common ancestor *C* resembles *A* more than it resembles *B*. Obviously degree of resemblance is difficult to measure, but we can have a clear idea of what it would, in principle, mean to measure it. We can think in terms of the methods of numerical "phenetic" taxonomy (Sneath and Sokal, 1973) and imagine applying them to an ancestor, were it available to us. Using our best guess as to what the common ancestor *C* was like, we could ask: Is the phenetic distance between *C* and *A* larger or smaller than that between *C* and *B*? If a clear answer emerges, we are entitled to use the relation "is more primitive than." See Figure 1.

I am prepared to guess that an earthworm is a more primitive animal than a crab because, although I have never seen the common ancestor, I suspect that the earthworm resembles it more than the crab does. But as we shall see, we are not entitled to assume that all, or even many, cases will be so clear cut.

I am not advocating the use of "higher and lower" terminology at all. But if you must use it, the relation "is more primitive than" is one of your better hopes of making it meaningful.

A is simpler than B. Many people think that they know what they mean by simple, and its opposite, complex, but there have been few attempts to define these terms precisely. J. W. S. Pringle (1951) uses information theory to clarify what is surely the dominant meaning in most minds. He first insists that "complex" makes reference to descriptions of things rather than to the things themselves. The complexity of an animal is the information content of a description of that animal. In practice we can hardly hope to measure the information content of a description of a lobster in bits, but here is an approximate rule of thumb. If you have a lobster and an earthworm and you wish to decide which is the more complex, proceed as follows. Write a book about the lobster, write another book about the earthworm, and count the number of words in the two books. The animal

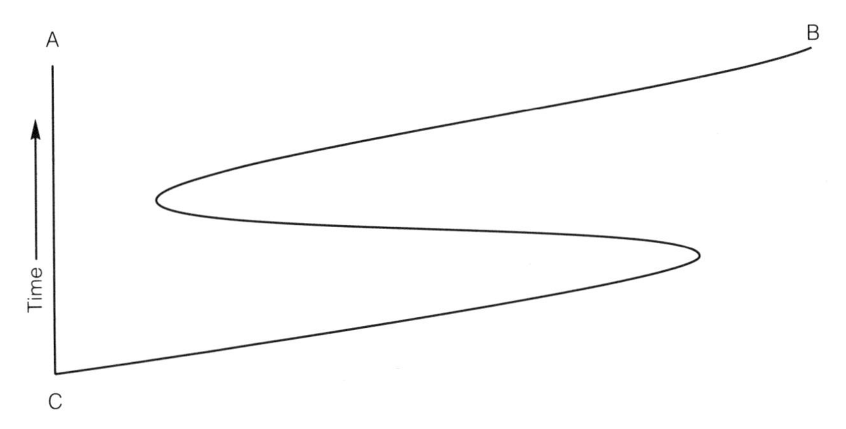

Figure 1 During the time since their common ancestor, *C*, *A* has evolved hardly at all while *B* has evolved a great deal. *A* is therefore more primitive than *B*.

that needs the larger book is the more complex. Obviously it is important that the two books should be roughly equally verbose, and should go down to equivalent levels of detail. You could easily spend more words on an amoeba than a man if you described every molecule of the amoeba but stuck to the gross anatomy of the man. But if you write the two books at an equivalent level of detail and with the same economy of style it is uncontroversial that man will emerge as more complex than amoeba, and lobster as more complex than earthworm. Probably the most economical way to describe an earthworm is to describe one segment in detail, then simply say "and repeat *n* times" where *n* is the number of segments. It would also be necessary to indicate such minor differences as there may be from segment to segment, but, essentially, when you've seen one segment you've seen them all. The lobster is also segmented, but here the segments are so different that you would need a new chapter for each, at least until you reached the abdomen. The lobster book would be considerably fatter, therefore the earthworm is a simpler animal than the lobster.

So the relation "is simpler than" can in principle be made meaningful. Even if we do not go to the trouble of actually counting words in descriptions, you now know what I mean if I say that in my opinion animal *A* is simpler than animal *B*. If you disagree with that opinion, the thought-experiment of the two books enables us at least to agree over what it is that we disagree about. This may not seem like much, but in the field of phylogenetic controversy it is a major achievement.

I am not going to discuss the remaining four candidate relations in detail, and will allow only a sentence or two to each. I hold no brief for the relation "is stupider than." It is notoriously difficult to compare mental abilities across species (MacPhail, 1982). It is hard enough within species (Kamin, 1974). I include the relation in my list simply for completeness; it undoubtedly lurks in the back of the mind of many people when they employ "higher/lower" terminology. Probably rather few people any more take "is lower than" to mean "is less well adapted than"; it has become sufficiently commonplace that rats, cockroaches, and bacteria are very successful species, while animals that we like to think of as "high," such as whales and gorillas, are threatened with extinction. Much the same applies to "is less adaptable than." "*A* resembles man less than *B* does" is still rather a common hidden meaning of "is lower than"; its objectionable speciesism (Ryder, 1975; Singer, 1976) hardly needs comment.

Are the candidate relations correlated? Let us for the moment allow that each of the seven relations could separately be made meaningful. Then in order for the general relation "is lower than" to stand usefully for a combination of more than one of the seven subsidiary relations, it is necessary that there should be a perfect, or at least a very high, association between rankings using the several relations. For example, if *A* is less complex than

B then *A* must also resemble the common ancestor more than *B*, obviously a highly dubious assumption to make. I know of no attempt systematically to measure the correlation between rankings on any one of the seven relations with rankings on any other. It might be worth doing, but only because of the light it might throw on evolutionary processes, not, in my opinion, because it would license us to talk about higher and lower animals. Meanwhile, the relation "is lower than" should never be used as a vague composite of subsidiary relations.

I turn now to some logical points that are often not appreciated, but that apply whichever particular relation you use to rank animals. For the sake of argument, I will talk in terms of only one of the relations, "is more primitive than," but the points are generally applicable to all seven of them.

Just because A is more primitive than B in one respect, it does not have to be more primitive than B in all respects. First we note that the relation—in this case "is more primitive than"—is being applied to parts of animals rather than complete animals. In principle there is nothing difficult about this. It necessitates a simple modification of the definition already given for the whole animals. To say that the brain of *A* is more primitive than the brain of *B* means that the common ancestor *C* had a brain that resembled *A*'s brain more than it resembled *B*'s brain. If this is true, it does not follow that the tail of *A* is more primitive than the tail of *B*. Yet this kind of mistaken deduction frequently appears in the literature.

In the following quotation, the very distinguished zoologist Sir Gavin de Beer (1954) was discussing the nature of our earliest multicellular ancestors, and supporting a then-heretical theory by J. Hadzi. The facts don't matter, for the point that I wish to criticize is a purely logical one. De Beer's premise is that "the Anthozoan plan of construction is more primitive than that of the Scyphozoa and that the Anthozoa were descended from bilaterally symmetrical ancestors" (1954, p. 26). Never mind whether this is true, or even what it means. What is remarkable is de Beer's deduction from it: "It follows . . . that the polypoid person, which is the only one represented in the Anthozoa, is more primitive than the medusoid person found in Scyphozoa and Hydromedusae" (1954, p. 26). His statement reduces to "*A* is more primitive than *B* in one respect (bilateral symmetry); therefore *A* is more primitive in all respects (for example in being a polyp rather than a medusa)." When it is put like this, anyone can see that it is a logical blunder of the most elementary sort. The biological evidence provided by the author for his premise was purely concerned with bilateral as against radial symmetry. The conclusions finally drawn concerned the polyp form as against the medusa. There is simply no logical connection between premise and conclusion at all.

Any lingering doubts about this can quickly be dispelled by considering the following perfectly reasonable alternative, compatible with de Beer's premise. The ancestral coelenterate could have been a bilaterally symmetrical medusa. From this the Scyphozoa evolved by acquisition of radical symmetry and the Anthozoa by changing into a polyp. The reason this possibility did not occur to the author after he had established his premise about bilateral symmetry was, I suspect, as follows. A bilaterally symmetrical medusa would not be classifiable in any of the modern classes of coelenterates; therefore it could never have existed! The concealed assumption is that "ancestors" should be sought among modern animals. This neglects the likelihood that in the remote past animals might have existed that would not fit into any modern pigeonhole. I shall return to the point under the next heading.

Whether *A* and *B* are whole animals or parts of animals, to say "*A* is more primitive than *B*" in the sense defined above is equivalent to saying: "Since the common ancestor *C*, a greater quantity of evolutionary change has taken place along the line to *B* than along the line to *A*." Given that the time available for evolution along the two lines has been the same, statements about primitiveness are really statements about rates of evolution. By the same token, the statement that, because *A* is more primitive than *B* in one respect, *A* must be more primitive than *B* in all respects, makes an assumption about rates of evolution. It implicitly assumes that rates of evolution in different parts of the body within any one lineage are constant, but rates of evolution in different lineages vary. Usually people who make this implicit assumption do not realize that they are doing so. If an animal is primitive from the waist down, we tend to assume that it is primitive from the waist up too. Maybe this is true; certainly one can think of good functional reasons why it often might be. But it is an interesting point to argue about, not something to be assumed without even being made explicit.

Just because one of the two animals A and B may *be more primitive than the other, it does not follow that either one* must *be more primitive than the other.* It is possible that for any given stage in our ancestry there is a living fossil still with us. Living fossils may exist, but some phylogenists write as if we were somehow entitled to expect them to exist, as a right! This assumption is implicit in most speculative writing about remote ancestry where fossils are not readily available, for example, about the origin of the metazoa and the origin of the vertebrates. The vertebrates have been traced "back" to various different groups, for instance, the Tunicates, the Echinoderms, and the Nemerteans. It is entirely reasonable to trace vertebrate ancestry back to an animal that would not have been classified as a vertebrate. But to assume that this remote primogenitor must necessarily have been classifiable in some named taxon or another of the modern animal kingdom is simply illogical. If there were no vertebrates in those far-off days, why should there have been tunicates or nemerteans either? To be sure, there must have been ancestors of tunicates and nemerteans, but these need not have been classifiable under those names any more than the remote ancestors of vertebrates were classifiable as vertebrates. Tunicate and nemertean ancestors have had just as long to evolve into something else since those times as vertebrate ones have.

The assumption I am attacking can be represented diagrammatically (see Figure 2). The bold main line of evolution surges ever upward toward *D* (often man), and every now and then it throws off a living fossil. *A*, *B*, and *C* successfully "arrive on the scene" and then stop evolving. This is a misleading picture. There is no such thing as a main line of evolution. Lines of evolution are simply branching lines of descent from living parent to living offspring. After an evolutionary branch, further change is presumably just as likely to occur in one daughter line as in the other. Figure 3 redraws the same phylogeny less contentiously in two equivalent ways.

In terms of the true information that they convey, Figures 2 and 3 are identical, and so are both versions of Figure 3. The additional connotations of Figure 2 are uncalled for. So is all talk of "side branches" and "main lines" and of lines coming off the "main line" earlier or later than others.

I am not, of course, saying it is never true that some lines of descent stagnate while their cousins evolve. I am simply objecting to the uncritical assumption that this will necessarily be the case. If, as Ernst Mayr (1969) avers, "a new group almost invariably buds off from the parental taxon

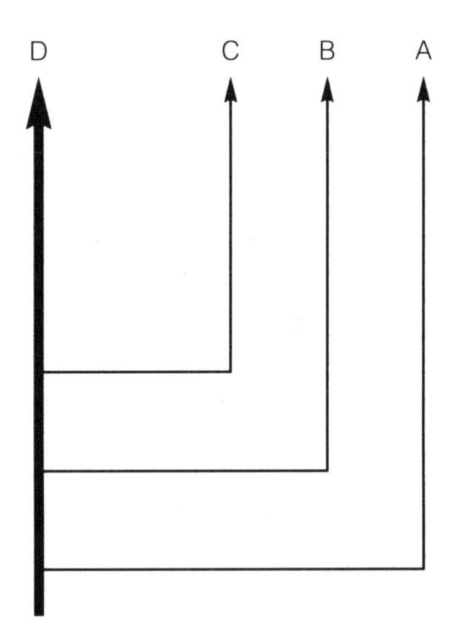

Figure 2

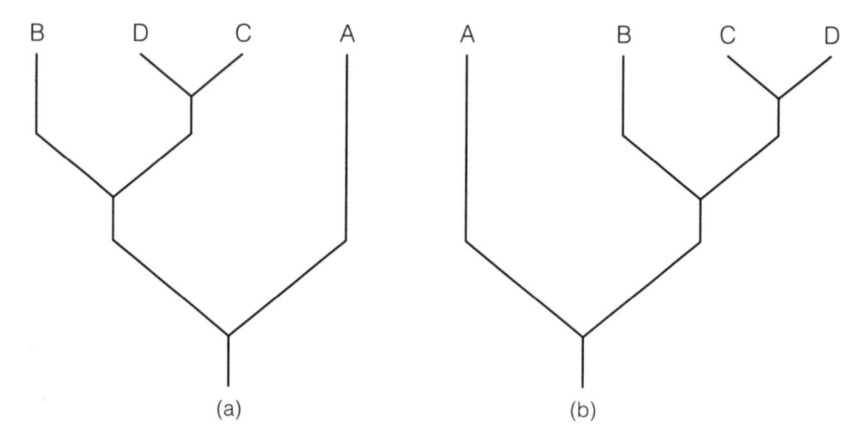

Figure 3

which continues to exist with very little change, sometimes for more than 100 million years," then let us see the evidence, and let us hear discussion about why such a remarkable fact should be. Let us not make assumptions without realizing that we are making them.

Just because one member of taxon X is more primitive than one member of taxon Y, it does not follow that all members of X are more primitive than all members of Y. Primatologists often refer to nonprimate mammals as "subprimate." For the sake of argument we may charitably assume that "sub" means something, namely, "more primitive than." Accepting this, examine the logic of using it to prefix the word "primate." It is evidently of this form. Some nonprimates are more primitive than some primates; therefore all nonprimates are more primitive than all primates. For example, if we can agree that a weasel is more primitive than a man, it is deduced that all carnivores are more primitive than all primates, and therefore that cats are more primitive than lemurs. I can place no more charitable interpretation on the widespread usage of "subprimate," "submammalian," and similar expressions.

Figure 4 correctly expresses the phylogenetic relationships of four mammals. *X* is the common ancestor of all four, *Y* is the common ancestor of the two primates, and *Z* the common ancestor of the two carnivores. Following our definition, the most primitive of the four animals is the one that most resembles *X*. Suppose for the sake of argument that this is the lemur. What this means is that relatively little evolutionary change has occurred along the line from *X* to *Y* and from *Y* to Lemur. Suppose that the least primitive of the four is man. This must mean—since we now know that *Y* was primitive—that in the line from *Y* to man a great deal of

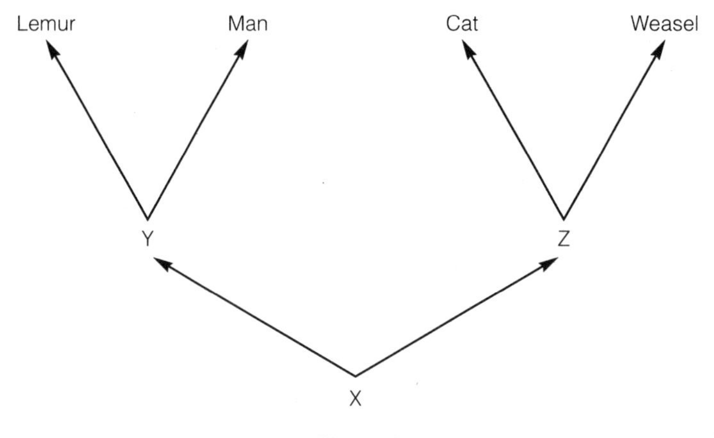

Figure 4

evolution has taken place. Without violating any laws of logic, and without stretching biological credulity, we can rank the four animals in order of primitiveness as follows: primate, carnivore, carnivore, primate. Lemur and man undoubtedly are closer cousins to each other than either is to cat or weasel. Yet they may well be farthest apart in terms of their resemblance to the common ancestor *X*. In terms of resemblance to the common ancestor, all possible rank orderings among modern animals can easily be obtained simply by manipulation of rates of evolution along the various lineages. "Subprimate" and like phrases are nonsensical. More generally we must beware of making sweeping statements about the primitiveness (complexity, etc.) of whole taxa on the basis of information about smaller included taxa. If we make such statements we are, in effect, assuming that rates of evolution in some taxa (such as the Order Primates) are characteristically higher than rates of evolution in other taxa (such as the Order Carnivora). This may be so, but there is no particular reason to expect it to be, and it is certainly not something to assume uncritically.

The assumptions underlying the "higher and lower animals" way of thinking are unconscious. I have tried to make them explicit so we can see clearly whether there is any biological basis for them or not. These unjustified assumptions may be summarized as follows.

1. There is a necessary connection between (at least) seven different ways of defining the relation "is lower than."
2. A lineage that evolves rapidly in certain characteristics will tend to evolve rapidly in all characteristics.

3. When a lineage branches into two, one of the branches will continue to evolve, while the other will stagnate.
4. Rapid evolution in one lineage within a taxon will tend to be associated with rapid evolution in all lineages within that taxon.

I make no attempt to assess the validity of these four assumptions, though I must admit that when I see them spelt out explicitly they do not strike me as very plausible. What I have argued is that they have been smuggled clandestinely into our thinking, wrapped up so as to be unrecognizable, disguised, by muddle rather than by mischief, to look more plausible than they really are. I recommend that evolutionary writers should no longer, under any circumstances, use the adjectives "higher" and "lower."

Random Drift

John Beatty

Random drift is a heterogeneous category of evolutionary causes and effects, whose overall significance relative to other modes of evolution (especially evolution by natural selection) has been greatly disputed. What most of the phenomena so designated have in common is one or another biological form of random or indiscriminate sampling, and consequent sampling error. But beyond that, the collective phenomena are very different. Moreover, there are phenomena sometimes included in the category of random drift that have nothing to do with random sampling.

The person who did the most to articulate and defend the importance of "the" phenomenon of random drift was Sewall Wright (see, e.g., Wright, 1931, 1932, 1940, 1948, 1949a, 1955, 1968–1978; Provine's 1986 biography of Wright is the most important source on the history and conceptual foundations of theories about random drift). But Wright entered an ongoing discussion. For instance, R. A. Fisher (1922) had already criticized a form of random drift that he attributed to A. L. and A. C. Hagedoorn (1921) and dubbed the "Hagedoorn effect." By 1950, though, Wright's name was so closely associated with random drift that Fisher and E. B. Ford directed their criticisms instead to the "Sewall Wright effect" (adding the first name, perhaps, in order to avoid the self-fulfilling sense conferred by the term "Wright effect"; Fisher and Ford, 1950). Wright, in response, insisted that the honor of priority belonged to J. T. Gulick (1872; Wright 1951, p. 452), and indeed a variety of random drift is sometimes called the "Gulick effect" (e.g., Roe and Frederick, 1981, p. 110). Interestingly, if not surprisingly, the priority-conscious Hermann J. Muller once laid his own claim to random drift; in this instance, Muller credited Wright only with the naming of the phenomenon (Muller, 1940, p. 216; see the reference to Muller, 1918, p. 481).

Varieties of random drift were first articulated and defended in the late nineteenth and early twentieth centuries in opposition to the "neo-

Darwinian" notion that natural selection is an all-sufficient agent of evolution. Evidence against the overarching significance of natural selection included many traits that had no apparent adaptive value, and many polymorphisms of apparently equally fit (or equally useless) traits. Various mechanisms involving large-scale, mass, and directed mutation were invoked to account for these phenomena (see, e.g., Kellogg, 1907). Varieties of random drift (or rather what came to be known simply as "random drift") were also proposed as appropriately nonadaptive mechanisms for explaining these phenomena.

Actually, as early as the first edition of *On the Origin of Species,* Charles Darwin acknowledged the significance of one sort of phenomenon that would later be included in the category of random drift: fluctuations in frequency of variations that have no adaptive significance or are otherwise equally fit (Darwin, 1859b, p. 81).

Darwin did not elaborate. Why would frequencies of equally fit variations fluctuate? Although not (explicitly) responding to Darwin, Gulick had an answer. He described a similar mode of evolution that he attributed to "indiscriminate destruction or failure to propagate of a part of the individuals of an intergenerating section of a species" (Gulick, 1889, in 1905, p. 209). Only a very discriminating cause, Gulick reasoned, could guarantee the maintenance of the "average character" of a breeding group. If the causes of differential survival and reproduction among members of a breeding group are entirely indiscriminate, then there is no guarantee that the frequencies of the various types in one generation will be repeated in the next generation.

Any cause of differential survival and reproduction among equally fit types would necessarily be indiscriminate with regard to fitness differences. But even among differentially fit forms, there can be indiscriminate causes of survival and reproduction. Gulick invoked the (by now familiar) thought example of a natural catastrophe (a volcano eruption, in Gulick's example) that kills all the nearby members of a species indiscriminately, resulting by chance in a change in frequency of the various types within the species.

Incorporating Mendelian insights, the Hagedoorns articulated another sort of evolutionary change that has since also been incorporated into the category of random drift; they viewed it as "the most important gain in knowledge which we owe to Mendel's work" (Hagedoorn and Hagedoorn, 1921, pp. 103–139, quotation p. 120). They showed that genetic variation in a population can be reduced by means other than natural selection. Given that heterozygous parents may leave homozygous offspring, and may by chance leave only offspring homozygous for one of the alleles in question, then in a finite population, an allele can be accidentally lost. This loss of variation is what Fisher called the "Hagedoorn effect," and criticized as being an insignificant agent of evolutionary change (Fisher, 1922, pp. 328–331).

Instead of focusing on the ultimate loss of variation that results from Mendelian inheritance in a finite population, one can focus instead on the unpredictable changes in frequency of alleles that eventuate in their loss or fixation. Given that a heterozygous organism, *Aa*, produces 50% *A* gametes and 50% *a* gametes only in the long run, then the finite number of *Aa* parents in any one generation may by chance contribute more *A*'s than *a*'s to their offspring. The frequencies of *A* and *a* may thus change from generation to generation until one or the other is eliminated.

In the early thirties, Wright began to refer to such evolutionary changes in terms of the "random drifting of gene frequencies" (e.g., 1931, p. 151). He attributed random drift to either or both of two different kinds of "accidents of sampling" (1932, p. 360). These two sources of evolutionary change correspond to those already discussed with respect to Darwin and Gulick, on the one hand, and the Hagedoorns, on the other. As Wright summarized, "If the population is not indefinitely large, another [evolutionary] factor [i.e., in addition to natural selection, mutation, etc.] must be taken into account: the effects of accidents of sampling among those that survive and become parents in each generation and among the germ cells of these" (1932, p. 360).

Following Dubinin and Romaschoff's (1932) explication, Theodosius Dobzhansky (1937, p. 129) popularized this notion of randomly drifting gene frequencies in terms of the classic means of illustrating chance processes: the blind drawing of colored balls from an urn. This has become, in turn, the classic means of illustrating this form of evolution. The balls in this case are genes for alternative traits. The alternative gene-balls are different colors, but they are otherwise indistinguishable by a blindfolded sampling agent—blindfolded to reflect indiscriminate causes of survival and reproduction, and/or the indiscriminate nature of Mendelian segregation. One urn of balls represents one generation of alternative genes—a finite number, characterized by particular gene frequencies. The frequencies of the next generation of genes are determined by a blind drawing of balls from the urn. This second generation of genes fills a new urn, blind drawings from which determine the frequencies of genes in the third generation, and so on. The frequencies of genes will inevitably differ from urn to urn—generation to generation—as a result of the fact that frequencies of otherwise indistinguishable balls sampled by blind drawings will inevitably not be representative of the frequencies in the urns from which the samples were drawn. The probability of drawing a representative sample from a population of a given finite size is easy to calculate—the smaller the population, the smaller the probability.

So construed, the possibility of random drift is part of population genetic theory. The Hardy-Weinberg "law," the core principle of population genetic theory, states that *in an infinitely large population,* gene and genotype frequencies will (after one generation) remain stable as long as there is no selection, mutation, or migration. There has been no disagreement about whether population size should be included as a variable in population genetic theory. There has, however, been considerable dispute about the relative importance of this particular variable—in other words, about how often population sizes and selection coefficients are sufficiently small that sampling error will make a substantial evolutionary difference. As for the real disagreements, as Wright once put it, "It is a question of the relative values of certain coefficients" (1948, p. 291).

Disputes about the relative importance of random drift will be considered in more detail shortly. But first let us consider some rather broader notions of random drift. Gulick was especially interested in another type of evolutionary phenomenon often included in the category of random drift: the divergent evolution of recently isolated breeding groups (Gulick, 1872, 1889, 1890, 1905; see also Lesch, 1975). One way of accounting for the divergence of two such groups was to suppose that they had been split "indiscriminately," that is, indiscriminately with respect to the inheritable variations of their members. A newly split-off group, especially if small, would be unlikely to have all the inheritable variations—and certainly not in the same proportions—as the original group. As Gulick explained, "It is evident that when the separated sections are small there is more likely to be *diversity* in the average character of the sections and that, roughly stated, the probability of divergence from this cause will be in direct proportion to the variableness of the species and in inverse proportion to the size of the different sections" (1889, in 1905, p. 186). Ernst Mayr would later call something very akin to this phenomenon the "founder effect," and would champion its importance in speciation (1942, p. 237).

Mayr himself does not consider the founder effect to be an instance of random drift, but Wright did; Dobzhansky was not sure (Mayr, 1963, pp. 204ff. and p. 534n; Wright, 1959, p. 86; Dobzhansky, 1959, pp. 85–86). On the one hand, the founder effect is indeed different from the varieties of random drift discussed previously insofar as it has to do primarily with divergence between units of evolutionary change, while parental and gametic sampling error have to do primarily with the evolution of one unit (and only consequently with divergence between units). On the other hand, the founder effect *is* a case of indiscriminate sampling and sampling error. The usual ball-and-urn illustration of random drift can easily be extended to cover the founder effect.

Moreover, the founder effect is, like the other cases of sampling error discussed, and in contrast to evolution by natural selection, a nonadaptive mode of evolution. For instance, Gulick (1872, 1905) relied on indiscriminate divisions of breeding groups, together with indiscriminate causes of survival and reproduction, to account for the differences that distinguish the more than two hundred species of the snail family Achatinellidae on the island of Oahu. He considered evolution by natural selection incompetent in this regard, inasmuch as there were no apparent environmental differences correlated with the species differences, and often no environmental differences at all. Over time, Wright extended the category of random drift even further to include phenomena that are not instances of random sampling error. By the late forties, he was distinguishing "systematic" from "random" changes in gene frequency, the former being determinate in direction, the latter indeterminate in direction and determinate only with respect to variance (Wright, 1949a, p. 369). The category of random changes included those due both to "accidents of sampling" and also to random fluctuations in selection pressure, mutation rate, and migration rate. By the mid-fifties, Wright was referring to the category of random changes as cases of "random drift," in contrast to cases of systematic change or "steady drift" (1955, pp. 17–19). The most important aspect of this change is that random drift came to be construed as including some cases of evolution by natural selection (cases where selection pressures change randomly).

By broadening the category of random drift in this way, Wright dissolved the distinction between random drift and natural selection per se. But reinforced as it was by ongoing controversies over the relative evolutionary significance of random drift versus natural selection, that distinction was firm in the minds of many other biologists.

During the 1930s and 1940s, random drift was frequently invoked to account for many apparently nonadaptive patterns of variation (Dobzhansky and Queal, 1938; Diver, 1940; Wright, 1940). But the tide turned, as many of the more celebrated cases of random drift were reinterpreted in terms of evolution by natural selection (Dobzhansky, 1943; Wright and Dobzhansky, 1946; Cain and Sheppard, 1950, 1954; Sheppard, 1951, 1952; Clarke, 1961; see also, e.g., Gould, 1983; Provine, 1986a, pp. 404–456; Beatty, 1987). These turnarounds seemed to many to constitute overwhelming support for the overwhelming importance of natural selection. In the 1950s and 1960s, proponents of the all-importance of evolution by natural selection repeatedly trotted out these successes as evidence of the shortsightedness of proponents of the importance of random drift (e.g., Mayr, 1963, pp. 203–214). Selectionists of this period began to wonder whether any purported case of drift would stand up to rigorous investigation. Invocations of drift seemed to them to be just admissions of ignorance regarding the subtle selective mechanisms actually at work in the cases at issue. A. J. Cain, perhaps the most vocal of these selectionists, derided the logic of the proponents of the importance of drift: "This is the real basis for every postulate of . . . genetical drift. The investigator finds that he, personally, cannot see any [evidence of selection], and concludes that, therefore, there is none" (1951a, p. 424). It is unfortunate, he added, that drift explanations owe their only success to the "failure of the investigator" to find evidence of selection (1951b, p. 1049).

This change of attitude toward random drift is paradigmatic of what Stephen Jay Gould has called the "hardening" of the evolutionary synthesis. According to Gould, the architects of the evolutionary synthesis were originally genuinely pluralistic with regard to the importance of the various agents of evolution (including random drift). But during the forties, fifties, and sixties, they and their students came to emphasize more and more the importance of natural selection, so that all other agents came to be seen as unimportant (Gould, 1983).

These disputes over the relative significance of natural selection versus random drift reflected the generally held view that natural selection and random drift are distinctly different agents of evolution. No wonder, then, that Wright was criticized, for example by Mayr (1963, pp. 204–205), for broadening the notion of random drift so far as to include some cases of natural selection (where selection pressures vary at random). Mayr suspected Wright of attempting to guarantee the importance of random drift simply by construing it so broadly, but Mayr certainly exaggerated when he complained that Wright extended the domain of random drift to cover *all* evolutionary changes.

The immediate occasion for Wright's expansion of the category of random drift was indeed suspicious. Fisher and Ford had investigated fluctuations in gene frequency underlying coloration of the moth *Panaxia dominula* (Fisher and Ford, 1947). The sizes of the fluctuations, together with the sizes of the populations studied, made sampling error a very unlikely cause; they cited Wright as a proponent of the importance of sampling error. They proposed instead, as the cause of the fluctuations, randomly changing selection pressures, which Wright then promptly incorporated into the domain of random drift! A good example of the resulting confusion about the meaning of random drift—and hence its overall evolutionary significance—is the discussion by the selectionists Cain and Currey (1963, pp. 57ff). Cain and Currey note the difficulties of deciding whether such phenomena as the founder effect and random changes in selection pressure do or do not belong to the category random drift. In a form of compromise, they urge adoption of more precise terms such as "sampling drift" to make clear what is being referred to in invocations of random drift.

Wright's expansion of the category of random drift may seem questionable in certain respects, but it is quite reasonable in others. Consider the role played by random drift in Wright's "shifting balance" theory of evolution. This theory laid out the conditions under which the adaptive response (or "lability") of species was maximized (Wright, 1931, 1932, 1948, 1949a; Provine, 1986a, pp. 277–326). According to Wright, a species is best able to respond adaptively to a particular set of selection pressures when there is plenty of genetic and genotypic variation present for natural selection to act upon. Such variation is more likely to be available, he argued, in species divided into small, partially isolated subpopulations. Ideally each subpopulation would be small enough that parental and genetic sampling error would keep them genetically and genotypically different, but not so small and not so isolated that the same sorts of sampling errors would lead to loss of variation within each subpopulation.

Sampling error due to indiscriminate subdivisions of the species would further contribute to genetic diversity between subpopulations, and hence within the species. Random changes in selection pressure would also help maintain variation within the species. In other words, all the various factors that Wright included in the category random drift had the effect of increasing the evolutionary lability of species.

Wright's apparent disregard for a clear-cut distinction between natural selection and random drift can also be attributed to his conception of random drift as facilitating the process of evolution by natural selection. He regarded evolution by random drift not as a strict alternative to evolution by natural selection but, rather, as a principal component of evolution by natural selection.

Provine (1986b, pp. 404–456) argues that Wright himself instantiated, in a way, the hardening of the evolutionary synthesis. That is, Wright originally invoked random drift to explain the prevalence of nonadaptive patterns of variation. In time, random drift came to be viewed by Wright first and foremost as a means of facilitating adaptive evolution. It is in part because of Wright's early views, according to Provine, that Wright is so commonly portrayed as a pro-drift anti-selectionist.

Since the late fifties and early sixties, there has been a slight reversal of the hardening of the synthesis. For instance, while Dobzhansky's selectionist account of chromosome inversions in *Drosophila* has pretty well stood the test of time, the selectionist accounts of banding and color pattern in *Cepaea* and of blood groups in humans have not. Lamotte (1959) has argued forcefully that selection alone is not responsible for *Cepaea* color patterns—that drift is also largely responsible. And Cavalli-Sforza (1969) has successfully accounted for differences in blood group frequencies in northern Italy in terms of drift alone.

Other biological considerations have also contributed to the recent rise of interest in random drift. Molecular biological findings have played a large role in the founding of a school that emphasizes the prevalence of "neutral" (equally fit) variation. Beginning in the fifties and sixties, it became clear that different DNA sequences can code for the same amino acid. It was realized that these so-called synonymous genetic variations are prime candidates for the sorts of selectively neutral variations whose frequencies would be expected to be entirely a matter of random drift. Theoretical considerations also suggested to molecular biologists of the sixties that proteins could differ in some of their constituent amino acids—the result of nonsynonymous genetic variations—and still be functionally equivalent, and hence selectively neutral. The frequencies of such variations in a population would be a matter of random drift, not natural selection. As King and Jukes (1969) expressed these basic ideas in their classic position paper, "Natural selection is the editor, rather than the composer, of the genetic message. One thing the editor does *not* do is to remove changes which it is unable to perceive" (p. 788).

Motoo Kimura combined this line of reasoning with another consideration, this one having to do with the enormous amount of intrapopulational genetic variation discovered by gel electrophoretic methods in the mid-sixties. If this variation represented differentially fit genotypes, it would indicate a large amount of relatively less fit variations and hence constitute an enormous "genetic load." This load might even threaten the perpetuation of the population. One way of reconciling such large amounts of intrapopulation variation with the perpetuation of populations, it is sometimes argued, is to suppose that the variations in question are largely neutral (Kimura and Crow, 1964; Kimura, 1968; see also Lewontin, 1974a, pp. 189–271).

Another line of reasoning that played a large role in the rise of neutralist thought in the late sixties and the seventies concerned findings of constant evolutionary rates across different taxa. Evidence was gathered suggesting that the evolution of even a functionally very important protein such as hemoglobin proceeded at the same rate (where rate is measured in terms of amino acid changes per unit time) in very diverse groups of mammals. If the evolution of hemoglobin in each group of mammals were mainly a matter of natural selection in favor of appropriate hemoglobin forms for each taxon, under the changing environmental conditions faced by that taxon, then it would be very unlikely for the rate of hemoglobin evolution to be so similar from group to group. If most of the amino acid changes in question were selectively neutral, however, there would be no reason to expect them to proceed faster in one group of mammals than another (see, e.g., Kimura, 1969).

Members of the neutralist school have, from time to time, conceived the notion of neutrality rather broadly. Some suggest extending the meaning of "neutral" from strict selective neutrality to "near" or "effective" selective neutrality. By "nearly" or "effectively neutral" genes is meant genes that behave evolutionarily just like strictly neutral genes—that is, their frequencies drift randomly to the same extent that the frequencies of strictly neutral genes would. Technically, effectively neutral genes are those whose relative selective advantage or disadvantage is small in comparison with the reciprocal of the effective population size. As Kimura acknowledges, the neutral theory is not really what it appears to be: "The essential part of the neutral theory is not so much that molecular mutants are selectively neutral in the strict sense as that their fate is largely determined by random genetic drift" (1983, p. 34, also p. xii; see also Crow, 1982). Merrell has complained that the notion of near or effective neutrality reminds him of the notion of being slightly pregnant (Merrell, 1981, p. 326).

It is sometimes remarked, for instance by Kimura himself, that the random drifting of frequencies of neutral genes is a process that accounts for variation at the "molecular" level (e.g., gene and protein polymorphisms) while natural selection is more important in explaining variation at the level of the "phenotype" (Kimura, 1983—note the title of the book). Whether this distinction will withstand serious conceptual and empirical scrutiny remains to be seen.

An interesting indication of the changing fortunes of random drift vis-à-vis natural selection is the methodological priority that some evolutionary biologists now give it. In the fifties and sixties, when natural selection seemed to be all important, it was argued that random drift could properly be invoked only at the end of a sufficiently long attempt to find a selectionist account of the pattern of variation in question. Mayr expressed this attitude: "[The evolutionary biologist] must first attempt to explain biological phenomena and processes as the product of natural selection. Only after all attempts to do so have failed, is he justified in designating the unexplained residue tentatively as a product of chance" (1983, p. 326). But it has been argued of late that evolutionary investigations should begin with considerations of random drift hypotheses, inasmuch as these hypotheses represent proper "null hypotheses." Natural selection, the argument continues, should be invoked only when the null, random drift hypothesis has been rejected (e.g., Selander, 1985, pp. 87–88).

Random drift may be construed to cover different kinds of sampling error, or to cover random fluctuations in selection pressure and mutation and migration rates. (For related discussions of the role of stochastic processes in genetics and evolutionary biology, see Gigerenzer et al., 1989, pp. 132–162.) Alternatively, its significance may be reflected in patterns of "phenotypic" variation or in patterns of "molecular" variation. Clearly, the conceptual and empirical domains of random drift are far from settled.

Resource

Peter Abrams

There is general agreement that a resource of a given individual or population has the property of increasing survival or reproduction—and hence that resources are a major determinant of natural selection. Beyond this basic agreement, however, there is considerable disagreement on what a more precise definition ought to include.

What is a resource and what is not a resource? It is useful to begin with a list of some of the many different entities that are commonly referred to as resources. Foods are resources for all heterotrophic animals. Appropriate types of shelter or space have been termed resources when they have an effect on an individual's probabilities of birth and death. Light, mineral nutrients, water, pollinators, and seed dispersal agents have all been considered to be resources for plants. Within a species, members of one sex have been referred to as resources for members of the other sex.

Are there properties that all of these examples have in common, that can be used to distinguish an entity that is a resource from one that is not? We may consider three properties that have been proposed as characteristics of resources, and consider whether these allow an unambiguous classification.

The first property concerns the *effect on survival and reproduction.* For an entity to be a resource, it should, at least potentially, have an effect on the probability that an individual will survive and reproduce. Ricklefs' (1979, p. 878) ecology textbook defines a resource as a "substance or object required by an organism for normal maintenance, growth, and reproduction." This implies that an organism will not be represented in the next generation if it is unable to obtain a sufficient amount of any specific resource. Others have less stringent criteria for effects on population parameters; Ehrlich and Roughgarden (1986, p. 630) merely require that a resource be "consumed or occupied . . . during . . . growth, maintenance, or reproduction." This may be too broad; for example, it includes inadvertently ingested nondigestible material along with food as a resource. The effect on survival and reproduction need not occur under all circumstances or conditions; Wiens (1984, p. 401) states his belief that resources need only "potentially influence individual fitness." Under this definition, a nonpreferred food would still be a resource in an environment that contains sufficient amounts of preferred foods, and foods in general remain resources even when they are superabundant and have no fitness effects. Some resources may become toxic when their concentrations are too high. Resources are generally termed "nonlimiting" when a small change in their availability does not affect a consumer's probabilities of dying or reproducing. Limiting resources are by definition more relevant to population biology and evolution.

The second property concerns *consumption or use.* Not all factors or entities that have effects on probabilities of survival and/or reproduction are resources. Components of an organism's environment are usually considered to be resources only when they are consumed, used, or occupied. Tilman (1982, p. 11) includes in his definition that a resource must be "consumed," Wiens (1984, p. 401) suggests that they must be "directly used," and Ehrlich and Roughgarden (1986) require consumption or occupation. These are all nebulous terms, but seem to be intended to exclude those factors that are not affected by the organism's presence. Temperature and other aspects of weather are excluded from being resources by this requirement. "Use" is the broadest of the terms; birds that disperse seeds by eating fruit are usually not thought of as being "consumed" or "occupied" by the plant, but may perhaps be "used."

The third property involves *reduction in the resource's availability to other individuals.* The ecology textbook by Begon, Harper, and Townsend (1986) devotes an entire chapter to resources, which they distinguish from "conditions" by the property that resources are quantities that can be reduced in availability by the activity of the organism, while conditions cannot be reduced by such activity. Andrewartha (1961) drew a similar distinction between resources and other aspects of an organism's environment.

There seems to be general agreement that resources used by one individual or group must be, at least temporarily, less available to be used by other individuals or groups. Thus, although heat can certainly be used by poikilotherms and increases their expected reproductive output, the effect of the heating of any lizard population on the ambient temperature is so minuscule that heat is not made less available to others. If there is a limit to the number of places with suitable temperature and to the number of lizards that can occupy any such place, then the sites themselves would be considered to be resources.

Similarly, oxygen is consumed by animals, and is required for life, but it is seldom referred to in discussions of resources. In most environments, the consumption of oxygen does not affect the amount available any more than the heat absorbed by a lizard affects the temperature. In addition, the spatial or temporal variation in oxygen availability in most environments is very small, and does not affect probabilities of surviving or reproducing. In practice, those resources of most interest to evolutionary biologists and ecologists are those that actually do have an effect on birth and death rates in some circumstances and whose availability is at least occasionally reduced by their use.

There is an additional question about what part of the utilized or consumed entity constitutes the resource. Wiens (1984) has raised this question, which seems particularly relevant to food. Is the food itself the resource, or is it only those nutritional components of the food that contribute to survival and reproduction? The relationship between resources and population dynamics may be influenced by the answer to this question.

Although there are many unanswered questions in the consideration of these three properties, there seems to be agreement about the status of most aspects of an organism's environment as resource or nonresource. The questionable cases primarily involve entities (e.g., oxygen for terrestrial vertebrates) that do not contribute to differences between individuals in their representation in future generations and are therefore unlikely to be of interest to evolutionary biologists. Nevertheless, there has been less awareness of the definitional issues associated with this term than there might have been. Many textbooks contain no definition of resource, even though there are extensive discussions of derivative concepts such as resource partitioning, resource overlap, and resource competition (e.g., Futuyma, 1986).

What distinguishes one resource from another? The amount of competition between types or species depends in part upon the amount of overlap in resource use (Levins, 1968; Pianka, 1981; Schoener, 1983). Measuring overlap depends on the ability to distinguish different resources. Many predictions of competition theory depend on the manner in which resources are distinguished. The question, well known to ecologists (Hutchinson, 1978), of whether or when two or more species can coexist while using a single limiting resource, requires that a single resource be distinguished from a set of two similar resources. Almost all organisms use more than a single type of resource. It is clear, for example, that food and nesting sites are different entities, but are nesting sites that differ in their height in a tree different resources? Are nesting sites that differ only in location different resources?

This question seems to have first received a detailed treatment in Haigh and Maynard Smith (1972). They suggested that resources were distinct if there did not exist a functional relationship between them—that is, if the availability of one did not have a one-to-one relationship with the availability of the other. Their rationale for this definition was that it was required if the proposition that two species could not coexist on a single resource (Volterra, 1928) were to be true. Armstrong and McGehee (1980) later showed that this proposition held only for coexistence at fixed densities. Two different parts of a single organism could constitute distinct resources to a species that fed on that organism, because the amount of one part need not be functionally determined by the amount of the other (Haigh and Maynard Smith, 1972). In any resource that has a spatial distribution with finite movement rates, the amount of the resource at one point is not entirely determined by its abundance elsewhere. As a result, a single homogeneous substance may represent a practically infinite number of resources. Maynard Smith (1974) and Abrams (1988) have both stressed that the enumeration of resources requires some consideration of the abilities of the consumer to distinguish them. Abrams (1988, p. 226) proposed that distinct resources should be recognized by two properties: "(a) the population level of one is not a function solely of the population level of the other, and (b) there exist consumers and environments such that either consumer fitness or total amount of resource consumed depends on the relative amounts of the two resource types." Condition (b) is an operational definition of distinguishability. It implies that a mouse population is likely to represent more resources to its parasites than to its avian predators.

Nevertheless, it is common for the distinction between resources to be made on chemical or taxonomic grounds alone (Tilman, 1982; Wiens, 1984)—that is, that a single chemical element or a single biological species represents a single resource. As early as 1960, Cole had objected that defining resources so species using the same resource could not coexist would make the competitive exclusion principle circular and untestable. Hutchinson (1978), among others, has pointed out that the principle was not meant to be tested in the field, but this objection continues to resurface (e.g., den Boer, 1986). In addition, the optimum manner for distinguishing resources from the standpoint of competition theory may not be optimal for other purposes. It is unlikely that the final word on how to distinguish resources has appeared.

SEXUAL SELECTION: HISTORICAL PERSPECTIVES

Helena Cronin

IF YOU WERE asked to invent an irksome challenge to Darwinian theory, you could get a long way with a peacock's tail. And if you were asked to think up a solution to the challenge that would disconcert Darwinians, you would need go no further than Charles Darwin's own theory of sexual selection. Natural selection is a sober force, economical, workmanlike, and strictly utilitarian. It should abhor the peacock's tail—gaudy, ornamental, a burden to its bearer. Darwin took the view that natural selection would indeed frown upon such flamboyance. It had been concocted, he decided, by female preference. Peahens prefer to mate with the most lavishly tailed males. The longest tails will then be best represented in the next generation. And in the next. And so on and on, through evolutionary time, the females exerting an ever-demanding, relentless pressure, their mates responding with ever-greater splendor, until the peacock's finery escalates to the exuberance that we see today. And "peacock," of course, stands for all extravagant male display.

Darwin elaborated his theory of sexual selection in *The Descent of Man* (1871). Another aspect of his theory was aggressive competition between males for mates. But such competition was thought to be uncontentious, because it seemed to call for the kinds of characteristics—muscles, claws, a quick response—that natural selection would approve of anyway. The fortunes of this notion took an entirely different course, typical of the history of natural selection, and we shall not examine it here.

But how contentious the idea of female choice proved to be! According to Darwin, females choose their mates solely on aesthetic grounds: "a great number of male animals . . . have been rendered beautiful *for beauty's sake*"; "the most refined beauty may serve as a charm for the female, and *for no other purpose*"; "that ornament and variety is the *sole object*, I have myself but little doubt" (Darwin, 1959, p. 371; 1871, vol. 2, pp. 92, 152–153; emphasis added). Other Darwinians didn't like this "good taste" theory of female choice. One reaction was to try to explain mate choice as more level-headed, more practical—as "good sense." The other reaction was to deny that females choose their mates at all and to explain nature's apparent ornaments instead by the standard forces of natural selection. Changes in the Darwinian understanding of sexual selection amount to the careers of these three views (Cronin, 1992).

Nothing but natural selection. For sexual selection's first hundred years, orthodoxy proclaimed it unofficially dead. Ordinary natural selection was wheeled out to account for almost all the luxuriant glamour, the ornamental flourishes that Darwin attributed to female choice. Beginning with Alfred Russel Wallace (e.g., 1889, pp. 187–300, 333–337), most Darwinians turned their attention to sexual selection only to demolish it. They aimed to dispose of the theory entirely and install everyday natural selection in its place. Gaudy coloration became warning colors. A bright spot or flash mark became species recognition. A ruff or crest? Threat signals. A song or dance? Territorial defense. This program was vigorously promoted by an unbroken line of leading Darwinians through the nineteenth century right up to the 1970s. Around the turn of the century, for example, we find E. B. Poulton, a foremost specialist on coloration, maintaining that sexual selection was "relatively unimportant" in evolution and that "its action has always been entirely subordinate to natural selection" (1896, pp. 79, 188). In the 1920s, O. W. Richards, a leading expert on insects (insects and birds were Darwin's prime examples), concluded: "It has become obvious since Darwin wrote that display-characters are probably acquired most often as a result of Natural rather than Sexual Selection" (1927, p. 300). In the 1930s, Julian Huxley, then regarded as a major authority on sexual selection—perhaps *the* major authority—was declaring ex cathedra that Darwin "persistently attached too much weight to the view that bright colors and other conspicuous characters must have a sexual function"; "it has now become clear that the hypothesis . . . is inapplicable to the great majority of display characters"; and "Darwin's original contention will not hold" (Huxley, 1938a, pp. 1, 20–21, 33; see also 1938b). In the 1960s, an influential text by Ernst Mayr urged that the "song of the nightingale belongs here [with natural selection] and so does the strutting of the peacock" (1963, p. 96). Even as late as the centenary celebrations for *The Descent of Man*, natural selection was still the front-runner. Mayr's contribution to a commemorative volume reflects the consensus: "It is now evident . . . that there are three major . . . selection pressures which favor . . . sexual dimorphism, without requiring sexual selection" (Mayr, 1972b, p. 96); he cited epigamic selection, isolating mechanisms, and males and females utilizing different niches. Darwinians in this tradition did not always rule out sexual selection entirely. But they allowed it at best a meager role.

"Good sense" female choice. Now to the "good sense" alternative. On this view, females choose their mates for vigor or health, for territory size or nest quality—the sort of sensible characteristics that natural selection would be choosing anyway. Male flamboyance comes along as part of the package, but it is not what interests the females. This idea, too, was suggested by Wallace (the only concession that he did make to female choice). But he didn't develop it. It foundered on the obvious problem of why good sense qualities just happen to coincide with ornamentation—and, indeed, why males evolve ornaments at all. The Darwinism of Wallace's day could deal happily with female choice of the strongest or the swiftest, but it stopped short at the very puzzle that Darwin's theory was designed to explain.

Recently, however, taking advantage of new lines of explanation, good sense theories have flourished. Combine, for example, the notion of an arms race within a species and the notion that male embellishment can act as an indicator of sensible qualities, and one route to ornamental escalation immediately offers itself. Consider the elaborate strutting that male sage grouse use to court females (see discussion of Vehrencamp in Krebs and Harvey, 1988). A male beats his wings, makes popping and whistling sounds, and inflates his chest with two orange air sacks set in white feathers. So conspicuous a display is very energy-costly, and males vary enormously in how much they strut. Females prefer mates that strut the most. It seems that they are picking males that are most able to sustain themselves (perhaps the most proficient foragers). Display is probably serving as a readily visible, reliable marker of more elusive sensible qualities. Presumably, over evolutionary time, there has been an arms race among the males, a symmetric arms race, with both sides competing at doing the same thing, so that the best strutter is the one that goes that bit further, however far the rest have already gone. And there would also be an arms race between males and females, this time an asymmetric one, males trying for low-cost, high-impact advertising and females scrutinizing their display ever more stringently so as not to be caught out by dishonesty. Thus, locked into the strategic logic of escalation, females, in strict pursuit of sensible qualities, could end up preferring the most extravagant of displays. One good sense theory along these lines, suggested by W. D. Hamilton, is that females choose males for their hereditary resistance to parasites (Hamilton and Zuk, 1982). Loss of color is an early sign of parasitic debilitation, so females can use brightness as diagnostic of immunity. Males will then want to look brightly hued in order to advertise that they are parasite free, females will try to police for cheating—and so on, down the generations, the females' utilitarian stethoscope gradually burgeoning into a brilliant kaleidoscope. There is even a view—the "handicap" theory, first suggested by Amotz Zahavi—that flagrant excess is not an unfortunate side effect of female choice but the very criterion on which females are choosing (Zahavi, 1975, 1981). If a male is encumbered with a hefty tail, conspicuous colors, and a daily song and dance routine to be completed before breakfast, and yet still manages to get through to the mating season, he must indeed be of sterling quality. The greater the burden, the more severe the test. So females should prefer the most handicapped of males—as long as they are still alive and capable of mating! Upside-down as this seems, some modern theorists have recently moved far toward Zahavi's position (e.g., Anderson, 1986; Pomiankowski, 1987b; see also Dawkins' discussion of Grafen in the second edition of *The Selfish Gene* [1989]).

Empirical investigations over a wide range of species suggest that females are, indeed, often sensible (e.g., hanging flies: Thornhill, 1980a; mottled sculpins: Brown, 1981; bowerbirds: Borgia et al., 1987). So Wallace has at last been vindicated. Sensible choice can end up with males looking so defiantly ornamental that the choice appears intuitively to make no sense at all. "Good sense" choice has now established itself as a legitimate interpretation of "sexual selection."

"Good taste" female choice. Now to Darwin's own "good taste" theory of female choice. It's all very well to explain the peacock's tail as pandering to the peahen's fancies, but this only raises the equally difficult problem of why on earth the peahen indulges in such costly predilections. How can it be adaptive—how, indeed, can it be anything but downright deleterious—to choose a mate who drags himself down by his own tail, and whose sons and grandsons, and all future generations, will suffer the same fate? Darwin had no answer. And neither, for half a century, did anyone else. It was R. A. Fisher, the pioneer of statistics and population genetics, who eventually came to the rescue (Fisher, 1915; 1930b, pp. 143–156, especially pp. 151–153).

Fisher argued that choosing a mate for his attractiveness alone can be adaptive because he will have attractive sons. If there is a majority preference for long tails or, indeed, for anything whatsoever, no matter how arbitrary, a female's best move is to go along with the fashion, because her sons will inherit their father's attractive feature while the next generation of daughters will inherit their mothers' preference for this feature. We can think of it in the following way. Imagine that you are a peahen in a population of peahens in which the majority prefer long-tailed males. (Never mind, for a moment, how the majority came to be that way.) You could decide to ignore the majority preference and mate with a male with a sensibly short tail. But think what would happen in the next generation. Your sons would inherit their father's short tail. But the daughters of the peahens around you would inherit their mothers' preference for long-tailed males (and their sons would have the attractive tails). So, although your son would be better equipped than most for day-to-day living, it would be to no evolutionary avail because he would be unlikely to get a mate. Selection would soon write off both genes for short tails and genes for a preference for them.

What fuels this process—and gives rise to the typical escalatory, runaway quality of the peacock's tail—is an immensely powerful self-reinforcing mechanism, a positive feedback. It is generated by a connection between preference genes and tail-length genes, the measure of the connection being known as the coefficient of linkage disequilibrium. This is how the mechanism works. Suppose again that you are a peahen, this time with a preference for long tails. When you mate with a long-tailed male, your sons will inherit genes not only for their father's long tail but also for your long-tail preference, although the preference genes won't, of course, express themselves phenotypically; similarly, your daughters will have their father's genes for long tails (although not phenotypically expressed) as well as your long-tail preference genes. So your union has soldered a connection between the two kinds of genes. And the same will happen in subsequent generations. Thus the more that females exercise a fashionable preference for long tails, the more the fashion is reinforced, each choice of long-tailed mate automatically selecting in favor of copies of genes for that very choice. We can now see how readily this escalation could get off the ground. Fisher suggested that it might start with a "sensible" choice—slightly longer tails being better, say, for flying; once the preference gene had established itself through natural selection, it could take off under its own steam, reinforcing its success in inexorable escalation. But the process could start with a characteristic that has no initial "sensible" function. All that it needs is a preponderance of one preference over the rest. Even small chance fluctuations could be enough to trigger the unstable runaway. And incidentally this could, of course, consequently go in any direction; tails could run to becoming ever shorter, ever duller, ever simpler—Fisherian taste building the plumage of a sparrow just as readily as the beauty of a peacock. Fisher showed, then, that—at least in principle—male ornament and female taste, however apparently absurd, can evolve in tandem, united in explosive runaway, pressing ever onwards to spectacular heights, reaching the typical immoderation of the peacock's tail. He showed that Darwin could be right: selection could favor "beauty for beauty's sake." At the same time, he met Wallace's demand that female choice should be adaptive: if good taste shows fashion sense then good taste can, after all, make adaptive good sense.

Fisher's vindication could have been the start of a brilliant new career for Darwin's theory of sexual selection, which had, after all, languished, almost universally rejected, for half a century. But it took another half century before Fisher's contribution, which he merely sketched, was taken up by population geneticists and filled out in what has now become a variety of formal models (e.g., O'Donald, 1962, 1980; Lande, 1981; Kirkpatrick, 1982; Seger, 1985; see also Dawkins, 1986, pp. 195–215).

Do any species go for good taste? If the Darwin-Fisher runaway process is at work, then natural selection will be clamping down on sexual selection's reckless spiral. So females should harbor a latent preference for more than nature ever offers them. When female long-tailed widowbirds were given the choice, they exhibited just such a hitherto unmanifested preference (Andersson, 1982, 1983). A group of males had their tails clipped down to almost a quarter of their lavish, mating-season, length; the feathers were glued onto the tails of another group, extending them to twice their normal size. The females overwhelmingly preferred to mate with the super-tailed males, even though they could never have satisfied their taste without the experimenter's unnatural interference. Super-preferences are not exclusive to the Darwin-Fisher runaway process. Nevertheless, long-tailed widowbirds could perhaps be Fisherian followers of fashion.

Darwin's good taste theory is now a universally accepted meaning of "sexual selection." Nevertheless, some Darwinians still regard it as an oddity. Their discomfort stems mainly from their notion of adaptation. For Darwinism's first hundred years, Darwinians standardly thought of adaptations as utilitarian and economical, the result of brisk, no-nonsense engineering, of straightforward benefit to their bearers and perhaps to the bearer's offspring. During the period of good-for-the-species thinking, which became common from the 1920s until the mid-1960s, adaptations were also thought to be of benefit to the species as a whole. Eyes and fur, foraging and nest building—these were the stuff of adaptation.

Viewed in this light, sexual selection seemed to flout all the rules. The peacock's frivolous, costly tail, the female's whimsical preference for it—how could they possibly be good for their bearers or for the species? The males of sexually selected species were sometimes seen as selfishly furthering their own reproductive success to the detriment, even to the eventual extinction, of the species as a whole. Julian Huxley waxed indignant about what he called this "biological evil"—that "display characters confined to one sex could be . . . useless or even deleterious to the species" (Huxley, 1942, p. 484; 1938a, p. 13). And Konrad Lorenz said that female choice "may give rise to bizarre physical forms of no use to [and] . . . quite against the interests of the species . . . which may easily result in destruction" (Lorenz, 1966, pp. 31–33; see also, e.g., Grant, 1963, pp. 242–243; Simpson, 1950, p. 223). Even nowadays some Darwinians make a distinction between natural selection's adaptations, which are "beneficial to all members of the species as well as to the individual" and sexual selection's adaptations, which are "beneficial to the individual but harmful to other members of the species" (Bajema, 1984, pp. 11, 113; see also pp. 110,

262). The idea that sexually selected characteristics are not typical adaptations is not confined to species-level thinking. Lorenz remarked that sexually selected characteristics were often "nonadaptive . . . and irrelevant, if not positively detrimental to survival" (1966, p. 31); the evolution of the Argus pheasant, for example, "has run itself into a blind alley . . . these birds will never reach a sensible solution and 'decide' to stop this nonsense at once" (1966, p. 32). Mayr has stated that sexual selection "may produce changes in the phenotype that could hardly be classified as 'adaptations'" (1983, p. 324). Today's leading experts on sexual selection standardly contrast what they call "adaptive" sexual selection (by which they mean good sense, Wallacean sexual selection) with what they call "Fisherian" or even "maladaptive" (good taste) sexual selection (e.g., Kirkpatrick, 1982, p. 10; Andersson and Bradbury, 1987, pp. 2–4). Some experts argue for preserving the traditional distinction between natural and sexual selection because "structures that confer mating success may hinder the male in the struggle for survival: sexual selection and natural can be opposing processes" (Arnold, 1983, p. 70; see also pp. 68–71).

But there are other ways of looking at Darwinian theory. Beginning with the work of Fisher, J. B. S. Haldane, and Sewall Wright in the 1930s, but particularly since the 1970s, Darwinians have increasingly looked on it as dealing with the replication of genes. On this view, adaptations are for the benefit of the genes of which those adaptations are the phenotypic effects. Adaptations may not be prosaic or worthy; genes can further their destiny as much by apparent ostentation as by strict austerity. And they may not be of benefit to their bearers, let alone to the species as a whole. Adaptation can be as aptly epitomized by a kin-selected ground squirrel heroically calling warnings to her sisters, or a hapless host manipulated to suicide by its parasite, as by the swift's wing or the chameleon's camouflage. Viewed afresh in this light, the distinction between "adaptive" natural selection and "nonadaptive" sexual selection dissolves. Certainly, male ornament involves a trade-off between mating advantage and survival. But predation involves a trade-off between eating and, say, getting eaten. All adaptations are costly and all incur opportunity costs.

Aesthetic, intersexual, and coy: Some terminological curiosities. Finally, a few terminological curiosities. Many Darwinians are still embarrassed at good taste choice being called "aesthetic"—as Darwin insisted on calling it. At one time this description may indeed have been farfetched. But when applied to Fisherian runaway, the term is entirely apt. It nicely captures the notion that escalation can proceed without any concessions to traditional concepts of "good sense," fashion alone giving female choice its selective advantage, taste alone being the driving force.

Another curiosity is the habit that has crept in of referring to female choice as "intersexual" selection. What could this mean? Consider intraspecific and interspecific competition. Intraspecific competition means competition between individuals within a species—sibling rivalry or a scramble for prey. Interspecific competition means competition between species—rabbits and foxes or antelopes and lions. Now consider intrasexual and intersexual competition. Intrasexual does indeed mean reproductive competition between members of the same sex—males fighting with males or singing the loudest or growing the showiest tail. "Intersexual," then, if it meant anything, should mean reproductive competition between the two sexes—males and females competing for the privilege of being the sex that does the mating! A very odd idea indeed—and not sound on biological intuition. In fact, of course, so-called intersexual selection is still about males competing with other males. Certainly, they are competing for females. But this does not make their competition "intersexual." Competition among cats for mice is not interspecific competition, even though the mice are a different species. The terminology perhaps arose, as Jerram Brown (1983) suggests, because Julian Huxley coined the term "intrasexual" selection to cover the other half of sexual selection (aggressive competition between males) and, in some muddled way , "intersexual" might have seemed like the appropriate alternative (though Huxley called it "epigamic").

A final oddity from the vocabulary of mate choice. When females choose their mates, they are standardly called "coy" (and males "eager") (e.g., Bradbury and Andersson, 1987, p. 4). If males were choosy, would they be coy—or would they be discriminating, judicious, prudent? (And would females then be eager—or would they be wanton, brazen, flighty?)

For a hundred years, evolutionists wrote regular obituaries to sexual selection. Go back to the time of Darwinism's great twentieth-century synthesis and you will find that sexual selection barely scraped into the indexes of most of the major texts (e.g., Dobzhansky, 1937; Huxley, 1942; Mayr, 1942; Simpson, 1944, 1953; Rensch, 1959). And then, in the late 1970s, the peacock's tail started to rise, phoenix-like, from its own ashes. The theory was at last being assimilated within orthodox Darwinism (e.g., Blum and Blum, 1979; Searcy, 1982; Bateson, 1983; Thornhill and Alcock, 1983). Some Darwinians feel that it has now found its place. But for others, the final accolade will be a new invisibility, the recognition that sexual selection is not, after all, a thing apart, separate and different from other selective forces, but a natural part of natural selection itself.

Sexual Selection: Contemporary Debates

Hamish G. Spencer and Judith C. Masters

The concept of sexual selection was first described by Charles Darwin in *On the Origin of Species* (1859a) and later elaborated in *The Descent of Man* (1871). Darwin's view was that sexual selection was that form of selection arising from "the advantage which certain individuals have over other individuals of the same sex and species, in exclusive relation to reproduction" (1871, vol. 1, p. 256). Sexual selection is thus not about the struggle for existence (natural selection); rather it concerns the struggle to reproduce. Several authors (e.g., Ghiselin, 1974a; Arnold, 1983) have emphasized Darwin's distinction between these two evolutionary forces. Others insist that sexual selection is a subset or category of natural selection, albeit a useful concept (e.g., Bateman, 1948; Thornhill and Alcock, 1983). Karlin (1978) viewed sexual selection as a special case of nonrandom mating that, unlike most forms of nonrandom mating, was asymmetrical in the sexes.

Darwin (1871) saw sexual selection as comprising two processes, usually referred to as intrasexual and intersexual selection. The former arises from competition among individuals of the same sex for mates. For example, males of many mammalian species (e.g., elephant seals; see Le Boeuf, 1974) fight for control of a harem of females, the successful males obtaining most (if not all) of the subsequent matings. If the victorious males differ in a heritable way from those that are defeated, then the conditions for sexual selection are satisfied. This process is also known as "male-male competition," because in most of the proposed examples it is the males which compete.

Intersexual selection—also known as epigamic selection—occurs when members of one sex choose as mates individuals of the other sex whose phenotypes, again, differ in some heritable way. This form of sexual selection thus falls within the ambit of the behavioral phenomenon known as mate choice—in most cases demonstrated as female choice—and has proved a much more controversial idea than intrasexual selection (Ghiselin, 1969a). The demonstration of mate choice contains many methodological difficulties (Halliday, 1983), which are greatly aggravated by the problems of identifying a heritable basis to the chosen characters. Lambert et al. (1982) showed that many mate choice experiments and field studies purporting to reveal female choice failed to eliminate other possible explanations (e.g., male-male competition, females choosing territories rather than males). Most of the argument today concerns this form of sexual selection, intrasexual selection being accepted by most workers in the field.

The widespread use of the terms "male-male competition" and "female choice" for intrasexual and intersexual selection obscures the few cases of female-female competition and male choice that have been reported (see, e.g., Halliday, 1983, for examples of the latter). These instances may provide crucial tests for hypotheses concerning the more commonly reported situations. The asymmetry in the roles of females and males in most purported examples of sexual selection is generally attributed to alternate "investment strategies" (Trivers, 1972). Trivers argued that where there is a marked differential in parental investment in offspring by the two sexes, their strategies for optimizing their respective reproductive outputs will differ. Mating competition should be most intense in the lower-investing sex, given that polygamy is possible. Similarly, mate choice should be most emphasized in the higher-investing sex, because mistakes will be relatively costlier. It is generally assumed that males are the lower-investing sex and females the higher, but this assumption may often be unwarranted: the cost of sperm, for example, may not be negligible (Nakatsuru and Kramer, 1982).

Trivers' reasoning brings us to one important criticism that has been leveled at sexual selection in general and at intersexual selection in particular. This criticism is that both in its general and specific applications, sexual selection theory is a prime example of the adaptationist program, easily bent to the generation of fascinating stories rather than useful explanations of observable phenomena (Lewontin, 1977; Lambert et al., 1982; Arnold, 1983; Lande, 1987). Consider, for example, the following excerpt from Trivers (1972, p. 170): "In dung flies, in which females must mate quickly while the dung is fresh, male courtship is virtually nonexistent . . . The male who first leaps on top of a newly arrived female copulates with her. This lack of female choice may also result from the *prima facie* case the first male establishes for his sound reproductive abilities. Such a mechanism of choice may of course conflict with other criteria requiring a sampling of the male population, but in some species this sampling could be carried out prior to becoming sexually receptive." In this quotation, the theory is quite obviously untestable: even when female choice is entirely absent, it is possible to explain the phenomenon in terms of female choice.

The distinction Darwin made between natural and sexual selection is not as clear in practice as it is in theory. The effects of fertility selection on a population, for example, may be identical to those of sexual selection. Many of the field studies that at first glance appear to be examples of sexual selection can, on closer examination, be better described by natural selection. For instance, a common observation in lek species is that females tend to mate with males at the center of the lek. Thornhill (1980) and Partridge and Halliday (1984) have suggested that this may have more to do with protection from predators during mating than with any superiority on the part of the central males. Even sexual dimorphism may not necessarily be the result of sexual selection (Mayr, 1972b). For example, that the males of many species are larger than the females may be due not to either male-male competition or female choice but to bioenergetic factors in reproduction; Downhower (1976) suggested that smaller female body size can be advantageous in fluctuating environments, because smaller females are more sensitive to environmental changes that indicate favorable conditions for breeding and hence are likely to breed sooner and more often than their larger counterparts.

In addition, Lambert et al. (1982) and Paterson (1989) have pointed out that many aspects of courtship behavior are under strong stabilizing selection, because deviants are less likely to be recognized as conspecific mates. The existence of such stabilized specific-mate recognition systems is not evidence for the origin of these systems via directional sexual selection, as recognition (conspecific/nonconspecific) does not equate with choice. Some confusion of mate recognition with mate choice arises from the ambiguity of the term "(female) preference." A preference to mate with conspecific individuals (recognition) does not necessarily involve preferences for different kinds of individuals within that species (mate choice).

The issue of whether or not sexual selection can be construed as a subclass of natural selection depends on the question being addressed. Both sexual and natural selection are manifested as differential reproductive success. If the causes of this differential fitness are not at issue, the treatment of sexual selection as a category of natural selection is quite appropriate. If we wish to investigate the consequences of different forces on reproductive success, then it can be revealing to maintain the separation of these concepts. For example, in several of the models to be discussed, the action of sexual selection is contrary to that of what we usually consider as natural selection.

One of the strongest advocates of sexual selection among the early neo-Darwinists was Fisher (1930). He noted that in discussing female choice,

Darwin failed to provide an explanation for the origin of female preference. Fisher argued that for sexual preference to exist the risk of remaining unmated must be outweighed by the advantage of choosing a mate. Today we extend the risk of not mating to include the cost of delaying mating, i.e., the cost of choice (Parker, 1982; Pomiankowski, 1987a). Fisher proposed a model not only for the origin of a female preference but also for evolutionary change via female choice. According to this model, termed the "runaway process," a female preference may arise if the preferred male character is also favored by natural selection. Once this preference is established its intensity will be increased, because the sons of the females with the stronger preference (and thus also sons of the males exhibiting the trait) will be preferred in the following generation. Concurrently the character itself will be selected to be expressed more strongly. The preference and the trait thus evolve together, even beyond the point where the trait is favored by natural selection, and the process only stops when the sexually selective advantage accruing to males with the character more strongly developed is balanced by some naturally selective disadvantage. Fisher's model obviously rests on a relative criterion for choice: females must always choose mates from that extreme of the distribution in which the trait is relatively more developed, in order for the process to bring about directional change. The model also requires that some males mate more often than others, and so does not apply to monogamous populations.

O'Donald (1980) numerically examined several mathematical versions of the runaway process, and concluded that they supported some of Fisher's verbal predictions but failed to validate others. O'Donald proposed a two-locus model for diploid individuals, in which one locus coded for variation in the male character and the other influenced the female preference. He demonstrated how the nonrandom mating aspect of Fisher's hypothesis could generate genetic coupling between the male trait and the female preference, but his results could not confirm the runaway aspect of the model. This failure he attributed in part to the nongenetic nature of Fisher's hypothesis, and in part to what he saw as a non sequitur in Fisher's argument that the females should continue to prefer more and more extreme males (O'Donald, 1983). The disagreement here is over whether female preference is relative (i.e., a continual preference for more extreme development of the male character) or absolute (i.e., a preference for a particular level of male-character development).

Other analytic models have provided support for the runaway process: haploid models have been proposed by Kirkpatrick (1982) and Seger (1985), while Lande (1981) modeled the male character and the female preference quantitatively. These three authors all found a curve of equilibrium points on a plot of the frequency of the male versus female alleles or characters. That is, they identified a range of equilibrium values for which the deleterious male trait and female preference for it are in balance, the negative forces of natural selection being exactly counteracted by positive choice. Depending on various parameters, the curve (or portions of it, in the case where females chose from a finite number of males [Seger, 1985]) was either stable or unstable, meaning that populations represented by points off the curve moved toward or away from the curve, respectively. Fisher's runaway process corresponds to an unstable curve: populations would move toward either fixation or elimination of the trait (and, respectively, increase or decrease the female preference). When the curve was stable, however, it was argued that drift along the curve might result in rapid speciation if and when the trait and preference became fixed (Lande, 1981).

In contrast to Fisher's hypothesis, these models emphasize the role of drift, and, because of this, do not require an initial selective advantage for the male character (provided the female preference with additive genetic variance already exists). Further, Lande's (1981) model is not dependent on relative female choice, as Fisher argued. Evolutionary change is generated by the constant genetic variances and covariances of the characters, something O'Donald (1983) claimed was unlikely to be true. Relaxing Lande's assumptions (see Lande, 1987) could destroy the runaway aspect of the model.

A fundamental feature of all these formulations of the runaway process is that there is no direct selection on the female's preference: selection operates indirectly via linkage disequilibrium or genotypic correlation with the male character. In Lande's model, for instance, this effect is manifested in the assumptions that all females mate, and that a female's expected number of offspring is independent of the mate chosen. One curious consequence of this assumption is that the curve of equilibria passes through points in which the male trait is fixed and yet the female preference is polymorphic (figures in Lande, 1981; Kirkpatrick, 1982; Seger, 1985; and Bulmer, 1989). If the female preference is assigned a positive cost, then the line of equilibria is destroyed (Pomiankowski, 1987a; Bulmer, 1989), to be replaced by one (or more) discrete point(s). Mutation affecting either the character or the preference would have the same effect (Bulmer, 1989). Under Fisher's model, therefore, costly female choice should not arise. Bulmer (1989, p. 205) concluded that a better way of viewing intersexual selection is to "imagine female preference finding its optimal level under forces of mutation, direct selection and genetic drift, dragging the male trait with it." He considered selection on the female preference, arising from the cost of delaying mating (or even never mating), to be the most important factor, overriding any runaway process.

The "sexy son" hypothesis of Weatherhead and Robertson (1979) is an extension of the runaway process. In this model, females which mate with attractive males make up for any possible initial loss of fertility by producing attractive sons (who, of course, carry their genes). Hence, under this model, there is an explicit cost assigned to female mating choice (Kirkpatrick, 1985): the expected number of offspring a female produces decreases if an attractive male is chosen. Consequently, there is no curve of equilibria, and the only nontrivial equilibrium point corresponds to the elimination of females who choose sexy males (Kirkpatrick, 1985). Kirkpatrick therefore considered the model to be untenable. Diploid models, however, can lead to the opposite conclusion (see Curtsinger and Heisler, 1989, and references therein).

Related to the "sexy son" hypothesis are the "good genes" approach of Trivers (1972), and its more precise version, the "handicap principle" of Zahavi (1975, 1977). In contrast to Fisher's suggestion that the female preference arises because of a positive correlation between the male trait and fitness, Zahavi claimed that the initial correlation should be negative. The argument thus states that greater expression of the trait is a handicap, but that those males which survive to mate must have compensated by being fitter in other respects. The trait, therefore, is an advertisement for "good genes." Females who prefer such males will thus gain generally better genes for their offspring. Their daughters will gain unconditionally if the trait is expressed only in the males; their sons will gain matings, but lose fitness in other respects by expressing the trait. O'Donald (1980) raised several criticisms against this model, one of them being that the male trait would quickly become disassociated from the higher viability genes. More recently, however, Michod and Hasson (1990) presented an interesting model involving a modifier locus, which does allow the trait to become a reliable indicator of viability. Maynard Smith (1985) recognized three nonmutually exclusive types of handicap: (1) "Zahavi's handicap," in which all males with the handicap genotype express the handicap equally, (2) the "revealing handicap," where the expression of the handicap accurately reflects the genetic makeup of the male, and (3) the "conditional handicap," where the degree of expression of the handicap depends on the phenotypic condition of the male. The revealing and conditional handicaps essentially make the handicap a more reliable basis for advantageous female choice, and do have some effects on the actions of the mathematical models discussed below (see, e.g., Pomiankowski, 1987b), but do not alter the basic conclusions.

If the handicap is not inherited (for example, because it is environmentally induced) then the model appears unproblematic (Maynard Smith, 1976, 1985), although it would not explain the rise of the handicap. If, as seems more likely, the male character has some additive genetic variance, then the outcome is not so clear: for a review of several apparently contradictory simulations and calculations, see Maynard Smith (1985) and

Kirkpatrick (1986). Kirkpatrick (1986) argued that the conflicting results arose from the multiple equilibria present in the various models. He went on to develop a quantitative model of the handicap principle, envisaging it as the addition of a viability trait to Fisher's (1930) model of a male trait and a female preference. (Fisher, however, did invoke a viability argument in his attempt to explain the origin of female preference, although this does not as yet appear to have been modeled.) In Kirkpatrick's analytical model the equilibrium point reached by a population depended on the initial conditions. The equilibria all fell on a curve (as in the models of the runaway process), and once the population reached a stable point on the curve no further evolution occurred. Kirkpatrick interpreted the lack of further change to mean that the handicap principle did not work, because there was no necessary increase in female preference for the viable but handicapped males. Moreover, female preference could maintain the handicap trait at very high levels in the population, a maladaptive rather than adaptive outcome.

Pomiankowski (1987b) strongly criticized Kirkpatrick's (1986) model, claiming that the reason the handicap principle did not work in this and other models was that the additive genetic variance in viability was quickly exhausted. Indeed, in Kirkpatrick's reanalysis of several authors' haploid allelic models the lower viability allele was eliminated, and the system reduced to a model of the runaway process: hence the existence of multiple equilibria. When variation in the viability trait is maintained, Pomiankowski (1987b) showed that once female preference exceeds a certain threshold, the handicap principle does cause an increase in the male handicap and the corresponding female preference (more easily with the revealing and conditional handicaps than with Zahavi's). The Fisherian equilibrium curve is destroyed by the additive genetic variance in viability, and the only equilibria correspond to the fixation or elimination of the handicap. If the female preference is less than a particular value (depending on various genetic parameters), the handicap is eliminated; above this value, it is fixed. The disappearance of the equilibrium curve also implies that genetic drift and mutation are less important than in the models of the runaway process (Pomiankowski, 1987b).

The continued maintenance of additive variance for viability is not a trivial assumption, however, in spite of Pomiankowski's (1987b) assertion to the contrary. Much population genetic theory suggests that such variation should be transient, although Pomiankowski (1987b) claimed that the variation of selection pressures over space and time (e.g., as in the host-parasite interactions, suggested by Hamilton and Zuk, 1982, but see criticisms of this in Charlesworth, 1987, and Lande, 1987) and mutation would provide enough fresh variation. In contrast, Turelli (1986) has argued that Lande's Gaussian genetic models, on which most of the above quantitative models of female choice are based, implicitly assume mutation rates several orders of magnitude higher than those observed in nature.

It should also be noted that, as in the runaway hypothesis, the handicap principle explicitly assigns no cost to female choice. Again it appears as though the inclusion of a positive cost would eliminate any newly arising female preferences (Pomiankowski, 1987a), leaving a negligible level of variation in female preference and suggesting that a different view of intersexual selection is more appropriate (Bulmer, 1989).

Ironically, Bulmer's conclusions about the cost of female choice bring us back to one of the problems Darwin (and indeed Fisher) failed to solve—that of the origin of female preferences (see also Maynard Smith, 1987). In most models the preference is assumed to exist at the start. As Paterson (1978, 1982) and Lambert et al. (1982) convincingly argued, the primary function of courtship in sexually reproducing animals is to facilitate syngamy, and hence the male-female communication system is subject to strong stabilizing selection. Unusual or fussy individuals (whether male or female) will be at a disadvantage, because they reject suitable mates or are themselves rejected. To our minds, the questions of highest interest in sexual selection theory are those concerning the conditions under which the stabilizing influences on the male-female communication system may be overcome and the mate recognition system altered. To this end, the elucidation of the conditions favoring the origin and maintenance of female preferences for unusual male characters is essential.

Sexual selection can only be a useful concept if it is defined rigorously and applied appropriately. Uncritical acceptance of its ubiquitous occurrence (e.g., as in Trivers, 1972) does no service to evolutionary theory. It is our contention that heritable variance in reproductive success, based on characters that are not favored by natural selection, must be demonstrable before sexual selection may be invoked unequivocally. Empirically, we need to know more about the frequency and strength of sexual selection, as well as the apparent sexual asymmetry. From a theoretical viewpoint, we need to focus more on the consequences of the structure of mate-recognition systems, for example, the cost of female choice and also costs to males.

SPECIES: HISTORICAL PERSPECTIVES

Peter F. Stevens

THE DEVELOPMENT OF species concepts is a complex story. Some taxonomists have insisted that the act of describing species affords no room for conceptualization; taxonomists simply describe nature, a matter not of theory but of direct observation. Others, perhaps the majority, have utilized some reproductive criterion in their species concept—either the species is not fertile when crossed with other species, or at least the characters used to distinguish the species are constant over successive generations. There has been general agreement that species must be readily recognizable. In practice, however, the absence of absolute criteria for distinguishing species has prompted recourse to authority and tradition, and these have been used to justify stasis. For Linnaeus, and for many taxonomists in the nineteenth century, there was near identity between taxonomic species and what may loosely be called the functional or ecological units of the natural world. But by the end of the century little congruence was seen between the two, yet both were still called species. Attempts were made, especially in some branches of zoology, to restore congruence or even identity between taxonomy and evolution or biology. Botanical and some zoological taxonomists developed distinctly different species concepts, although authors described most species without explicit mention of their concepts; still other concepts were used by biologists who were not taxonomists.

Further complicating the issue, many of the words commonly used in discussions of species and species concepts—"objective," "real," "natural," "arbitrary," "abstract," and the like—are among the most ambiguous in the English language (Gregg, 1950). Finally, the rules of nomenclature are generally believed to be value-free, yet our choice of names both reflects and conditions our perception of the world (La Vergata, 1987).

The possibilities for ambiguities and contradictions that this summary suggests have been more than amply realized. Unfortunately, there is no appropriate history of species concepts to guide one through an extremely difficult literature, although Mayr (1982a) fills some of this gap. Here I focus on the period 1750–1965.

Taxonomic practice. An understanding of how taxonomists see the practice of their discipline and how taxonomic knowledge is transmitted is essential to any discussion of the conceptual tensions surrounding species in taxonomy. The perceived relationship between theory and practice in such a discipline is likely to be especially problematic. Taxonomy has not infrequently been considered something of an art (e.g., Simpson, 1961); indeed, the description of an artistic genius in Kant's *Critique of Judgement* could stand for the description of a good taxonomist. Further insulating the act of taxonomic judgment from analysis is the long-standing feeling expressed by Asa Gray (1854) and Charles Lyell (see Wilson, 1970) that "intuitive perception" is involved in this judgment. The quality most commonly attributed to great taxonomists of the past, such as George Bentham, is "instinct."

Taxonomic knowledge has been transmitted through an apprenticeship system (A. J. Cain, in Simpson, 1961; Stevens, 1986), the neophyte attending to exemplar treatments by the great taxonomists. At the end of the last century detailed study of populations engendered new methods for analyzing and displaying data, but their utility in delimiting taxonomic species was sometimes questioned (Diels, 1932), and such methods were relatively seldom used. Rule of thumb and usually poorly articulated data analysis informed the decision to assign species rank to a group of organisms. The main requirements of an adequate definition of species were that species be readily recognizable by the layman; that the nomenclature be stable; and that there not be too many names to learn. Paradoxically, names have been readily changed for nomenclatural reasons, especially from the later nineteenth century onwards, yet the allegiance to a stable nomenclature may have perpetuated the use of biologically inappropriate names (cf. Gilmour, 1958).

Grouping and ranking. The critical distinction between grouping and ranking was made only in the middle of this century (Gregg, 1950; Mayr, e.g., 1963; Hull, 1965). This distinction is simple: one can recognize a group of organisms—cats, daisies, aardvarks, red maples—because the members all look similar. The rank at which to place that group is a separate issue. I can recognize the red maple as a species, *Acer rubrum,* and place it by itself in a separate genus or reduce it to a variety of another species of *Acer*; although the rank of the group changes, its circumscription does not. Arguments at the beginning of the century, however, in particular over whether species were "real" or simply "concepts" (e.g., Bessey, 1908) or whether the genus was more "natural" than the species (Anderson, 1940), remain obscure. If species are not "real," does this refer to the species as group, or to the rank of species itself, or both? Does the term "natural" refer to groups at one taxonomic level compared with each other, or to a comparison of the levels themselves? A related issue concerns the current use of terms such as "typological" or "nominalist" to refer to species concepts, and the attitudes of people with such concepts. Essentialists see species, both groups and rank, as 'real,' with members of an individual species being members of a class. Hull (1965) teased apart some elements involved in essentialism, however, and it is clear that to call people such as Charles Darwin or T. H. Morgan essentialists or typologists obscures rather than clarifies what their attitudes actually were. As will be evident below, the same is true of the term "nominalism."

Species as taxonomic groups and species as actors in nature. Although it is usual to discuss taxonomic species and species concepts by themselves, ambiguity results from tensions between conceptualizations of species as part of a general explanatory program and the definition of species as a classificatory category (Hodge, 1987a; Sloan, 1979, 1987). Linnaeus (e.g., 1751) provided an almost apocryphal definition of a species: "we reckon as many species as there were diverse forms created in the beginning." Note that this and some other early species concepts were occasionally rejected because they were not operational; even Adam was too late to see the actual creation of plants and animals, and genealogical connections were considered to be "metaphysical." Linnaeus believed that the plants and animals he placed in species were the identical descendants of their created progenitors because of their common essence. Species (and groups in higher ranks of the hierarchy) were "real" both as groups and as ranks, and variation below the rank of species was ignored. This enabled Linnaeus to reduce the plethora of species, based on inconstant and slight differences that were then recognized. In terms of Linnaeus' understanding of the relationships between organisms and environment, species were units that functioned both in ecology and in classification (Stevens and Cullen, 1990). Even Jean Baptiste de Lamarck, who was a nominalist about species, continued to see them as at least locally distinct (see Burkhardt, 1987); they also included a part of the order of nature.

Darwin was also a nominalist, but of a type quite different from Lamarck. Darwin's species were discretely bounded groups that existed in nature, whether nature was considered in a local or a more general context. Although the rank of species was a matter of tradition and had no unique properties, species (and other low-level taxa) were nature's actors as well as taxonomic units (see Darwin, 1859a). Darwin was very much an exception in denying reality to the rank of species, and in attaching little importance to the inability to hybridize as a distinction of rank. He argued that the diversity of species concepts current in the middle of the nineteenth century was in fact tantamount to negation of the reality of the rank of species (see also Beatty, 1985).

The rank of species maintained its privileged status throughout the nineteenth century, even for many taxonomists who believed in evolution. Some, such as W. T. Thistleton-Dyer (in Cock, 1977) and E. B. Poulton (1908), even interpreted Darwin's work as support for the reality of rank. George Bentham (1875) was exceptional in his forceful articulation of the taxonomic consequences of Darwin's theories. Some feared that the acceptance of evolutionary thought would lead to the description of numerous new species, because there could be no implicit recourse to an ultimate authority (God) when deciding what was or was not a species. Some other taxonomists perceived implications of evolution: species were no longer facts, or evidence of the categories of God's mind, but judgments, perhaps even psychological projections "made" by the taxonomist (Gray, 1860, 1880; Cattell, 1898; Agassiz, 1857, 1860). Although most authors conceded that taxonomic ranks above the species level were largely arbitrary, they felt that the rank of species was different.

Tension between evolutionary notions of species and species as taxonomic units. In general, taxonomists conceived of species as groups of similar organisms that maintained this similarity through successive generations and did not hybridize with other groups. Zoologists, having discussed differences between the modes of reproduction of plants and animals from the time of Buffon (1749) onwards, tended to consider the reproductive criterion especially important (see Gray, 1863). It was only toward the end of the nineteenth century that tensions developed between species conceptualized in evolutionary terms and species as taxonomic units. These tensions were resolved differently in the fields of ornithology and botany, with a deceptively similar eventual outcome: the acceptance of a broad species concept.

In ornithology, in particular, evolutionary considerations inspired the emphasis on species as being units separated by discontinuities, and subspecies as units with some intermediates (American Ornithologists' Union, 1886). Nomenclature changed to accommodate these findings through the introduction of trinomials, the name of a bird consisting of the generic, specific, *and* subspecific names. This trend culminated in the biological species concept, which was explicitly formulated to make discussion of the evolution of species possible (Mayr, 1942; cf. in part Beatty, 1982), and also to emphasize local—that is, nondimensional—variation patterns. "Species" here denotes groups of actually or potentially interbreeding natural populations in reproductive isolation from other populations. Species and subspecies remained conceptually dependent on larger theories of the evolution of diversity, and the ontological reality of the rank of species was reasserted. Ernst Mayr explicitly focused on the results of speciation rather than on the process; the results, ex definitio, emerged only in the nondimensional situation, that is, when two forms became able to persist together and to remain distinct in a particular locality. The taxonomist, concerned with a larger universe, was not satisfied. Mayr has often seen a continuity among the species Linnaeus (and Ray) recognized in the field, the species of "naturalists" and local groups of people generally (for the latter, see Atran, 1987), and his nondimensional species. There was little tension between field and the taxonomists' closet in Linnaeus' time, because the properties of the local species could legitimately be extrapolated to taxonomic species; the ability or otherwise to interbreed was a property of species, not of individuals, and taxonomy and biology were one. Tension, however, pervades Mayr's own work (e.g., Mayr, 1942, 1982a), because such extrapolation was no longer straightforward. That the early formulations of subspecies concepts reduced the "unwieldy" number of names at the rank of species (and were in part designed to do so) should not obscure the biological intent behind these formulations (cf. Cracraft, 1989). Ultimately the new nomenclature was justified on evolutionary, rather than pragmatic, grounds.

Lepidopterists and mammalogists also helped to develop the biological species concept (see Rothschild, 1895; K. Jordan, 1896; D. S. Jordan, 1905). The situation in botany was different. The nomenclatural polemics at the turn of the nineteenth century led to an explicit rejection of the "ornithological" position on trinomials, and a continued nomenclatural separation of the rank of species from infraspecific ranks. This separation reflected a different conceptualization of species, the nomenclature lending support to those who considered infraspecific variation to be less real than specific variation, and the rank of species, in consequence, more real than lower ranks in the hierarchy. The work of Hugo de Vries (1905), William Bateson (e.g., 1913), J. Clausen (1922), G. Turesson (1922), and others, however, suggested that units below the level of the taxonomic species existed and appeared to be active in evolution. Many workers, including Clausen, Turesson, G. H. Shull, H. M. Hall, F. E. Clements, N. Vavilov, E. Anderson, and E. B. Babcock murmured approval of Linnaean (taxonomic) species, but their work failed to show the relevance of Linnaean species to evolutionary studies. Such ambiguity is also very evident in Theodosius Dobzhansky (1951).

It was clear to botanists that speciation occurred in several ways, one of which, the doubling of chromosome numbers following hybridization, was at best very rare in animals. With the focus on how species evolved, it is not surprising that the overwhelming majority of botanists thought that there could be no unified species concept, and did not accept the biological species as the only allowable species concept. (Botanical evolutionists such as G. L. Stebbins [1950], P. H. Davis and V. H. Heywood [1963], and a few zoologists such as Dobzhansky [1937], Julian Huxley [1942], and Cain [1953] also inclined to this position.) Because barriers to the exchange of genes were rarely absolute, either in the wild or in an experimental setting, botanists often argued that absolute reproductive criteria had little to do with taxonomic species (Grant, 1957; cf. Mayr, 1959). Debate also continued over actual *versus* potential barriers (e.g., Müntzing et al., 1931); both were used to circumscribe taxa at the same hierarchical level. The more fluid nature of reproductive barriers in plants allowed full play in the ambiguities surrounding the long-admitted connection of "genealogy" with species. W. H. Camp (1951) even suggested that the basic units of the taxonomic hierarchy should each include all forms that could be crossed one way or another, a position similar to that adopted by W. Herbert (1837) over a century before.

In sum, botanists had biological reasons for rejecting the biological species concept, depending as it does on particular views of reproduction and speciation. The traditional basic pragmatic demands for a limited number of easily recognizable species could thus continue to play a dominant role in the botanists' notion of species.

Consequences of the use of different species concepts. What were the consequences of these disparate species concepts on taxonomic practice? Herbert considered his genera to be the conceptual equivalent of other people's species; both ultimately represented individuals created by God. This raises the issue of how species concepts relate to how species are circumscribed, sometimes called the debate about "splitting" and "lumping." "Splitters" adopt a narrow definition of species, under which relatively fine differences are sufficient to justify a new taxon; "lumpers" adopt a broader definition, with a consequent smaller number of species. But "splitting" and "lumping" species has little to do with particular species concepts, as M. P. Winsor (1979; cf. Mayr, 1959) has persuasively argued. For instance, the species that Herbert recognized are not clearly affected by his distinctive views on hybridization and creation (A. Meerow, pers. com.). It is often hard to see clear connections between species concepts and the species actually recognized. J. D. Hooker (see Huxley, 1918) noted with a little exaggeration that taxonomists tended to state their views without acting on them. Connections established in a taxonomist's mind seem to have to do more with the justification of taxonomic conclusions than the actual circumscription of species. The charged nature of the debate derived from the effects of a species concept on an individual worker's ontology of species. Thus Gray (1850) considered that there might be "aboriginal" varieties if one species were originally created in more than one locality; Louis Agassiz (1857) rejected the possibility of such varieties; others thought that species necessarily emerged in only a single locality.

Did the existence of man-made variation necessarily bear on species concepts? Charles Naudin, Alfred Russel Wallace, Darwin, and Hooker all thought it did; others just as emphatically denied the connection (see Rolfe, 1900, on the relevance of artificial hybridizations for taxonomic species).

What about comparable arguments about the taxonomic significance of characters of adaptive value at particular taxonomic ranks, or of the particular importance of characters taken from anatomy or cytology, or of a qualitative rather than a quantitative nature that raged in the literature, especially after about 1890? In the nineteenth century, taxonomists' opinions on the reality of infraspecific and generic ranks and on evolution displayed almost every possible permutation. Similarly, different aspects of "reality" may be assigned greater or lesser taxonomic relevance. Thus A. L. de Candolle (1855), T. V. Wollaston (1856), and Hooker and Thomson (1855) all conceded the "reality" of some or all infraspecific variation, but considered the cause(s) of this variation to be of a less fundamental nature than that which produced true species. For a variety of reasons, the rank of species maintained its preeminence.

With the advent of evolutionary thought, taxonomists such as the malacologist Bourguignat (see Dance, 1970) and many botanists (Briquet, 1899) and ornithologists (Stresemann, 1975) adopted narrow species concepts. J. Briquet, combating what he saw as the pernicious influence of the ideas of Anton Kerner von Marilaun (e.g., 1891), who emphasized the heritability of characters that separated species, insisted that species be sharply distinguished from one another. Kerner von Marilaun and other workers developed very distinctive ideas of how evolution occurred, ideas which were reflected in how he circumscribed species. Lyell's worst fears seem to have been realized: no one Pope (his term for special creation) who could decree the limits of species, but many little Popes (taxonomists believing in evolution), all busily describing species (see also Bessey, 1908).

Although the fear that acceptance of evolutionary theory would produce splitters was thus apparently justified, a simple identification between evolutionists and splitters is not so easily supported. Not all splitters are evolutionists; Alexis Jordan, vehement in his belief that God created species, described many hundreds of new species, mostly from France, beginning in the 1840s. His approach was widely opposed by Naudin, long a believer in evolution (Diara, 1987). Among botanists, Hooker (1853), A. P. de Candolle (1813), and A. L. de Candolle (1855) all argued that belief or otherwise in evolution should not affect the circumscription of species; a taxonomist described what did not change. Hence not all evolutionists are splitters. Darwin himself held that his evolutionary view had little effect on his circumscription of species taxa (see letters in F. Darwin, 1887; Burkhardt and Smith, 1988). Darwin and numerous other authors discussed why change, or variation, or both were in fact relatively restricted, in some way validating the taxonomic species. (Change and variation were hard to ignore when studying fossils. It was a paleontologist, G. G. Simpson, who developed the evolutionary species concept in which each species had its "own unitary evolutionary role and tendencies," and in which change became integrated into the species concept; see, e.g., Simpson, 1961.)

Splitters were not necessarily "typologists," rather, they emphasized aspects of variation patterns thought by some to be unimportant. No correlation exists between disbelief in evolution and belief in the reality of genera, species, and varieties as both groups and ranks; the same is true of any other combinations of beliefs that have existed. Babcock and Hall (1924) even suggested that if Asa Gray had access to their data, he would have reached the same taxonomic conclusions as they had. (Hill's [1988] analysis of species concepts and taxon circumscription in the *Rana pipiens* group can be usefully reinterpreted from this point of view.) New approaches to botanical systematics in the early twentieth century were seen as having little effect on the circumscription of species taxa (Hagen, 1983, 1984). Few studies, however, attempt to link circumscription of species to concept in the context of the information that was available to particular workers (but see also Dean, 1979); still fewer studies trace the development of concept and group circumscription during the life of an individual worker. In the case of Gray, the changes to his species concept that did occur during his long working life (1835–1888) are relatively slight, despite his acceptance of evolutionary ideas in the late 1850s.

A strong case can be made in botanical systematics that the broad species concepts became established by conventions and authority. Evolutionary ideas impinged little on taxonomic species, and so neither concepts nor taxon circumscription faced the same challenges in botany as in parts of zoology (see Bock and Farrand, 1980, for discussions of how species numbers changed with changing concepts in ornithology). Even by the middle of the nineteenth century it was evident that the numbers of species recognized might differ by two orders of magnitude, depending on how one drew species limits. Differing species circumscriptions could not be ignored, therefore, and authors such as Hooker (1853; see also Hooker and Thomson, 1855), Gray (1854), and A. L. de Candolle (1862) explicitly promoted the idea that species should be broadly circumscribed. This favored a worker in a large herbarium over the field botanist; the former could (and did) claim that the discontinuities found by the latter broke down on examination of more copious amounts of material from a wider geographical range. C. G. van Steenis (1948) provides a particularly striking, influential, and more recent example of this attitude; he wished to maintain Linnaeus' *concepts* for taxonomic species. Readily recognizable species have remained a desideratum, and indeed are obtained using the broad approach. Pragmatic considerations thus played a major role in the victory of a broad species in botany.

Use of taxonomic species concepts in evolutionary studies. It was not particular species concepts that led authors such as Buffon and Darwin to their distinctive views of the world (Beatty, 1985; Sloan, 1987; Hodge, 1987a); rather, their species concepts depended on their larger theories of the world. For many twentieth-century biologists, the same relationship has held, yet they have used species described by taxonomists in their work, and we have seen that the disciplinary exigencies of taxonomy have their own dynamics and connection. But those interested in using the results of evolution for comparative purposes have emphasized species concepts formulated in broad congruence with evolutionary theory in order to have units useful to evolutionary studies. Students of evolution have also appeared to find taxonomic species useful for comparative purposes, and have used a "statistical" approach based on larger samples. This is essentially the approach adopted by Darwin in his botanical tabulations. But for twentieth-century biologists interested in how evolution actually happened, taxonomic species concepts in particular were seen to be of limited use. Detailed analyses of the changing variation patterns actually found in nature mattered; the taxonomic species into which these variation patterns could be fitted was of lesser importance.

Hooker was one taxonomist who felt that there were problems inherent in the taxonomic approach. In a study of the biogeography of the arctic, he explicitly recircumscribed taxonomic species to represent evolutionary units, enabling him to compare distribution patterns of these evolutionary species (Hooker, 1861). Such an approach by a taxonomist is exceptional, however. In addition, Hooker did not change the names of the taxonomic species, and this is the crux of the issue. To oversimplify the situation, attempts to study evolutionary patterns and processes separately from a formal taxonomic system (e.g., Turesson, 1922; Gilmour and Gregor, 1939) could be ignored as long as names were not changed; if names were changed because of information gained from such evolutionary studies, there was likely to be an outcry (see Hagen, 1984).

The history of the circumscription of taxonomic species in botany is perhaps aptly interpreted in pragmatic and instrumentalist terms (Dean, 1979). As long as the overriding concern is that species be readily recognizable, not too numerous, and stable in their nomenclature, challenges to concepts are easy to ignore. Furthermore, the somewhat cynical (or simply flippant) definition of a species—a species is what a competent taxonomist says it is—gained fairly widespread currency (Regan, 1926; Dobzhansky, 1937; Gilmour, 1940; Camp, 1951); it differs little from the repeated exhortations to follow the practice of the masters of the discipline. Whatever taxonomists say that their species concepts are, it is what the taxonomists are perceived to *do* that matters. The biological species concept that became prevalent in ornithology, at least, had a different conceptual basis, although this is partly obscured because of the simultaneous use of a pragmatic justification of the concept. Developments in the last quarter of a century have caused the biological species concept, even in ornithology, to come under heavy fire. It is not clear to me that our evolutionary interests can be easily squared with the taxonomic species. We are interested in the whole process of evolution, not only in what "has evolved."

SPECIES: THEORETICAL CONTEXTS

John Dupré

SPECIES ARE, by definition, the lowest-level classificatory unit, or basal taxonomic unit, for biological organisms. Although further subdivisions into subspecies, races, or varieties are sometimes used, these are generally assumed to be of little theoretical importance. It is also often supposed that species are a uniquely significant level of division of biological kinds; although classification extends to much broader groupings (genera, families, etc.), these are often taken to be arbitrarily chosen in a way that species are not. (Certain kinds of evolutionary taxonomy, however, do attach more significance to higher taxa, as we shall see.)

The concept of a species gives rise to two major theoretical issues. The first, more strictly philosophical, question concerns the ontological nature of species: is a species a natural kind, a set, or, as is currently much argued, an individual? A more straightforwardly biological question concerns the criterion of membership in a species: is an individual to be assigned to a particular species on the basis of morphological characteristics, reproductive links, phylogeny, or what? These questions will be briefly discussed in turn.

The ontological nature of species. It has traditionally been assumed that names of species, being paradigmatically classificatory, were general terms, and hence referred to classes, sets, or kinds. Historically, it has perhaps been most common to suppose that species were natural kinds. That is to say, at least, that they were the sorts of kinds that exist in nature independently of their discovery, or naming, by humans. It is often supposed that the members of natural kinds are united by common possession of an essence, a property possession of which is both necessary and sufficient for belonging to that kind. It is, however, possible to conceive of natural kinds in ways uncommitted to this essentialist component (see, e.g., Dupré, 1986).

The starting point for contemporary discussions of the nature of species is the rejection at least of this essentialist aspect of the view of species as natural kinds (Hull, 1965; Mayr 1970). Although some very influential arguments in the philosophy of language have purported to show that species must have essential properties, it is debatable how much weight these can have against the biological considerations to the contrary (Dupré, 1981; but see Kitts, 1983). Most central among the latter are those stemming directly from belief in the theory of evolution. Because evolution assumes both variation among the contemporaneous members of a species and secular change in the overall distribution of characteristics in a species, it is difficult to see how evolution could be reconciled with a theory of fixed and unchanging essences.

Precisely the intent to reconcile our conception of species fully with evolutionary assumptions lies behind the recent claims that species should be considered not as kinds, classes, sets, or any such general category but, rather, as particular individuals (Ghiselin, 1974b; Hull, 1976). The most central argument for this claim is that species are what the theory of evolution is about: they are the entities that evolve. But, it is argued, a kind could not possibly do anything, let alone evolve. In addition, it is widely believed that there are no laws governing the members of particular species (since, as remarked above, these are typically quite variable); this is just what should be expected if species are individuals, but seems puzzling if they are, even in a nonessentialist interpretation, natural kinds.

Although a large number of biologists and philosophers have been persuaded by these—and other—arguments, others have not. It has been argued that the conception of species as sets, whether or not these are natural kinds, permits translation of any claims about the evolution of species we may want to make (Kitcher, 1984a; see Sober, 1984b, and Kitcher, 1984b, for further discussion). Moreover, though there is little doubt that no very profound laws hold all of and only the members of particular species, there might still be relatively superficial and perhaps probabilistic laws. Although general claims about overall patterns of macroevolution will very probably require treating species as units, it is possible that more local (e.g., ecological) studies might attribute modest laws to the members of particular species. Thus a possible compromise between the competing views under consideration might hold that species can best be viewed as either individuals or sets, depending on the type of biological problem at issue. A final possibility is that a proper understanding of species requires appeal to some ontological category different both from an individual and a set (or kind). Thus Ernst Mayr (1987) suggests that populations should be seen as an ontological category *sui generis*.

The criteria of membership. The other major problem concerning species is the question of what determines the membership of an individual in a particular species. It should be noted that if it is taken that species are sets or kinds, this is a question about membership; if species are individuals, it is a question about mereological inclusion, or part/whole relations. Although one's view on the individual/set question will influence the plausibility of various positions on the present question, it does not appear that the former will logically constrain the latter (though perhaps it would come close to logical incoherence for a pheneticist to hold that species were individuals). Although there are many views about the criteria of species membership currently discoverable in the technical literature, it is possible to group the most influential under three broad categories, which I will refer to in the following discussion as morphological, evolutionary, and pluralistic.

Pre-Darwinian conceptions of species characteristically assumed that membership of a species should be determined by morphological (or phenetic) characteristics. (Some of these might also have been considered essential.) The post-Darwinian recognition of the omnipresence of variation has tended to put such views in disrepute. More recently, morphologically based taxonomy has been revised in the form often referred to as numerical taxonomy (Sokal and Sneath, 1963; Sneath and Sokal, 1973). This movement aimed to ground taxonomy in objective measurements of overall similarity determined by computer analysis of very large numbers of features. Although aspects of its methodology have been incorporated into taxonomic practice, this approach has not been widely welcomed. This is in part because of philosophical difficulties with the concept of objective similarity, and in part because of a perceived lack of contact with evolutionary theory.

Despite this rather widespread disfavor of morphological taxonomy, it should be recalled that most practical classification depends on the determination of reliable morphological diagnostic criteria. Moreover, in large areas of biology where more theoretically favored approaches seem impracticable or inapplicable—especially microbiology and to a considerable extent botany—morphological conceptions of taxonomy remain more respectable. It is, however, quite widely accepted that some independent criteria for assessing the relative significance of morphological features are required.

The second, and dominant, theme in contemporary systematics is undoubtedly the intent to connect taxonomy as directly as possible with evolution. Within this tendency, there are two major subdivisions, biological and phylogenetic.

The so-called *biological species concept* has been perhaps the most widely discussed conception of species membership in recent years, owing in considerable part to the great influence of Ernst Mayr (1963, 1970). This takes a species to consist of a group of organisms connected to one another by actual or possible reproductive links, and reproductively isolated from other organisms. This conception is most centrally motivated by the thought that reproductive isolation is a necessary condition for two groups to evolve independently, and thus lies at the heart of the explanation of biological diversity. In Mayr's thought it is intimately connected with the broader conception of evolution that conceives of speciation as beginning with geographical separation and ending with the establishment of mechanisms of reproductive isolation sufficient to survive even the breakdown of geographical barriers. A recent development of this basic idea, the mate-recognition species concept, emphasizes the mechanisms by which organisms identify conspecifics as potential mates as the fundamental basis of reproductive isolation (Paterson, 1985).

The major difficulty with the biological species concept is its limited applicability. First, and most obviously, it has no apparent application to asexual organisms, every one of which is isolated from every other. (Ghiselin [1987] and Mayr [1987] both bite the bullet here and suggest that there are no species of such organisms; Mayr refers to them as forming "paraspecies" and Ghiselin as "pseudospecies"; for others—e.g., Cowan [1971]—the existence of asexual species constitutes grounds for rejecting the biological species concept. Within bacterial taxonomy, which is Cowan's primary concern, pluralism has long been an attractive option.) Second, and perhaps more seriously, in a great many actual cases—especially, but by no means only, among plants—reproductive isolation is fairly weak. This has led some biologists to a somewhat more radical dissatisfaction with the biological species concept, which is reflected in a questioning of the overriding role of isolation in divergent evolution.

Phylogenetic taxonomy aims at a more direct connection with the historical component of evolutionary theory, by starting from the principle that taxonomy should accurately reflect genealogy. Thus a necessary condition for a group of organisms to constitute a species is that they should share descent from some common set of ancestors. The condition is evidently not sufficient, given that it could apply to anything from a handful of siblings to, perhaps, the entire set of terrestrial organisms. Thus something more needs to be said about what makes a genealogically coherent set of organisms correspond to the rank of species.

Different answers have been given to this question. It is useful to consider the available options by comparison of the taxonomic hierarchy with a genealogical tree. The latter represents each divergence in the evolution of life with a branch. One, perhaps extreme, position, generally referred to as cladistic taxonomy or cladistics (deriving from Hennig's classic [1966]), proposes that taxonomy and genealogy should aim for complete convergence. All and only distinct branches of the genealogical tree should be given distinct names. Thus species are simply the smallest twigs of the tree. (It may be noted that on this view higher taxa are no less objectively real than species, because all are equally required to be monophyletic.) Strict cladistics will require some very elaborate nomenclature. Wiley (1979) offers a suggestion for making things a bit more manageable.

Probably a majority of those sympathetic to phylogenetic taxonomy are in fact committed only to the much weaker demand that classification not be inconsistent with the genealogical tree. Typically, this will involve the requirement that the members of a taxon be monophyletic, but not that the taxon necessarily includes all organisms descended from the relevant ancestral population. To take a standard example, the belief that birds are descended from (primitive) reptiles should, on strict cladistic grounds, commit us to classifying birds as a subgroup of the class Reptilia. The more tolerant and traditional classification of Reptilia and Aves as separate classes remains generally favored. In addition, the possibility of anagenetic speciation—speciation, that is, within a continuing undivided lineage—need not be excluded; sufficient change within such a lineage may make the insistence on the same name for all its temporal parts extremely confusing. The general motivation for such divergence from strict cladistics is the thought that judgments of similarity and difference should have some relevance to taxonomy independent of the desirability of recording phylogeny. These positions will thus require some appeal to criteria of speciation distinct from phylogenetic separation. Cracraft (1983, 1987), for example, proposes the conjunction of one or more heritable diagnostic characters with the reproductive cohesiveness of the group; Van Valen (1976) defends an ecological criterion; and Nelson and Platnick (1981) require a "minimal evolutionary novelty" (this position is generally referred to as "pattern cladism").

Such proposals lead naturally to a third category of views about the criteria of species membership: a moderate brand of pluralism. Subject to the constraint that species be coherent genealogical units, it has been suggested that we might adopt various different criteria for deciding to call such units species (Mishler and Donoghue, 1982; Donoghue, 1985). The crucial distinction at work is that between a criterion of grouping and a criterion of ranking (see Mishler and Brandon, 1987). A grouping criterion determines which organisms are even candidates for conspecificity, whereas a ranking criterion determines the actual extent of the species. This moderate pluralism insists on a monistic grouping criterion—monophyly—while allowing pluralism of ranking criteria: the latter may be selected according to what kinds of causal processes are of greatest relevance for maintaining the cohesion of a lineage in a particular case.

Given the insistence on this monistic grouping criterion, it is appropriate that Mishler and Donoghue (1982) refer to their view of species as (a variant of) the phylogenetic species concept. By contrast, Kitcher (1984a) has advocated a radically pluralistic conception of species. Kitcher argues that both historical (evolutionary) and structural (or functional) inquiries should be accorded equal weight in biology, and they may require different classificatory schemes, the latter in some cases demanding a morphological classification.

SPECIES: CURRENT USAGES

Mary B. Williams

TWO AS YET unresolved problems are significant sources of our present difficulty in grasping the nature of species. Because one of these problems concerns the nature of particular species and the other concerns the nature of the concept "species," we must first be clear about the difference between the following two questions: How is a species defined? How is "species" defined? A species, for example, *Homo sapiens,* is defined (or, perhaps more properly, is characterized) by a species description that consists of a set of traits that differentiate *Homo sapiens* from other species. "Species" is defined by a statement which tells what sorts of groups of organisms should be called species; this definition tells the meaning of the word "species." I focus here on two conflicts. One concerns the nature of the species concept that the various definitions of "species" attempt to capture: Is "species" a unitary or a pluralistic concept? The second involves the ontological nature of a species: Is a species (e.g., *Homo sapiens*) an individual or a set of organisms?

Is "species" unitary or pluralistic? To say that "species" is unitary is to say that there is a single unifying characteristic or set of characteristics that all groups that are properly called "species" have; for example, if Ernst Mayr's biological species definition is the correct definition of species, all species have the interbreeding characteristics specified in the definition. To say that "species" is pluralistic is to say that there is no such single unifying characteristic; for example, those who consider that asexually reproducing organisms form species might claim that the correct definition of "species" would be of the form: S is a species if and only if it is either a Mayrian biological species or a morphospecies.

None of the proposed definitions of "species" is completely satisfactory. To some extent this can be attributed to the fact that the meaning of "species" is still very much clouded by lack of knowledge about the nature of those groups that we intuitively recognize as species, and about their role in the biological world. But a number of biologists and philosophers of biology have begun to claim that the real source of this difficulty is that we have been trying to find a unitary definition for a pluralistic concept.

There are two important ways in which the species concept may be pluralistic. First, it may be that "species" has a different meaning for different taxonomic groups—for example, that "species" means something different for plants than for animals, or for asexual species than for sexual species. Second, it may be that "species" has a different meaning for evolutionary biology than it has for taxonomy, so that attempts to give a definition of "species" that captures both its evolutionarily important characteristics and its taxonomically important characteristics are doomed to failure.

Does "species" have a different meaning for different taxonomic groups? A major objection to many species definitions is that either some traditional species do not satisfy the definition or some groups that do satisfy the definition would not generally be considered to be species. On the one hand, if "species" is defined in terms of interbreeding within the group and lack of interbreeding between groups, then traditionally recognized species in asexually reproducing organisms would not be species. On the other hand, in some plant groups (e.g., oaks) there is extensive interbreeding among the traditionally recognized species in a genus; Mayr's interbreeding criterion would reject these traditionally recognized species and would identify some larger group as the species. Other species definitions have similar problems. It is possible that these problems indicate that "species" does have a different meaning for different taxonomic groups. But it is possible that the problems have a different source.

One source of the problems may be that the traditionally recognized set of species was simply wrong in including (or excluding) the groups in question; Ghiselin (1987) has used this type of defense for the biological species definition by asserting that asexually reproducing organisms simply do not have species. If the defenders of one of the unitary species definitions can convince biologists in general that their definition captures the true concept of species even though the set of species it recognizes is not exactly the same as the traditionally recognized set, then they will have shown that the appearance of pluralism was due to our having misidentified some groups as species.

Templeton's (1989) cohesion concept species definition offers a different way of explaining the appearance of pluralism: "The cohesion concept species is the most inclusive population of individuals having the potential for phenotypic cohesion through intrinsic cohesion mechanisms" (p. 12). The cohesion mechanisms referred to (e.g., isolating mechanisms, common descent, natural selection) are all mechanisms that promote genetic identity. Any traditionally recognized species that does not show any of the presently recognized cohesion mechanisms could be explained as having an as yet unrecognized cohesion mechanism. A defender of the cohesion concept could claim that cohesion is the unifying characteristic that all species have, and that the appearance of pluralism has been due to definitions of species based on a single cohesion mechanism.

It is not yet clear whether different taxonomic groups have different species concepts or whether one of these types of explanation is correct. If the other possible source of pluralism is correct, then the theoretical reason for believing that the species concept must be unitary loses its strength.

Does "species" have a different meaning in evolutionary biology than in taxonomy? Beginning with Darwin, biologists have recognized evolutionary theory as providing the theoretical basis for taxonomy. It seems to follow naturally that the basic unit of taxonomy (the T-species) coincides with the basic unit of evolution (the E-species). It has, consequently, generally been assumed that a single species concept that merges the taxonomic unit and the evolutionary unit will ultimately be found. While some (e.g., Ghiselin, 1987) claim that the biological species concept satisfactorily achieves this merger, others suggest new ways of defining "species" that explicitly merge the two units; for example, Cracraft suggests that we should "equate [taxonomic] species with evolutionary units. Accordingly, a species can be defined as an irreducible cluster of organisms, within which there is a parental pattern of ancestry and descent, and which is diagnosably distinct from other such clusters" (Cracraft, 1987, p. 341).

But this is not an issue which can be settled by fiat. Essential properties of the unit of evolution are specified by the theory of evolution, while essential properties of the unit of classification are specified by the nature of classification (e.g., a classification must partition the world into nonoverlapping groups). Thus the claim that "species" can be defined so that these two types of units coincide is a claim that the basic units of the evolutionary process satisfy the restrictions on units of classification.

However, challenges to this assumption have recently arisen from both biologists and philosophers; the following three assertions illustrate the positions taken by the challengers.

A. The "main criterion for blocking out species units in taxonomy is convenience and workability in practical classification, identification, and museum filing . . . Conflicts between taxonomic and population-based [evolutionary] systems of classification thus arise in several problem areas. In these situations each field must be true to its own objectives. The resolution is to recognize the legitimacy of both species concepts" (Grant, 1985, pp. 207–288).

B. We should consider the taxonomic species as merely a necessarily somewhat arbitrary naming convention (Holsinger, 1984), while making the taxonomic species as close to the evolutionary species as convenience allows.

C. The evolutionary unit is a unit with respect to a particular evolutionary process; consequently any organism is in many different evolutionary units, and the set of taxonomic species is a subset of the set of evolutionary units (Williams, 1989; see also Mishler and Donoghue, 1982).

If the basic unit of the evolutionary process (evolutionary species) is different from the basic unit of classification (taxonomic species), then the claim that evolutionary theory provides a theoretical foundation for taxonomy is undermined. Indeed, challenge A (and, to a lesser extent, challenge B) seems to imply that there is no fundamental theory uniting all taxonomic species, and that convenience for taxonomists is the only fundamental principle of taxonomy. This contradicts the deep intuitions that nature is the primary determinant of taxonomic classification and that Darwinian theory is about the natural process which determines it.

But the nonequivalence of evolutionary and taxonomic units does not *necessarily* doom these intuitions; if many evolutionary units are not taxonomic species but all taxonomic species are evolutionary units (as in challenge C), then evolutionary theory provides at least a partial theoretical basis for systematics. And if, for example, a general theoretical basis could be found for the concept "important" in Mishler and Brandon's definition, or for the concept "cohesive" in Templeton's definition, then this theoretical basis, together with evolutionary theory, could form a complete theoretical basis for systematics. So it is possible to deny the equivalence of evolutionary units and taxonomic units without destroying evolutionary theory as a theoretical basis for taxonomy.

If the challengers are correct, then much of the problematic nature of the species concept may be due to its having been used to mean two intrinsically different things. The species definitions that have been proposed are typically based in evolutionary theory, and are actually definitions of E-species, while the traditionally recognized species that serve as counterexamples have typically been T-species. If the attempt to find a definition that equates T-species and E-species has warped the search for a definition, then recognition of this fact may allow unproblematic definitions for each to be found.

Are species individuals or classes? The discussion so far has been about the nature of the category "species." Let us turn now to a discussion of the nature of the particular species in the category. The most important recent insight (or claim of an insight) is the claim that species are individuals rather than sets. (See Ghiselin, 1987, and the responses in the same volume for a discussion of this claim.) The idea that a species is a class (or set) of organisms is so intuitive that it needs no clarification here. The problem is to clarify the unintuitive claim that species are individuals.

Suppose that there is a very small (about the size of an atom) scientist who happens to be inside a baseball. This scientist sees his surroundings as consisting mostly of space, with molecules of various sizes inhabiting, and moving in all directions through, the space. Being a scientist, he has instruments that show him that in fact he is in a region of space that is more densely populated with atoms and molecules than the surrounding region of space; he is in a cloud of molecules. Because his life span is in microseconds, he has little opportunity to observe that, although the individual molecules move in all directions within the cloud, the cloud is moving as a unit through space. He has even less opportunity to observe that this unit obeys Newtonian laws.

Our microscientist sees the baseball as a set of molecules, just as we see the species as a set of organisms. But with respect to Newtonian laws, the baseball is an individual—that is, because of the cohesive forces that hold its molecules together, it acts as a unit with respect to these laws. The claim that species are individuals in evolutionary biology is a claim that species are held together by cohesive forces (e.g., common selection forces on a common gene pool) so that they act as units with respect to the laws of evolution. The importance of this claim, if true, is in the insight it gives us into the nature of the laws of evolutionary biology.

Suppose our microscientist has historical records showing the locations of the ball-cloud and of a bat-cloud over a period of time during which the bat (using our terminology) hit the ball, and suppose that he also has historical records of similar incidents. He is then in a position to discover the Newtonian laws. The important breakthrough that our microscientist has made is the recognition that, for his attempt to find laws governing the changes in position of clouds of molecules, the scientifically relevant properties are not properties of individual molecules but properties of clouds of molecules; this breakthrough is similar to Galileo's breakthrough in recognizing that the scientifically important property was not speed but acceleration. The recognition that species are individuals plays a similar role in evolutionary biology; it implies that the properties of species, not the properties of organisms, are the relevant properties for the laws of evolutionary biology.

On the one hand, for example, consider the following putative law of evolutionary biology (adapted from Williams, 1989).

Mullerian Mimicry Law: If S1 and S2 are species whose organisms

1. are unpalatable,

2. are structurally similar and occasionally have mutations increasing that similarity,
3. have in common major visually hunting predators who learn to avoid prey recognizably similar to previously encountered unpalatable prey,
4. are all connected by organism-predator interactions,

then both interspecific and intraspecific polymorphism in the prey recognition characters will decrease over evolutionary time.

Notice that, although properties of organisms play some role in this law, the change specified is a change in polymorphism, which is a species property.

On the other hand, if species are sets it should be possible to express this law wholly in terms of properties of organisms, with all mention of species eliminated. No definition of species that would allow this elimination has yet been offered. (One major problem is to find a definition that would satisfactorily treat the organisms during the initial process of splitting of one species into two. This is not a problem if the species is an individual, because in that case the exact status of such organisms is of no theoretical significance, just as the exact status of a molecule that escapes from the baseball-cloud is of no theoretical significance to our microscientist in his work with the Newtonian laws.)

A type of law that is possible if species are sets, but not be possible (as a part of evolutionary theory) if species are individuals, is a law of the form "If X is a chimpanzee of type A, then X has behavioral trait T," where "chimpanzee" could be replaced by any other species name. If species are individuals, then a species name (such as chimpanzee) is a proper name, the name of an individual; because laws must be universal in form and thus cannot contain proper names, there cannot be any laws about particular species. But if species are sets, the name of the species is merely an abbreviation for the defining properties of the set; thus in this case there could be laws about particular species.

Teleology

James G. Lennox

When we say that the wing coloration and pattern of one species of butterfly mimics that of another *in order to* avoid predators, what sort of claim are we making? When we read, in a discussion of the biochemistry of photosynthesis, "thus photoinduced cyclic electron flow has a real and important *purpose,* namely, *to* transform the light energy absorbed by the chlorophyll molecules in the chloroplast into phosphate bond energy" (Lehninger, 1971, p. 110), what meaning does the concept of purpose convey? In discussing physical traits, behavior patterns, or biochemical processes in this way, are we saying anything more than that each of them has predictable *effects*? Such questions set the context for the discussion of the concept of teleology.

Historical considerations. The word "teleologia" was coined to refer to explanations by final causes by Christian Wolff (*Logica,* chap. III, sec. 85) in 1728, and found its way into English as "teleology" in volume 41 of *Philosophical Transactions of the Royal Society* (Eucken, 1879, pp. 132–133; OED, p. 3251). The etymological roots on which Wolff was drawing, which stretch back to ancient Greece, suggest the idea of *giving an account of something by reference to an end or goal.* It became progressively more common throughout the nineteenth century, no doubt in part because of Immanuel Kant's adoption of Wolff's terminology in his *Critique of Judgement* (Lenoir, 1982, pp. 22–35).

Not surprisingly, the semantic roots of the concept also stretch back to classical Greece, and the concepts of teleology of Plato and Aristotle help illuminate certain presuppositions guiding contemporary discussions.

The concept, though not the word, "teleology" originates in autobiographical remarks of Socrates.

> One day, however, I heard someone reading from a book he said was by Anaxagoras, according to which it is, in fact, Mind that orders and is the

> cause for everything. Now this was a cause that pleased me; it seemed to me, somehow, to be a good thing that Mind should be the cause of everything. And I thought that, if that's the case, then Mind in ordering all things must order them and place each individual thing in the best way possible; so that if anyone wanted to find out the reason why each thing comes to be or perishes or exists, this is what he must find out about it: how is it best for that thing to exist or to act or be acted upon in any way? (*Phaedo* 97b8–d1; Gallop trans. slightly modified)

In the *Timaeus,* aiming to succeed where Anaxagoras had failed, Plato pictured the natural world as the end-product of a divine craftsman who looked to the world of eternal being for his model of the good and then created a natural order that was as good as it could possibly be.

Teleological explanation on the Platonic model thus involves the following components (Furley, 1987, chaps. 1, 12; Lennox, 1985).

a. The identification, by a rational agent, of some state as good, or at least as better than other alternatives.
b. Actions taken by that agent which are sufficient, and perhaps necessary, for that entity's being, or coming to be, in that state because it is seen by the agent as good.

This model is the origin of what is sometimes referred to as "external teleology" (Goudge, 1961, p. 193; Ayala, 1970b, p. 11; Hull, 1973, p. 55). The "externality" involved is twofold: (a) *the agent* whose goal is being achieved is external to the object that is being explained teleologically, and (b) *the value* aimed at is the agent's value, not the object's. It is not surprising that in contemporary discussions this term is used to describe the application of teleological language to artifacts, for the Platonic model quite explicitly treats the natural world as the production of a divine craftsman. This vision of the physical universe as the result of a rational agent acting to achieve the good, first systematically presented in Plato's *Timaeus,* is also the origin of what is sometimes referred to as a "cosmic teleology," the idea that the entire cosmos is ordered, usually by a rational agent, toward one end or kingdom of ends (Mayr, 1982a, p. 50).

Aristotle's approach to teleology is embedded within a very different theory of causality and explanation. We have scientific understanding of something when we can correctly answer the question "Why?" about it, and such answers involve the identification of its causes *(aitia).* But for any one fact to be explained there may be a number of different answers, reflecting different kinds of cause (*Physics* II.3). Among these kinds is "the things which stand to the rest as their end and good; for what the other things are for tends to be best and their end" (*Physics* II.3, 195a23–25). Aristotle devotes two chapters of the *Physics* (II.8, 9) and most of the first chapter of *Parts of Animals, I,* to justifying his view that certain natural changes take place, and certain natural attributes exist, *for the sake of some end.* Not only do these changes and attributes contribute to an end—they take place and exist in part *because* they contribute to an end.

It is important, both historically and philosophically, that Aristotle's naturalistic teleology be set in contrast with Plato's. First, Aristotle's argument for the operation of ends as causes in nature in no way depends on the actions of a rational agent. Thus his arguments for natural teleology are not part of the tradition leading toward the natural theology of the seventeenth-nineteenth centuries, based as that tradition is on the assumption that the universe is the product of intelligent design. The classical roots of that tradition are quite explicitly Platonic. For this reason, Aristotle is often identified as the origin of what is referred to as "immanent" or "internal" teleology (Goudge, 1961, p. 193; Ayala, 1970b, pp. 14–15; Hull, 1973, pp. 55–56). This notion carries two distinct connotations, parallel to those of "external teleology," which should not be confused. The notion of "immanence" may simply stress that the goal or function involved is a goal or function of the individual organism under consideration, rather than of an "external" designer. But it may also carry connotations of a "quasi-conscious" agent inside natural objects, so to speak. The teleological explanations associated with certain forms of vitalism in the nineteenth and twentieth centuries occasionally invoke immanent intentionality of this sort. Aristotle's teleology is immanent in the first, but not the second, sense (cf. Balme, 1987, pp. 279–281).

Second, the explanatory relevance of Aristotle's regular identification of the end as good has to be thought through carefully. If a rational evaluator plays no necessary role in the production of goal-directed processes, it is difficult to see how the fact of the goal's goodness is causally relevant. It is also necessary to rethink the basis of the valuation of the goal. For Aristotle, the end which a natural process is directed toward, or for which an attribute exists, is simply the being (in fact, the life) of the natural object in question.

Third, and arising from the first two points, Aristotle appears to restrict full-blown natural teleology to the biological domain. His *Meteorology,* for example, explains all sublunary nonbiological phenomena without once appealing to a "final cause." Even within the biological realm, many features of living things are not there for the sake of anything (cf. *Generation of Animals* V.1). This seems to be a consequence of the fact that natural teleology for him depends on the capacity of living things to maintain themselves and, through the act of reproduction, to initiate a process directed toward there being another organism of the same form (Gotthelf, 1987). Self-maintenance can thus be extended beyond the individual, allowing it to partake in the eternal and divine, as Aristotle tells us in two important passages (*de Anima* II.4, 415a23–415b8; *Generation of Animals* II.1, 731b18–732a2). The end of biological development (the form of an actual living thing) can thus be identified as a cause of what contributes to that end. Except for the organism's form, its capacity for self-maintenance, none of the parts that contribute to the organism's life would come to be or exist (Balme, 1987; Cooper, 1987; Gotthelf, 1987).

This naturalistic approach to teleology allows Aristotle to offer teleological explanations of organic parts and behavior that sound remarkably like modern "adaptational explanations." If a part comes to be because of its contribution to its organism's life, that sanctions its identification as *there for the sake of* that contribution. Taking its continued existence as good for it, those of the animal's parts and behavior that develop to contribute to its life in various ways can then be described as for the sake of *the good* or *the valuable.* Thus the use of evaluative language in the Aristotelian teleological tradition doesn't depend, as it does in the Platonic tradition, on the end being *seen as* valuable by a rational agent. The accompanying table sums up the implications for teleological concepts of these contrasting models.

Both the Platonic and Aristotelian traditions found their way into the Renaissance. Aristotelian naturalism was alive and well in Renaissance Padua (Randall, 1961; Schmitt, 1983), and can be seen most straightforwardly in the work of William Harvey (O'Malley et al., 1961; Pagel, 1967, 1976; Lennox, 1981).

There was no common attitude to teleology among the so-called Mechanical Philosophers, although Plato's influence was strong. In his *Disquisition upon the Final Causes of Natural Things,* Robert Boyle defends a teleology that sees nature as the product of a "divine craftsman," but finds evidence for such a craftsman primarily in the way living things are constructed (Lennox, 1983, pp. 40–43). Descartes' opposition to teleology rests on skepticism about knowing God's aims, which again implicitly rests on a Platonic understanding of teleology. With thinkers such as

Semantic Element	Platonic Model	Aristotelian Model
a. Causation	Rational agency	Inherent potential for form
	Intentionality required	Intentionality not required
b. Valuation	External value	Immanent value
c. Extension	Entire cosmos	Living structures and processes

Thomas Hobbes and Francis Bacon, there emerges an opposition to teleology of the sort that was championed by Epicureans such as Lucretius—the only legitimate sort of cause is an efficient cause; ends are not efficient causes; so explanation by appeal to ends is illegitimate (Lennox, 1983, pp. 38–40). This appears to be the basis of Francis Bacon's famous dismissal of final causes as "barren Virgins" (on which more later).

The Natural Theology of people such as William Paley, Thomas Malthus, or John Herschel, in which Charles Darwin was schooled, is a Christian form of the Platonic tradition of external teleology. (In fact the line of descent from Boyle and Newton to these thinkers is direct; cf. Hull, 1973, pp. 57–66.) Animals are structured as they are and behave as they do as a result of being designed for a purpose by a benevolent Creator. It is part of the naturalist's role to discover divine purpose by the careful study of adaptation. Likewise, the application of value concepts to nature is based on the theological assumption that the Creator acted with good intentions. When necessary, this approach allows one to account for apparent flaws in the natural order, such as the excessive reproductive powers of certain species, by appeal to a "greater good" that the Creator had in mind. This is a move not open to a thinker in the Aristotelian tradition, who must restrict teleological explanation to contributions to the individual natures of the animals in question.

It would be surprising if Charles Darwin did not at some point come face to face with the concept of teleology. By his own admission he admired the argument from design in Paley's work while at Cambridge, and was apparently ready to accept creation by design as the most reasonable explanation for adaptation when the *Beagle* sailed. Less than ten years later he was confident of the main outlines of a theory that explained adaptation, and the creation of new species, by references to natural causes, and was an avowed agnostic.

Throughout his Species Notebooks Darwin uses the concept of "final cause" in a consistent way (cf. Notebook B, pp. 5, 49; C, p. 236; D, pp. 114, 135, 167; E, pp. 48–49, 146–147; M, p. 154). When the central question being considered is "What is this for?" Darwin refers to the answer as the final cause. By contrast, when he thinks it is reasonable to suppose that the fact in question has no function, he denies that it has a "final cause." The most interesting passage on Darwin's attitude to teleology is found in his notes on John Macculloch's *Proofs and Illustrations of the Attributes of God from the Facts and Laws of the Physical Universe, being the Foundation of Natural and Revealed Religion* (Barrett et al., 1987, pp. 631–641). Darwin is systematically testing natural selection against Macculloch's Creator in explaining adaptations. At 58r we find the following comment on page 234 of Macculloch's work: "The Final cause of innumerable eggs is explained by Malthus,—[is it anomaly in me to talk of Final causes: consider this!—] consider these barren Virgins" (Barrett et al., 1987, p. 637).

The "barren Virgins" reference is, of course, to Francis Bacon's famous negative assessment of final causes (Bacon, 1665), a passage cited by William Whewell in his Bridgewater treatise of 1833 (pp. 355–356). The parenthetical note to himself indicates that Darwin is concerned whether, once one gives up the designs of an intelligent, benevolent Creator as the productive cause of an adaptation, one should also give up the language of final cause. He apparently decided that the answer was no. He continued using the term "final cause" in *On the Origin of Species* (1859b, pp. 216, 435, 448) and after (see his descriptions of Bates' selectionist explanation of mimicry as giving the final cause in Barrett, 1977, vol. 2, p. 89). More important, he consistently claimed that natural selection acts *for the good of each being,* and that its products are present *for* various functions, purposes, and ends (Darwin 1859b, pp. 149, 152, 224, 237, 451). Such claims are scattered liberally throughout his many botanical publications. In June of 1874 Asa Gray noted, in *Nature,* "Darwin's great service to Natural Science in bringing back to it Teleology: so that instead of Morphology versus Teleology, we shall have Morphology wedded to Teleology." Darwin quickly responded: "What you say about Teleology pleases me especially and I do not think anyone else has ever noticed the point" (F. Darwin, 1892, p. 367).

It seems historically inaccurate, then, to say that "Darwin conceived of teleological explanations as presupposing that there has been foresight and deliberate action on the part of some agent to fit the organ to its function" (Ghiselin, 1969a, p. 137). As far as the evidence goes, it seems that Darwin faced the issue of the compatibility of natural selection and teleology, and decided they were compatible. At the same time, his notebooks show him increasingly impatient with the invocation of the plans of a Creator to explain adaptation. In the years after the publication of the *Origin,* Darwin debated long and hard with Gray over the compatibility of natural selection and "special design," but the primary issue being debated was whether Darwin's "chance variations" did not in fact evidence divine preordination (Moore, 1979, chaps. 11, 12). Darwin's Calvinist followers insisted on interpreting teleology along Platonic, natural theological lines. The idea of a natural teleology of a more Aristotelian variety was not what they had in mind.

The debate over the scientific status of teleology also raged in the study of development and physiology (Appel, 1987, chap. 8; Lenoir, 1982). Major successes in these disciplines resulted from the application of the experimental methods of analytical chemistry to biological processes. The question thus arose to what extent living things were in fact simply complex chemical systems, fully explicable by reference to physical and chemical processes. The debate, variously influenced by Kant's *Critique of Judgement*, German *naturphilosophie*, and French positivism, often focused on the manifestly goal-directed organization of living systems. The central questions were whether organic systems could be understood without reference to forces peculiar to them, and without reference to goals. Under the influence of Blumenbach, Kant, Cuvier, and others, a school of research into the nature of development and physiology emerged that accepted, as a principle of research, the self-regulatory and goal-directed nature of living things. Near the end of his life, one of the most articulate spokesmen for this "teleo-mechanical" point of view, Karl Ernst von Baer, represented the debate as follows: "Nearly a century ago Kant taught that in an organism all the parts must be viewed as both ends and means [Zweck und mittel] at the same time. We would rather say: goals and means [Zeile und Mittel]. Now it is announced loudly and confidently: Ends do not exist in nature, there are in it only necessities; and it is not even recognized that precisely these necessities are the means for reaching certain goals. Becoming [ein Werden] without a goal is simply unintelligible" (quoted in Lenoir, 1982, p. 271).

This remark was made as part of a critique of Darwinism—seen as a theory that reduced the explanation of living phenomena to the interaction of chance and necessity. In the recommended replacement of "Zeile" for "Zweck" we see that teleology in this research tradition was of the Aristotelian, "internal" variety—there was no association of teleology with the purposes of a rational agent in this defense. Rather the debate was over whether this reference to a goal—the life of the organism—was required in order to achieve a full understanding of biological development and functioning.

Thus in the nineteenth century we see representatives of both the Platonic and the Aristotelian traditions in teleology ranging themselves against various aspects of Darwinism. Darwin, however, seemed to see natural selection as a new underpinning for teleological explanation.

This historical background clearly conditions the ambivalent attitude one finds toward teleology among leading neo-Darwinians (cf. Simpson, 1964, chaps. 5, 10, 11; Mayr, 1982a, pp. 47–51). It is their clear sense that the concept of teleology carries with it unshakable metaphysical commitments that are theistic, vitalistic, or both. The following remark from Simpson is typical: "Darwin had substituted a scientific teleology for a philosophical or theological one. The redefinition did not take. The older meanings of the word *teleology* were ineradicable, and they brought a certain scientific (although not necessarily philosophical) disrepute to the whole subject" (Simpson, 1964, p. 102).

Contemporary discussions. Contemporary discussions of teleology have thus been marked by a tension that needs to be understood against the historical development of the concept. On the one hand, most biologists and philosophers of science acknowledge the apparent value of identifying various features of living things in terms of goals, ends, functions, design, and so on. On the other hand, there is anxiety that the use of such language implies one or more of the following illicit features.

1. Backward causation (the future goal state causally influences the events leading up to it).
2. Anthropomorphism (i.e., a reference to conscious, typically divine, purpose).
3. Reference to "internal" vital forces beyond the reach of empirical investigation.

Discussions of teleology among biologists and philosophers of science during the last forty years have focused on providing an account of teleological concepts and explanations within biology that does not commit the person using these concepts to any of these implications. Indeed, avoiding the twin specters of "natural theology" and "vitalism," inherited from different nineteenth-century traditions, was part of the motivation behind the attempt, by leading neo-Darwinians, to replace the word "teleology" with "teleonomy" (Huxley, 1960; Simpson, 1964; Mayr, 1965a, 1982a; Williams, 1966).

Contemporary philosophical analyses have tended to be based on one of two distinct biological contexts, corresponding to the focus of the two nineteenth-century backgrounds we have discussed—goal-directed activity on the one hand, adaptation and selection on the other. We will begin with goal-directed behavior.

The analysis of organic activity in terms of negative feedback models seems to stem from behaviorism in psychology (Perry, 1921; Tolman, 1932; Hoffstader, 1941) and cybernetic technology (Rosenbueth, Wiener, Bigelow, 1943). A number of philosophers attempted to do justice to the claims of "organismic" biologists such as E. S. Russell—that the phenomena of biological development, physiological homeostasis, and animal behavior demanded teleological characterization and explanation—by extending this analysis to these phenomena. According to this analysis, a teleological system is one in which a complex sequence of changes, involving a number of correlated variables of a relatively closed system, tends toward either the achievement of, or the maintenance of, a certain goal-state of that system. (The phrase "tends toward" captures the idea that the system will, within definable limits, adjust the values of its variables in ways required for achieving or maintaining the goal-state [Sommerhoff, 1950; Braithwaite, 1953; Nagel, 1961, chap. 12; Beckner, 1959, chaps. 6–9; Ruse, 1973, chap. 9]. The behavior involved in predator/prey or mating interactions, homeostatic mechanisms in physiology and the regulation of clutch size in bird populations have all been cited as examples of such systems. A brief, clear statement of this approach is Beckner, 1971.)

Beckner has nicely summarized the common ground of these very similar discussions: "they agree unanimously that no nonphysical agents or causes need be active in teleological behavior and that no unique method of treatment is required in its explanation. In fact, the teleological system is simply a special case of a physical system in the ordinary sense of that term, and can be described, *qua* teleological, in a vocabulary suited for the description of nonteleological systems" (Beckner, 1959, pp. 132–133).

Furthermore, if such systems are ultimately, as proponents of the above analysis insist, to be analyzed in terms of necessary and/or sufficient conditions for the production of the goal-state, the facts that such systems can fail to achieve their goal, can display such behavior in the complete absence of the goal, and can achieve the goal in the absence of one or the other of relevant changes are embarrassing.

An alternative analysis of goal-directed behavior has been suggested by the work of Charles Taylor (1964) and Larry Wright (1976). What makes such behavior *teleological* is that the behavior or process in question can be shown to occur *because* it is required for the achievement of the goal state. On the Taylor/Wright analysis, only so long as the best explanation for a type of behavior or process is its tending to bring about a certain effect, is it teleological. This approach aims to account for the appearance of cause/effect reversal in teleological language without invoking intentionality. Andrew Woodfield (1976, 1979), however, has insisted that the very concept of a goal is intentional, implying something like a cognitive perspective on the part of such systems.

The directed behavior paradigm is difficult to apply to explanations of why types of animals have the traits they do, and yet these have always been among the standard list of facts requiring teleological explanation. A number of biologists and philosophers have focused on the facts of adaptation in thinking about teleology (Ayala, 1970; Brandon, 1981; Simpson, 1964; Williams, 1966; Wimsatt, 1972). Among philosophers, these explanations are often termed "functional," indicating that propositions of the form "the *function* of T is E" carry much of the same semantic content as "O has T *in order to (for the sake of)* E." (I believe some confusion has arisen from Ernst Mayr's characterization of domains of biology where *nonteleological* inquiry dominates "functional biology," whereas philosophers have typically viewed functional explanation as teleological; see Mayr, 1982a, pp. 68–71.)

The central issue at stake regarding the context of selective adaptation can be stated in a manner parallel to the basic question regarding goal-directed behavior: When I say a behavior pattern is *designed to* attract mates, or that a chemical reaction occurs *in order to* regulate serum cholesterol levels, is this merely an anthropomorphic holdover from natural theology? Am I really just saying that these processes have these effects; or do I mean that having this effect is (at least part of) the reason why these processes are taking place? Similarly, when I say that a certain character trait is *an adaptation for* predator avoidance, does this simply mean that avoiding predators is one consequence of having that trait; or does it mean that this trait's leading to predator avoidance is (at least part of) the reason why that trait is present in the populations that have it? A variety of philosophers and biologists, coming to this issue from various standpoints, have insisted that it means the latter (Ayala, 1970, p. 12; Brandon, 1981, p. 103; Taylor, 1964, p. 9; Wainwright et al., 1976, p. 1; Williams, 1966, p. 261; Wimsatt, 1972, p. 70; Woodfield, 1979, p. 416; Wright, 1976, p. 81).

Each of these writers stresses that the appropriateness of teleological ascriptions of a functional or adaptational sort depends on identifying a certain effect or consequence of a trait as causally relevant to its presence. In an evolutionary context, it is the fact that certain character traits have advantageous consequences for their possessors in certain environments that accounts for their being selectively favored. Natural selection is a consequence-oriented force, and thus provides design without the need for an intelligent designer.

Attempts to give a unified account of teleology, given these different paradigmatic starting points, have taken two forms: assimilation and abstraction. Some philosophers and biologists have attempted to show that the teleology imbedded in functional explanation is simply the teleology of a goal-directed system. (For a prime example, see Beckner's [1959, pp. 167–172] attempt to assimilate selection theoretic explanations to those invoking goal-directed systems.)

An alternative strategy is to seek a unified account by identifying some abstract feature or features of the two types of explanations in virtue of which they possess their teleolgical character (Wimsatt, 1972; Wright, 1976). For example, Wimsatt ties teleology to a broad concept of selection, while on Wright's view what transfers from one explanatory context to another—and indeed from one historical context to another—is *causation by consequence*. Another approach might be to see explanations in both contexts as answers to the same sort of question, that is "What is X [a trait, a behavior] for?" Providing an abstract characterization of the force of teleological explanation can help explain not only the semantic similarities between explanations in different theoretical contexts but also the fact that such explanations can retain their power through major changes in biology's theoretical foundations.

Unit of Selection

Elisabeth A. Lloyd

For at least a decade, some participants in the "units of selection" debates have been aware that more than one question concerning the unit involved in the process of natural selection is at stake. Richard Dawkins, for instance, introduced the terms "replicator" and "vehicle" to stand for different roles in the evolutionary process (1978, 1982a,b). He proceeded to argue that the units of selection debate should be not about vehicles, as it had formerly been, but about replicators. David Hull, in his influential article "Individuality and Selection" (1980), suggested that Dawkins' "replicator" subsumes two distinct functional roles, and the separate categories of "replicator," "interactor," and "evolver" were born. Robert Brandon, arguing that the force of Hull's distinction had not been appreciated, analyzed the units of selection controversies further, claiming that the question about interactors should more accurately be called the "levels of selection" debate, in order to distinguish it by name from the dispute about replicators, which he allowed to keep the "units of selection" title (1982).

My purpose here is to delineate further the various different questions asked by Dawkins (1978, 1982a,b, 1986), Wright (1980), Hull (1980), and Brandon (1982, 1990) under the rubric "units of selection." In the context of considering what a unit of selection is, four quite distinct questions have, in fact, been asked.

My analysis does not resolve any of the conflicts about which research questions are most worth pursuing. I do not attempt to decide which of the questions or combinations of questions should be considered "*the* units of selection question." Although I have argued elsewhere that the interactor question is the primary question for evolutionary genetics, that claim is intended as historical and descriptive; most evolutionary genetics models that address any version of the units of selection question have focused on which level of interaction must be represented in the model in order to make it dynamically and empirically adequate (Lloyd, 1988; see especially chaps. 5, 6). Furthermore, the mere persistence of the three other questions attests to both their importance and their general interest.

Several terms need clarification. The term "replicator," originally introduced by Dawkins, but since modified by Hull, is used to refer to any entity of which copies are made. Dawkins classifies replicators using two orthogonal distinctions. A "germ-line" replicator, as distinct from a "dead-end" replicator, is "the potential ancestor of an indefinitely long line of descendant replicators" (1982a, p. 46). For instance, DNA in a chicken's egg is a germ-line replicator; DNA in a chicken's liver is a dead-end replicator. An "active" replicator is "a replicator that has some causal influence on its own probability of being propagated"; a "passive" replicator is never transcribed, and has no phenotypic expression whatsoever (1982a, p. 47). Dawkins is especially interested in *active germ-line replicators*, "since adaptations 'for' their preservation are expected to fill the world and to characterize living organisms" (1982a, p. 47).

Dawkins also introduced the term "vehicle," which he defines as "any relatively discrete entity . . . which houses replicators, and which can be regarded as a machine programmed to preserve and propagate the replicators that ride inside it" (1982b, p. 295). According to Dawkins, most replicators' phenotypic effects are represented in vehicles, which are themselves the proximate targets of natural selection (1982a, p. 62).

Hull, in his introduction of the term "interactor," observes that Dawkins' theory has replicators interacting with their environments in two distinct ways: they produce copies of themselves, *and* they influence their own survival and the survival of their copies through the production of secondary products that do ultimately have phenotypic expression. Hull suggests the term "interactor" for entities that function in this second process. An "interactor" denotes that entity which interacts, as a cohesive whole, directly with its environment in such a way that replication is differential—in other words, on which selection directly acts (Hull, 1980, p. 318). The process of evolution by natural selection is "a process in which the differential extinction and proliferation of interactors cause the differential perpetuation of their replicators that produced them" (Hull, 1980, p. 318; cf. Brandon, 1982, pp. 317–318).

The interactor question. In its traditional guise, the interactor question is "What units are being selected in the process of natural selection?" As such, this question is involved in the oldest forms of the units of selection debates (Darwin, 1859b; Haldane, 1932; Wright, 1945; I found nearly two hundred references to books and papers by biologists and philosophers that treat this question, and these represent just a fraction of the literature on the topic—see Lloyd, 1988). In his classic review paper, Richard Lewontin's purpose was "to contrast the levels of selection, especially as regards their efficiency as causes of evolutionary change" (1970, p. 7).

L. B. Slobodkin and A. Rapaport assumed that a unit of selection is something that "responds to selective forces as a unit—whether or not this corresponds to a spatially localized deme, family, or population" (1974, p. 184).

Questions about interactors focus on the description of the selection process itself—that is, on the interaction between entity and environment, and on how this interaction affects evolution; it does not focus on the outcome of this process (see Wade, 1977; Vrba and Gould, 1986). The interaction between some interactor and its environment is assumed to be mediated by "traits" that affect the interactor's expected survival and reproductive success. (An interactor might be a group, an organism, a chromosome, a kin group, or a gene.) In other words, the expected fitness of the interactor is directly correlated with the "value" of the trait in question. The expected fitness of the interactor is commonly expressed in terms of genotypic fitness parameters, that is, in terms of the fitness of replicators; hence, interactor success is assumed to be reflected in and counted through replicator success. There are several methods available for expressing such a correlation between trait and (genotypic or organismic) fitness, including regression and variances, and covariances. There are also a number of models available for representing interactors; in all of these, the interactor's trait is correlated with replicator fitness values, and the component of the replicator fitnesses attributed to the interactor is not available or reproducible from a lower level of interactor. (See, e.g., Li, 1967; Crow and Kimura, 1970; Price, 1972; Hamilton, 1975; Wade, 1978, 1980, 1985; Uyenoyama and Feldman, 1980; Wimsatt, 1980, 1981; Colwell, 1981; D. S. Wilson and Colwell, 1981; Arnold and Fristrup, 1982; Crow and Aoki, 1982; Lande and Arnold, 1983; Ohta, 1983; D. S. Wilson, 1983a; Heisler and Damuth, 1987; Damuth and Heisler, 1988. See discussion in Lloyd, 1988.)

Much of the interactor debate has been played out through the construction of mathematical genetical models. The point of building such models is to determine what kinds of selection, operating on which levels, might be effective. But not all discussion of which levels of selection are causally efficacious has been quantitative. Many authors have attempted to determine what levels of selection must or should be taken into account through qualitative descriptions of interactors. (e.g., Lewontin, 1970; Wimsatt, 1980, 1981; Colwell, 1981; Sober, 1981, 1984a; Arnold and Fristrup, 1982; Brandon, 1982; Sober and Lewontin, 1982; Wade, 1985; Lloyd, 1986, 1988, 1989; Heisler and Damuth, 1987; Damuth and Heisler, 1988).

Note that what I am calling the "interactor question" does *not involve attributing adaptations or benefits to the interactors.* Interaction at a particular level involves only the presence of a trait at that level with a special relation to genic or genotypic expected success that is not decomposable into fitness components at another level. (See the models already cited for various technical approaches to expressing this special relation between fitness and trait.) The claim about interaction indicates only that there is an evolutionarily significant interaction occurring at the level in question; it says nothing about the existence of adaptations at that level. The most common error made in interpreting many of the genetical models is that the presence of an interactor at a level is taken to imply the existence of an adaptation at that level.

The replicator question. Starting from Dawkins' view, Hull refined and restricted the meaning of "replicator," which he defined as "an entity that passes on its structure directly in replication" (1980, p. 318). From here on I shall use the terms "replicator" and "interactor" in Hull's sense.

Hull's definition of "replicator" corresponds more closely to a long-standing debate in genetics about how large or small a fragment of a genome ought to count as the replicating unit (that is, as something that is copied, and which can be treated separately) (see especially Lewontin, 1970). This debate revolves critically around the issue of linkage disequilibrium, and led Lewontin, most prominently, to advocate the usage of parameters referring to the entire genome rather than allele and genotypic frequencies in genetical models (Lewontin, 1970, 1974a; Franklin and Lewontin, 1970; Templeton, Sing, and Brokaw, 1976; see discussion in Wimsatt, 1980; Brandon, 1982). The basic point is that with a great deal of linkage disequilibrium, individual genes cannot be considered as replicators, because they do not behave as separable units during reproduction. Although this debate remains pertinent to the choice of state space for genetical models, it has been eclipsed by concerns about interactors in evolutionary genetics.

The beneficiary question. What benefits from a process of evolution by selection? There are two predominant interpretations of this question: "Who benefits ultimately, in the long term, from the selection process?" and "Who gets the benefit of possessing adaptations as a result of a selection process?"

On the issue of the ultimate beneficiary, there are two obvious answers—two alternate ways of characterizing the long-term *survivors* of the evolutionary process. One might say that the species or lineages (Hull's and M. B. Williams' "evolvers"), or even living things in general, are the beneficiaries of the evolutionary process. Alternatively, one might say that the lineages characterized on the genic level, that is, the surviving alleles, are the long-term beneficiaries. I have not located any authors holding this first view, but, for Dawkins, the latter possible interpretation is the primary fact about evolution. In order to arrive at this conclusion, Dawkins adds the requirement of *agency* (cf. Hampe and Morgan, 1988). To

Dawkins, a beneficiary does not simply passively accrue credit in the long term; it must function as the initiator or causal source of a biochemical causal pathway. Under this picture, the beneficiary is causally responsible for all of the various effects that arise further down the pathway, independently of which entities might reap the long-term rewards.

A second interpretation of "benefit" involves the notion of adaptation. If something is a unit of selection, then the selection process "benefits" that level of entity through producing *adaptations* at that level (G. C. Williams, 1966; Maynard Smith, 1976b; Eldredge, 1985; Vrba, 1984). This is a distinct question from the identity of the ultimate beneficiaries of a selection process. One can think (and Dawkins does) that there is organismic adaptation without thinking that individual organisms are the "ultimate beneficiaries" of the selection process. I shall therefore treat this second sense of beneficiary as a separate issue, discussed in the next section.

The owner-of-adaptations question. At what level do adaptations occur? Elliott Sober puts this question as follows: "When a population evolves by natural selection, what, if anything, is the entity that does the adapting?" (1984a, p. 204).

As already mentioned, the presence of adaptations in a given level of entity is sometimes taken to be a *requirement* for something to be a unit of selection. Sewall Wright recognized that the combination of the interactor question with the question of what entity had adaptations created a problem in the group selection debates. He distinguished group selection for "group advantage" from group selection per se (1980). The identification of a unit of selection with the owner of an adaptation at a certain level has, I submit, caused a great deal of confusion.

Some of this confusion is a result of a very important but neglected duality in the meaning of "adaptation" (in spite of related discussions in Brandon, 1978; Burian, 1983; Sober, 1984a). Sometimes "adaptation" is taken to signify *any trait at all* that is a direct result of a selection process at that level. On this view, any trait that arises directly from a selection process is, *by definition,* an adaptation (see, e.g., G. C. Williams, 1966, p. 25; Arnold and Fristrup, 1982; Sober, 1984a; Brandon, 1990). Other times, "adaptation" is reserved for traits that are "good for" their owners, that provide a "better fit" with the environment, and that intuitively fit some notion of "good engineering" (G. C. Williams, 1966; Bock, 1980; Gould and Lewontin, 1979; Dunbar, 1982). These two meanings, which I call the "product of selection" and "engineering" definitions, respectively, are distinct, and in some cases, incompatible.

G. C. Williams, in his influential book *Adaptation and Natural Selection,* tacitly advocated an engineering definition of adaptation (1966). He believed that it was possible to have evolutionary change result from direct selection favoring a trait *without* having to consider that changed trait an *adaptation*. Consider, for example, his discussion of Waddington's (1956) genetic assimilation experiments. Williams interprets the results of Waddington's experiments in which latent genetic variability was made to express itself phenotypically because of an environmental pressure (1966, pp. 70–81; see the discussion in Sober 1984a, pp. 199–201). Williams considers the question of whether the bithorax condition (resulting from direct artificial selection on that trait) should be seen as an adaptive trait, and his answer is that it should not. Williams instead sees the bithorax condition as "a disruption . . . of development," a failure of the organism to respond (1966, pp. 75–78). Hence Williams draws a wedge between the notion of a trait that is a direct *product* of a selection process, and a trait that fits his stronger, *engineering* definition of an adaptation (see Sober 1984a, p. 201; cf. Dobzhansky, 1956).

This distinction between product of selection and engineering views of adaptation is far from established; note, for example, that Williams also says, "natural selection would produce or maintain adaptation as a matter of definition" (1966, p. 25; cf. Mayr, 1976). This comment conflicts with the conclusions Williams draws later in his discussion of Waddington, though the tension between his comments remains unacknowledged.

The engineering notion of adaptation is also at work in the long dispute over the relationship between natural and sexual selection. Many evolutionists, starting with Darwin, rejected the idea that the products of a sexual selection process should be considered *adaptations*. In fact, analysis of the process of sexual selection is sometimes motivated by the drive to find an explanation for the presence of "maladaptive" traits. Even in recent discussions, the engineering notion of adaptation plays an important role. Mark Kirkpatrick, for instance, uses a notion of adaptedness based on mean survival values in his argument that sexual selection does not always produce adaptations (1987).

Consider for a moment the two schools of sexual selection theory. The "good genes" school claims that mate choice evolves under selection for females to mate with ecologically adaptive genotypes. The assumption here is that even though it appears that the females are basing their mate choice on a nonadaptive character, the character is actually an indication of the male's adaptedness (see, e.g., Vehrencamp and Bradbury, 1984). The "nonadaptive" school claims that "preferences frequently cause male traits to evolve in ways that are not adaptive with respect to their ecological environment" (Kirkpatrick, 1987, p. 44). In other words, female preferences do not correspond with the kinds of males favored by natural selection. The result is a compromise between natural and sexual selection, the final state being one "that is maladaptive with respect to what natural selection acting alone would produce" (Kirkpatrick, 1987, p. 45). Fisher developed mathematical models showing how preferences for maladaptive males could evolve (1930b; see discussion in Lande, 1980; see also SEXUAL SELECTION).

But an alternate concept of adaptation is available—the sexually selected traits that are advantageous to mating can still be seen as adaptations, once the meaning of "adaptation" is adjusted. In this school of thought, the notion of "adaptation" should be broadened to include traits that contribute exclusively to reproductive success, even though the more traditional definition is in terms of survival (vs. Bock, 1980; see Kirkpatrick, 1987).

On my analysis, a great deal of the heat in the debates about group selection has arisen because one set of researchers, including Maynard Smith, Dawkins, and G. C. Williams, attribute group selection only when there is a group adaptation in the engineering sense, while others, including Wade, D. S. Wilson, and Uyenoyama and Feldman, do not see such an adaptation as necessary in order to establish group selection (Dawkins, 1982b, 1989; Lloyd, 1988; Lloyd and Gould, 1992; Maynard Smith, 1976b; Uyenoyama and Feldman, 1981; Wade, 1985; G. C. Williams, 1966; D. S. Wilson, 1983a; Wright, 1980). More recently, however, Maynard Smith has moved in the direction of Wright (1980), and agreed that the question of group-level adaptation in (what I call here) the engineering sense is supplementary to the question of whether the group is an interactor, or whether group selection is producing adaptations in the product of selection sense (1987a). In sum, when asking the question about whether a given level of entity has acquired adaptations as a result of a selection process, it is necessary to state not only the level in question, but also the notion of adaptation—either product of selection or engineering—being used.

REFERENCES · INDEX

REFERENCES

Abrams, P. A. (1983). "The theory of limiting similarity." *Annual Review of Ecology and Systematics* 14: 359–376.

——— (1988). "How should resources be counted?" *Theoretical Population Biology* 33: 226–242.

Adams, Mark B., ed. (1990). *The Wellborn Science: Eugenics in Germany, France, Brazil, and Russia.* New York: Oxford University Press.

Agassiz, L. (1857–66). *Contributions to the Natural History of the United States*, 4 vols. Boston: Little, Brown.

Alberch, P.; Gould, S. J.; Oster, G. F.; and Wake, D. B. (1979). "Size and shape in ontogeny and phylogeny." *Paleobiology* 5(3): 296–317.

Alexander, R. D. (1974). "The evolution of sexual behavior." *Annual Review of Ecology and Systematics* 5: 325–383.

——— (1979). *Darwinism and Human Affairs.* Seattle: University of Washington Press.

Allee, W. C. (1955). *Cooperation among Animals, with Human Implications.* New York: Shuman.

Allen, Garland (1968). "Thomas Hunt Morgan and the problem of natural selection." *Journal of the History of Biology* 1(1): 113–139.

——— (1969). "Hugo de Vries and the reception of the Theory." *Journal of the History of Biology* 2(1): 55–87.

Alvarez, L. W.; Alvarez, W.; Asaro, F.; and Michel, H. (1980). "Extraterrestrial cause for the Cretaceous-Tertiary extinction." *Science* 208: 1095–1108.

American Ornithologists' Union (1886). *The Code of Nomenclature and Check List of North American Birds Adopted by the American Ornithologists' Union.* New York: American Ornithologists' Union.

Anderson, E. (1940). "The concept of the genus, II. A survey of modern opinion." *Bulletin of the Torrey Botanical Club* 67: 363–369.

Anderson, V. L.; and Kempthorne, O. (1954). "A model for the study of quantitative inheritance." *Genetics* 39: 883–898.

Andersson, M. (1982). "Female choice selects for extreme tail length in a widowbird." *Nature* 299: 818–820.

——— (1983). "Female choice in widowbirds." *Nature* 302: 456.

——— (1986). "Evolution of condition-dependent sex ornaments and mating preferences: Sexual selection based on viability." *Evolution* 40: 804–816.

Andersson, M. B., and Bradbury, J. W. (1987). "Introduction." In *Sexual Selection: Testing the Alternatives*. Report of the Dahlem Workshop on Sexual Selection, Life Sciences Research Report 39, Dahlem Konferenzen, Berlin, ed. W. Bradbury and M. B. Andersson, 1–8. Chichester: John Wiley.

Andrewartha, H. G. (1961). *Introduction to the Study of Animal Populations*. Chicago: University of Chicago Press; London: Methuen.

Andrewartha, H. G., and Birch, L. C. (1984). *The Ecological Web*. Chicago: University of Chicago Press.

Anonymous (1670). "Review of Historia Generalis Insectorum." *Philosophical Transactions of the Royal Society* 5: 2078–80.

Anonymous (1989). "How to get and control a market niche." *Agency Sales Magazine* 19(1): 5–7.

Anstey, R. L. (1987). "Astogeny and phylogeny: Evolutionary heterochrony in Paleozoic bryozoans." *Paleobiology* 13(1): 20–43.

Antonovics, J.; Clay, K.; and Schmitt, J. (1987). "The measurement of small-scale environmental heterogeneity using clonal transplants of *Anthoxanthum odoratum* and *Danthonia spicata*." *Oecologia* 71: 601–607.

Antonovics, J.; Ellstrand, N. C.; and Brandon, R. N. (1988). "Genetic variation and environmental variation: Expectations and experiments." In *Plant Evolutionary Biology*, ed. L. D. Gottlieb and S. K. Jain, 275–303. London: Chapman and Hall.

Appel, Toby A. (1987). *The Cuvier-Geoffroy Debate: French Biology in the Decades before Darwin*. Oxford: Oxford University Press.

Armstrong, R. A., and McGehee, R. (1980). "Competitive exclusion." *American Naturalist* 115: 151–170.

Arnold, S. J. (1983). "Sexual selection: The interface of theory and empiricism." In *Mate Choice*, ed. P. Bateson, 67–107. Cambridge: Cambridge University Press.

Arnold, S. J., and Fristrup, K. (1982). "The theory of evolution by natural selection: A hierarchical expansion." *Paleobiology* 8: 113–129.

Arnold, S. J., and Wade, M. J. (1984a). "On the measurement of natural and sexual selection: Theory." *Evolution* 38: 709–719.

——— (1984b). "On the measurement of natural and sexual selection: Applications." *Evolution* 38: 720–734.

Arthur, W. (1988). *A Theory of the Evolution of Development*. New York: Wiley-Interscience.

Ashlock, P. (1971). "Monophyly and associated terms." *Systematic Zoology* 20: 63–69.

Atchley, W. R., and Woodruff, D. S., eds. (1981). *Evolution and Speciation: Essays in Honor of M. J. D. White*. Cambridge: Cambridge University Press.

Atran, S. (1987). "The early history of the species concept: An anthropological reading." In *Histoire du concept de l'espèce dans les sciences de la vie*, ed. S. Atran et al., 1–36. Paris: Singer-Polignac.

Austin, M. (1983). "Continuum concept, ordination methods, and Niche Theory." *Annual Review of Ecology and Systematics* 16: 39–61.

——— (1987). "Models for the analysis of species' response to environmental gradients." *Vegetation* 69: 35–45.

Axelrod, R. (1981). "The emergence of cooperation among egoists." *American Political Science Review* 75: 306.

——— (1984). *The Evolution of Cooperation*. New York: Basic Books.

Axelrod, R., and Dion, D. (1988). "The further evolution of cooperation." *Science* 242: 1385–89.

Axelrod, R., and Hamilton, W. D. (1981). "The evolution of cooperation." *Science* 211: 1390.

Ayala, F. A. (1968). "Biology as an autonomous science." *American Scientist* 56: 207–221.

——— (1970a). "Competition, coexistence and evolution." In *Essays in Evolution and Genetics in Honor of Theodosius Dobzhansky*, ed. M. K. Hecht and W. C. Steere, 121–148. New York: Appleton-Century-Crofts.

——— (1970b). "Teleological explanations in evolutionary biology." *Philosophy of Science* 37: 1–15.

——— (1985). "The theory of evolution: Recent successes and challenges." In *Evolution and Creation*, ed. E. McMullin, 59–90. Notre Dame: University of Notre Dame Press.

Babcock, E. B., and Hall, H . M. (1924). "*Hemizonia congesta*, a genetic, ecological and taxonomic study of the hay-field tarweeds." *University of California Publications in Botany* 13: 154–200, pl. 1–7.

Bacon, Francis (1631). *Sylva Sylvarum; or a Naturell Historie in Ten Centuries*, 3rd ed. London: William Lee.

——— (1665). *Opera Omnia*. Frankfort: Joannis Baptistae Schonwtteri.

Baer, Karl Ernst von (1828). *Entwickelungsgeschichte der Thiere: Beobachtung und Reflexion*. Königsberg: Bornträger.

——— (1986). *Autobiography of Dr. Karl Ernst von Baer*, ed. Jane Oppenheimer, trans. H. Schneider. Canton, Mass.: Science History Publications.

Bajema, C. J., ed. (1984). *Evolution by Sexual Selection Theory: Prior to 1900*. Benchmark Papers in Systematic and Evolutionary Biology 6. New York: Van Nostrand Reinhold.

Baldwin, J. M. (1902). *Development and Evolution*. New York: Macmillan.

Balme, D. M. (1972). *Aristotle's De Ortibus Animalium I and De Generatione Animalium I*. Oxford: Clarendon Press.

——— (1987). "Teleology and necessity." In *Philosophical Issues in Aristotle's Biology*, ed. Allan Gotthelf and James G. Lennox, 275–286. Cambridge: Cambridge University Press.

Barnes, J., ed. (1984). *The Complete Works of Aristotle: The Revised Oxford Translation*. Princeton, N.J.: Princeton University Press.

Barrett, P. H., ed. (1977). *The Collected Papers of Charles Darwin*. Chicago: University of Chicago Press.

Barrett, P. H.; Gautrey, P. J.; Herbert, S.; Kohn, D.; and Smith, S., eds. (1987). *Charles Darwin's Notebooks, 1836–1844*. Ithaca, N.Y.: Cornell University Press.

Bateman, A. J. (1948). "Intra-sexual selection in *Drosophila*." *Heredity* 2: 349–368.

Bates, H. W. (1862). "Contributions to an insect fauna of the Amazon Valley." *Transactions of the Linnean Society of London* 23: 495–566.

Bateson, P. P. G., ed. (1983). *Mate Choice.* Cambridge: Cambridge University Press.

Bateson, W. (1894). *Materials for the Study of Variation: Treated with Especial Regard to the Discontinuity in the Origin of Species.* London: Macmillan.

——— (1902). *Mendel's Principles of Heredity: A Defense.* Cambridge: Cambridge University Press.

——— (1913). *Problems in Genetics.* New Haven: Yale University Press.

Baum, B. R. (1988). "A simple procedure for establishing discrete characters from measurement data, applicable to cladistics." *Taxon* 37: 63–70.

Beadle, G. W., and Tatum, E. L. (1941). "Genetic control of biochemical reactions in *neurospora.*" *Proceedings of the National Academy of Sciences* 27: 499–506.

BEAR (1956). *The Biological Effects of Atomic Radiation.* Washington, D.C.: National Academy of Sciences–National Research Council.

Beatty, J. (1982). "What's in a word? Coming to terms in the Darwinian revolution." *Journal of the History of Biology* 15: 215–239.

——— (1985). "Speaking of species: Darwin's strategy." In *The Darwinian Heritage,* ed. D. Kohn, 265–281. Princeton: Princeton University Press.

——— (1987). "Dobzhansky and drift: Facts, values, and chance in evolutionary biology." In *The Probabilistic Revolution,* vol. 2, ed. L. Krüger et al. Cambridge, Mass.: MIT Press.

Beckner, Morton (1959). *The Biological Way of Thought.* Berkeley: University of California Press.

——— (1971). "Teleology." In *Man and Nature,* ed. Ronald Munson, 92–101. New York: Dell Publishing.

Beer, G. (1983). *Darwin's Plots: Evolutionary Narrative in Darwin, George Eliot and Nineteenth-Century Fiction.* London: Routledge and Kegan Paul.

Begon, M.; Harper, J. L.; and Townshend, C. R. (1986). *Ecology: Individuals, Populations, and Communities.* Sunderland, Mass.: Sinauer.

Bell, G. (1982). *The Masterpiece of Nature: The Evolution and Genetics of Sexuality.* Berkeley: University of California Press.

Bentham, G. (1875). "On the recent progress and present state of systematic botany." *Report of the British Association for the Advancement of Science* 1874: 27–54.

Benzer, S. (1957). "The elementary units of heredity." In *The Chemical Basis of Heredity,* ed. W. D. McElroy and B. Glass, 70–93. Baltimore: Johns Hopkins University Press.

Bernstein, H.; Byerly, H. C.; Hopf, F. A.; Michod, R. A.; and Vemulapalli, G. K. (1983). "The Darwinian dynamic." *Quarterly Review of Biology* 58: 185–207.

Bessey, C. E. (1908). "The taxonomic aspect of the species problem." *American Naturalist* 42: 218–224.

Birch, L. C. (1957). "The meanings of competition." *American Naturalist* 91: 5–18.
Black, Max (1962). *Models and Metaphors*. Ithaca: Cornell University Press.
——— (1979). "More about metaphors." In *Metaphor and Thought*, ed. Andrew Ortony. Cambridge: Cambridge University Press.
Blum, M. S., and Blum, N. A., eds. (1979). *Sexual Selection and Reproductive Competition in Insects*. New York: Academic Press.
Bock, W. (1963). "Evolution and phylogeny in morphologically uniform groups." *American Naturalist* 97: 265–285.
——— (1974). "Philosophical foundations of classical evolutionary classification." *Systematic Zoology* 22: 375–392.
——— (1977). "Foundations and methods of evolutionary classification." In *Major Patterns in Vertebrate Evolution*, ed. M. K. Hecht, P. C. Goody, and B. M. Hecht, 851–895. New York: Plenum.
——— (1980). "The definition and recognition of biological adaptation." *American Zoologist* 20: 217–227.
Bock, W. and Farrand, J. (1980). "The number of species and genera of recent birds: A contribution to comparative systematics." *American Museum Novitates* 2707: 1–29.
Bock, W., and von Wahlert, G. (1965). "Adaptation and the form-function complex." *Evolution* 19: 269–299.
Boerhaave, Hermann (1744). *Praelectiones Academicae*, vol. 5, part 2. Edited with notes by Albertus Haller. Göttingen: Bandenhoeck.
Bonnet, Charles (1762). *Considerations sur les corps organisés*, 2 vols. Amsterdam: Marc-Michel Rey.
——— (1769). *La Palingénésie philosophique, ou Idées sur l'etat passé et sur l'etat futur des etres vivans*, 2 vols. Geneva: Philibert and Chiroi.
Bookstein, F. L., et al. (1985). *Morphometrics in Evolutionary Biology*. Special Publication No. 15. Philadelphia: The Academy of Natural Science.
Borgia, G.; Kaatz, I.; and Condit, R. (1987). "Female choice and bower decoration in the satin bowerbird *Ptilonorhynchus violaceus*: A test of hypotheses for the evaluation of male display." *Animal Behaviour* 35: 1129–39.
Boucher, D. H. (1985a). "The idea of mutualism, past and future." In *The Biology of Mutualism: Ecology and Evolution*, ed. D. H. Boucher, 1–28. London: Croom Helm.
——— (1985b). "Lotka-Volterra models of mutualism and positive density-dependence." *Ecological Modeling* 27: 251–270.
——— (1985c). *Mutualism*. Oxford: Oxford University Press.
Bowler, P. J. (1975). "The changing meaning of evolution." *Journal of the History of Ideas* 36: 95–114.
——— (1983). *The Eclipse of Darwinism*. Baltimore: Johns Hopkins University Press.
——— (1984a). "E. W. MacBride's Lamarckian eugenics and its implications for the social construction of scientific knowledge." *Annals of Science* 41: 245–260.
——— (1984b). *Evolution: The History of an Idea*. Berkeley: University of California Press.

——— (1985). "Lotka-Volterra models of mutualism and positive density-dependence." *Ecological Modelling* 27: 251–270.

——— (1986). *Theories of Human Evolution: A Century of Debate, 1844–1944.* Baltimore: Johns Hopkins University Press; Oxford: Basil Blackwell.

——— (1988). *The Non-Darwinian Revolution: Reinterpreting a Historical Myth.* Baltimore: Johns Hopkins University Press.

——— (1989). *The Mendelian Revolution: The Emergence of Hereditarian Concepts in Modern Science and Society.* London: Athlone Press; Baltimore: Johns Hopkins University Press.

Boyd, R., and Richerson, P. J. (1990). "Group selection among alternative evolutionary stable strategies." *Journal of Theoretical Biology.* 145(3): 331–342.

Boyden, A. (1943). "Homology and analogy: A century after the definitions of 'homologue' and 'analogue' of Richard Owen." *Quarterly Review of Biology* 18: 228–241.

——— (1973). *Perspectives in Zoology.* Oxford: Pergamon Press.

Bradburd, R. M., and Ross, D. R. (1989). "Can small firms find and defend strategic niches? A test of the Porter Hypothesis." *Review of Economics and Statistics (Netherlands)* 71: 258–262.

Bradbury, J. W., and Andersson, M. B., eds. (1987). *Sexual Selection: Testing the Alternatives.* New York: Dahlem Workshop Reports and Wiley Interscience.

Bradshaw, A. D. (1965). "Evolutionary significance of phenotypic plasticity in plants." *Advances in Genetics* 13: 115–156.

Braithwaite, R. B. (1953). *Scientific Explanation.* Cambridge: Cambridge University Press.

Brakefield, P. (1988). "What is the progress towards understanding the selection webs influencing melanic polymorphisms in insects?" In *Population Genetics and Evolution,* ed. G. de Jong, 148–162.

Brandon, R. N. (1978). "Adaptation and evolutionary theory," *Studies in the History and Philosophy of Science* 9: 181–206.

——— (1981). "Biological teleology: Questions and explanations." *Studies in the History and Philosophy of Science* 12: 91–105.

——— (1982). "The levels of selection." In *PSA 1982,* vol. 1, ed. P. Asquith and T. Nickles, 315–323. East Lansing, Mich.: The Philosophy of Science Association.

——— (1985). "Adaptation Explanations." In *Evolution at a Crossroads,* ed. D. J. Depew and B. H. Weber, 81–96. Cambridge, Mass.: MIT Press.

——— (1990). *Adaptation and Environment.* Princeton: Princeton University Press.

Brandon, R. N., and Burian, R. M. (1984). *Genes, Organisms, Populations: Controversy over the Units of Selection.* Cambridge, Mass.: MIT Press.

Bridges, C. B. (1916). "Non-disjunction as proof of the chromosome theory of heredity." *Genetics* 1: 1–52; 107–163.

Briquet, J. (1899). "Observations critiques sur les conceptions actuelles de l'espèce végétale au point de vue systematique." In *Flore des Alpes Maritimes,* ed. E. Burnat, 3(1): v–xxxvi. Geneva and Basel: Georg.

Brock, T. (1985). "Procaryotic population ecology." In *Engineered Organisms in the Environment: Scientific Issues,* ed. H. O. Halvorson, D. Pramer, and M. Rogul, 176–179. Washington, D.C.: American Society for Microbiology.

Brown, J. L. (1983). "Intersexual selection." *Nature* 302: 472.

——— (1987). *Helping and Communal Breeding in Birds.* Princeton: Princeton University Press.

Brown, L. (1981). "Patterns of female choice in mottled sculpins *(Cottidae, Teleostei).*" *Animal Behavior* 29: 375–382.

Buffon, Georges Louis Leclerc, Comte de (1749). *Histoire naturelle, genérale et particulière,* vol. 2. Imprimerie Royale.

——— (1766). "De la dégénération des animaux." In *Oeuvres complètes de Buffon,* vol. 4, ed. Pierre Fourens. Paris: Garnier, 1852–1855.

Bulmer, M. G. (1980). *The Mathematical Theory of Quantitative Genetics.* Oxford: Oxford University Press.

——— (1989). "Structural instability of models of sexual selection." *Theoretical Population Biology* 35: 195–206.

Burian, R. M. (1983). "Adaptation." In *Dimensions of Darwinism,* ed. M. Grene, 287–314. Cambridge and New York: Cambridge University Press.

Burian, R. M., et al. (1988). "The singular fate of genetics in the history of French biology, 1900–1940." *Journal of the History of Biology* 21: 357–402.

Burkhardt, F., and Smith, S. (1988). *The Correspondence of Charles Darwin, 1847–1850,* vol. 4. Cambridge: Cambridge University Press.

Burkhardt, Richard (1977). *The Spirit of the System.* Cambridge, Mass.: Harvard University Press.

——— (1987). "Lamarck and species." In *Histoire du concept de l'espèce dans les sciences de la vie,* ed. S. Atran et al., 161–180. Paris: Fondation Singer-Polignac.

Burks, B. S. (1928). "The relative influence of nature and nurture upon mental development: A comparative study of foster parent–foster child resemblance and true parent–true child resemblances." *27th Yearbook of the National Society for the Study of Education.* Bloomington, Ill.: Public School.

Bush, G. L. (1981). "Statispatric speciation and rapid evolution in animals." In *Evolution and Speciation,* ed. William Atchley and David Woodruff, 201–218. Cambridge: Cambridge University Press.

——— (1983). "What do we really know about speciation?" *Perspectives on Evolution,* ed. R. Milkman, 119–128. Sunderland, Mass.: Sinauer.

Buss, L. W. (1983). "Evolution and development, and the units of selection." *Proceedings of the National Academy of Sciences, USA* 80: 1387–91.

——— (1987). *The Evolution of Individuality.* Princeton: Princeton University Press.

Butler, Samuel (1879). *Evolution, Old and New.* London: Hardwicke and Bogue.

Byerly, H. C. (1983). "Natural selection as a law: Principles and Processes." *American Naturalist* 120: 739–745.

Cain, A. J. (1951a). "So-called non-adaptive or neutral characters in evolution." *Nature* 168: 424.

——— (1951b). "Non-adaptive or neutral characters in evolution." *Nature* 168: 1049.

——— (1953). "Geography, ecology and coexistence in relation to the biological definition of the species." *Evolution* 7: 76–83.
Cain, A. J., and Currey, J. D. (1963). "Area effects in *cepaea*." *Philosophical Transactions of the Royal Society of London B* 246: 1–81.
Cain, A. J., and Sheppard, P. M. (1950). "Selection in the polymorphic land snail *Cepaea nemoralis*." *Heredity* 4: 275–294.
——— (1954). "Natural selection in *cepaea*." *Genetics* 39: 89–116.
Camin, J., and Sokal, R. (1965). "A method for deducing branching sequences in phylogeny." *Evolution* 19: 311–326.
Camp, W. H. (1951). "Biosystematy." *Brittonia* 7: 113–127.
Campbell, B., ed. (1972). *Sexual Selection and the Descent of Man, 1871–1971*. London: Heinemann.
Candolle, A. P., de (1813). *Théorie elémentaire de la botanique*. Paris: Déterville.
Candolle, A. L., de (1855). *Géographie botanique raisonée*, vol. 2. Paris: Victor Masson.
——— (1862). "Etude sur l'espèce à l'occassion d'une révision de la famille des Cupulifères." *Bibliot. Univ. Arch. Sci. Phys. Nat.* 15: 211–237.
Cannon, W. F. (1961). "The basis of Darwin's achievement: A revaluation." *Victorian Studies* 5: 109–134.
Carlisle, E. Fred (1980). "Literature, science, and language: A study of similarity and difference." *Pre/Text* 1(1–2): 39–72.
Carlson, E. A. (1966). *The Gene: A Critical History*. Philadelphia: W. B. Saunders.
Carr-Sanders, A. M. (1922). *The Population Problem: A Study in Human Evolution*. Oxford: Clarendon Press.
Carter, G. S. (1951). *Animal Evolution: A Study of Recent Views of Its Causes*. London: Sidgwick and Jackson.
Castle, W. E. (1906). "Yellow mice and gametic purity." *Science* 24: 275–281.
——— (1914). "Pure lines and selection." *Journal of Heredity* 5: 93–97.
Caswell, H. (1983). "Phenotypic plasticity in life-history traits: Demographic effects and evolutionary consequences." *American Zoologist* 23: 35–46.
——— (1988). "Theory and models in ecology: A different perspective." *Bulletin of the Ecological Society of America* 69: 102–109.
Cattell, J. M. (1898). "The definition of species." *Science*, n.s. 7: 751–752.
Cavalli-Sforza, L. L. (1969). "Genetic drift in an Italian population." *Scientific American* 223(2): 26–33.
Cavalli-Sforza, L. L., and Bodmer, W. F. (1971). *The Genetics of Human Populations*. San Francisco: W. H. Freeman.
Cavalli-Sforza, L. L., and Feldman, M. W. (1973). "Cultural versus biological inheritance: Phenotypic transmission from parents to children." *American Journal of Human Genetics* 25: 618–637.
——— (1978). "Dynamics and statistics of traits under the influence of cultural transmission." In *Genetic Epidemiology*, ed. N. E. Morton and C. S. Chung, 133–144. New York: Academic Press.
Chambers, R. (1844). *Vestiges of the Natural History of Creation*. London: Churchill.
Charlesworth, B. (1970). "Selection in populations with overlapping generations;

1. The use of Malthusian parameters in population genetics." *Theoretical Population Biology* 1(3): 352–370.

——— (1987). "The heritability of fitness." In *Dahlem Workshop on Sexual Selection: Testing the Alternatives,* ed. J. W. Bradbury and M. B. Andersson, 21–40. Chichester: Wiley.

Child, C. M. (1915). *Individuality in Organisms.* Chicago: University of Chicago Press.

Christiansen, F. B. (1983). "The definition and measurement of fitness." In *Evolutionary Ecology: B. E. S. Symposium,* ed. B. Shorrocks, 23: 65–79.

Churchill, Frederick B. (1974). "William Johannsen and the genotype concept." *Journal of the History of Biology* 7: 5–30.

——— (1979). "Sex and the single organism: Biological theories of sexuality in mid-nineteenth century." *Studies in the History of Biology* 3: 139–177.

Clark, R. W. (1968). *JBS: The Life and Work of JBS Haldane.* London: Hodder and Stoughton.

Clarke, C. A. (1961). "Blood groups and diseases." *Progress in Medical Genetics* 1: 81–119.

Clarke, G. (1954). *Elements of Ecology.* New York: Wiley.

Clausen, J. (1922). "Studies on the collective species *Viola tricolor* L. II." *Bor Tidsschr.* 37: 363–416.

Clausen, J.; Keck, D. D.; and Heisey, W. W. (1948). "Experimental studies on the nature of species. 3. Environmental responses of clomatic races of *Achillea.*" *Carnegie Institution of Washington Publication* 581: 1–129.

Clements, F. E. (1905). *Research Methods in Ecology.* Lincoln, Neb.: University Publishing Company.

——— (1916). *Plant Succession: An Analysis of the Development of Vegetation.* Publication No. 242. Washington, D.C.: Carnegie Institution of Washington.

Clements, F. E., and Goldsmith, G. W. (1924). *The Phytometer Method in Ecology.* Publication No. 356. Washington, D.C.: Carnegie Institution of Washington.

Clements, F. E., and Shelford, V. E. (1939). *Bio-Ecology.* New York: Wiley.

Clements, F. E.; Weaver, J. E.; and Hanson, H. C. (1929). *Plant Competition: An Analysis of Community Functions.* Publication No. 398. Washington, D.C.: Carnegie Institution of Washington.

Cloninger, C.; Rice, J.; and Reich, T. (1979). "Multifactorial inheritance with cultural transmission and assortative mating. II. A general model of combined polygenic and cultural inheritance." *American Journal of Human Genetics* 31: 176–198.

Clutton-Brock, T. H., ed. (1988). *Reproductive Success: Studies of Individual Variation in Contrasting Breeding Systems.* Chicago: University of Chicago Press.

Clutton-Brock, T. H., and Harvey, P. H. (1979). "Comparison and adaptation." *Proceedings of the Royal Society of London, B* 205: 547–565.

Cock, A. G. (1977). "Bernard's symposium: The species concept in 1900." *Biological Journal of the Linnean Society* 9: 1–30.

Cole, L. C. (1960). "Competitive exclusion." *Science* 132: 348–349.

Coleman, William (1971). *Biology in the Nineteenth Century.* Cambridge: Cambridge University Press.

Colinvaux, Paul (1978). *Why Big Fierce Animals Are Rare*. Princeton: Princeton University Press.

Colless, D. H. (185). "On 'character' and related terms." *Systematic Zoology* 34: 229–233.

Collins, G. N. (1921). "Dominance and the vigor of first generation hybrids." *American Naturalist* 55: 116–133.

Colwell, R. K. (1981). "Evolution of female-based sex ratios: The essential role of group selection." *Nature* 290: 401–404.

——— (1985a). "Literature search: Life in the niches." *Planning Review* 13(5): 47.

——— (1985b). "Stowaways on the hummingbird express." *Natural History* 94(7): 56–63.

Colwell, R. K., and Fuentes, E. (1975). "Experimental studies of the niche." *Annual Review of Ecology and Systematics* 6: 281–310.

Colwell, R. K., and Futuyma, E. R. (1975). "On the measurement of niche breadth and overlap." *Ecology* 52: 567–576.

Conner, R. (1986). "Pseudoreciprocity: Investing in mutualism." *Animal Behavior* 34: 1652–54.

Conrad, M. (1983). *Adaptability: The Significance of Variability from Molecule to Ecosystem*. New York: Plenum Press.

Cooper, J. M. (1987). "Hypothetical necessity and natural teleology." In *Philosophical Issues in Aristotle's Biology*, ed. Alan Gotthelf and James G. Lennox, 243–274. Cambridge: Cambridge University Press.

Cooper, W. S. (1984). "Expected time to extinction and the concept of fundamental fitness." *Journal of Theoretical Biology* 107: 603–629.

Cope, E. D. (1876). "The theory of evolution." Reprint from *Proceedings of the Academy of Natural Science Philadelphia*. In *The Origin of the Fittest: Essays in Evolution*, by E. D. Cope (1887). New York: Macmillan.

——— (1887). *The Origin of the Fittest: Essays in Evolution*. New York: Macmillan.

Cornell, J. F. (1984). "Analogy and technology in Darwin's vision of nature." *Journal of the History of Biology* 17: 303–344.

Cornford, F. M. (1937). *Plato's Cosmology*. London: Routledge and Kegan Paul.

Corsi, Pietro (1988). *The Age of Lamarck*. Berkeley: University of California Press.

Cowan, S. T. (1971). "Sense and nonsense in bacterial taxonomy." *Journal of General Microbiology* 67: 1–8.

Cowell, A. (1988). "Nile gives abundantly as doomsday recedes." *New York Times* (December 20).

Cox, D. (1980). "A note on the queer history of 'niche'." *Bulletin of the Ecological Society of America* 61: 201–202.

Cracraft, J. (1967). "Comments on homology and analogy." *Systematic Zoology* 16: 355–359.

——— (1981). "The use of functional and adaptive criteria in phylogenetic systematics." *American Zoologist* 21: 21–36.

——— (1983). "Species concepts and speciation analysis." *Current Ornithology* 1: 159–187.

——— (1987). "Species concepts and the ontology of evolution." *Biology and Philosophy* 2: 329–346.

——— (1989). "The empirical consequences of alternative species concepts for understanding patterns and processes of differentiation." In *Speciation and Its Consequences*, ed. D. Otte and J. A. Endler, 28–59. Sunderland, Mass.: Sinauer.

Cravens, H. (1978). *The Triumph of Evolution: American Scientists and the Heredity-Environment Controversy, 1900–1914*. Philadelphia: University of Pennsylvania Press.

Crombie, A. C. (1947). "Interspecific competition." *Journal of Animal Ecology* 16: 44–73.

Cronin, H. (1992). *The Ant and the Peacock*. Cambridge: Cambridge University Press.

Crovello, T. J. (1970). "Analysis of character variation in ecology and systematics." *Annual Review of Ecology and Systematics* 1: 55–98.

Crow, J. F. (1958). "Some possibilities for measuring selection intensities in man." *Human Biology* 30: 1–13.

——— (1985). "The neutrality-selection controversy in the history of evolution and population genetics." In *Population Genetics and Molecular Evolution*, ed. T. Ohta and J. Aoki, 1–18. Tokyo: Japan Scientific Societies Press.

Crow, J. F., and Aoki, K. (1982). "Group selection for a polygenetic behavioral trait: A differential proliferation model." *Proceedings of the National Academy of Sciences, USA* 79: 2628–31.

Crow, J. F., and Kimura, M. (1970). *An Introduction to Population Genetics Theory*. New York: Harper and Row. Reprinted (1977), Minneapolis: Burgess.

Curio, E. (1973). "Towards a methodology of teleonomy." *Experientia* 29: 1045–58.

Curtsinger, J. W., and Heisler, I. L. (1990). "On the consistency of sexy-son models: A reply to Kirkpatrick." *American Naturalist* 134: 978–981.

Cuvier, G.. (1812). *Recherches sur les ossemens fossiles de quadrupedes, ou l'on rétablit les characteres de plusiers especes d'animaux que les revolutions du globe paroissent avoir détruits*. Paris: Deterville.

Damuth, J. (1985). "Selection among 'species': A formulation in terms of natural functional units." *Evolution* 39: 1132–46.

Damuth, J., and Heisler, I. L. (1988). "Alternative formulations of multilevel selection." *Biology and Philosophy* 3: 407–430.

Dance, S. P. (1970). "'Le Fanatisme du Nobis': A study of J. R. Bourguignat and the 'Nouvelle Ecole.'" *Journal of Conchology* 27: 65–86.

Darden, L. (1977). "William Bateson and the promise of Mendelism." *Journal of the History of Biology* 10: 87–106.

Darden, L., and Cain, J. (1989). "Selection type theories." *Philosophy of Science* 56: 106–129.

Darlington, P. J. (1980). *Evolution for Naturalists: The Simple Principles, the Complex Reality*. New York: John Wiley and Sons.

——— (1983). "Evolution: Questions for the modern theory." *Proceedings of the National Academy of Sciences, USA* 80: 1960–63.

Darwin, C. (1841). "Notes on the habits of bees." Unpublished manuscript, dated Maer, June 1841. In Darwin Ms. 46.2, Cambridge University Library.

——— (1842). "Sketch of 1842." In *Evolution by Natural Selection,* by C. Darwin and A. Wallace, ed. G. De Beer. Cambridge: Cambridge University Press.

——— (1842, 1844). *The Foundations of the Origin of Species: Two Essays Written in 1842 and 1844 by Charles Darwin,* ed. Francis Darwin. Cambridge: Cambridge University Press, 1909.

——— (1859a). *On the Origin of Species by Means of Natural Selection, or the Preservation of Favoured Races in the Struggle for Life.* London: John Murray.

——— (1859b). *On the Origin of Species, A Facsimile of the First Edition* with an introduction by Ernst Mayr. Reprint (1964), Cambridge, Mass.: Harvard University Press.

——— (1859c). *On the Origin of Species.* Reprint (1968), Baltimore: Penguin Books.

——— (1868). *The Variation of Plants and Animals under Domestication,* 2 vols. New York: Orange Judd.

——— (1869). *On the Origin of Species,* 5th ed. London: John Murray.

——— (1871). *The Descent of Man and Selection in Relation to Sex.* London: John Murray.

——— (1876). *The Effects of Cross and Self-fertilisation.* London: John Murray.

——— (1881). Letter to Karl Semper. In *More Letters of Charles Darwin,* I, ed. F. Darwin and A. C. Seward, 391. London: John Murray.

——— (1909). *The Foundations of the Origin of Species: Two Essays Written in 1842 and 1844 by Charles Darwin,* ed. Francis Darwin. Cambridge: Cambridge University Press.

——— (1959). *On the Origin of Species by Charles Darwin: A Variorum Text,* ed. M. Peckham. Philadelphia: University of Pennsylvania Press.

——— (1969). *The Autobiography of Charles Darwin,* ed. Nora Barlow. New York: Norton.

——— (1975). *Charles Darwin's Natural Selection,* ed. R. Stauffer. Cambridge: Cambridge University Press.

——— (1987). *Charles Darwin's Notebooks, 1836–1844,* transcribed and edited by P. H. Barrett et al. Cambridge: Cambridge University Press.

Darwin, Erasmus (1796). *Zoonomia or the Laws of Organic Life,* 2nd ed., 2 vols. London: Johnson.

Darwin, F., ed. (1887). *Life and Letters of Charles Darwin,* 2 vols. London: John Murray.

——— (1892). *The Autobiography of Charles Darwin and Selected Letters.* Reprint (1958), New York: Dover Publications

Darwin, F., and Seward, A. C. (1903). *More Letters of Charles Darwin,* 2 vols. London: John Murray.

Davis, B. D. (1984). "Science, fanaticism, and the law." *Genetic Engineering News* 4(5): 4.

Davis, P. H., and Heywood,V. H. (1963). *Principles of Angiosperm Taxonomy.* Edinburgh: Oliver and Boyd.

Dawkins, R. (1976). *The Selfish Gene.* 2nd ed., 1989. Oxford: Oxford University Press.

——— (1978). "Replicator selection and the extended phenotype." *Zeitschrift fur Tierpsychologie* 47: 61–76.

——— (1980). "Good strategy or evolutionarily stable strategy?" In *Sociobiology: Beyond Nature/Nurture*, ed. G. W. Barlow and J. Solverberg, 331–367.

——— (1982a). "Replicators and vehicles." In *Current Problems in Sociobiology*, ed. King's College Sociobiology Group, Cambridge, 45–64. Cambridge: Cambridge University Press.

——— (1982b). *The Extended Phenotype*. San Francisco: W. H. Freeman.

——— (1983). "Universal Darwinism." In *Evolution from Molecules to Men*, ed. D. S. Bendall, 403–425. Cambridge: Cambridge University Press.

——— (1986). *The Blind Watchmaker*. London: Longman; New York: Norton.

——— (1989). "The evolution of evolvability." In *Artificial Life, Studies in the Sciences of Complexity*, ed. C. Langton, 201–220. Reading, Mass.: Addison-Wesley.

Dayhoff, M. O.; Barker, W. C.; and Hunt, L. T. (1983). "Establishing distant homologies in protein sequences." *Methods in Enzymology* 91: 524–545.

Dean, J. (1979). "Controversy over classification." In *Natural Order: Historical Studies in Scientific Culture*, ed. B. Barnes and S. Shapin, 211–230. Beverly Hills, Calif.: Sage Publications.

DeAngelis, D. L., and Waterhouse, J. C. (1987). "Equilibrium and non-equilibrium concepts in ecological models." *Ecological Monographs* 57: 1–21.

DeBary, A. (1879). *Die Ersheinung der Symbiose*. Strasburg: Karl J. Trubner.

De Beer, G. R. (1930). *Embryology and Evolution*. Oxford: Clarendon Press.

——— (1938). *Evolution: Essays on Aspects of Evolutionary Biology, Presented to Professor E. S. Goodrich on His Seventieth Birthday*. Oxford: Clarendon Press.

——— (1940). *Embryos and Ancestors*. 3rd ed., 1958. Oxford: Clarendon Press.

——— (1954). "The evolution of Metazoa." In *Evolution as a Process*, ed. J. S. Huxley, A. C. Hardy, and E. B. Ford, 24–33. London: Allen and Unwin.

——— (1971a). "Teleology." In *Man and Nature*, ed. Ronald Munson, 92–101. New York: Dell Publishing.

——— (1971b). *Homology, an Unsolved Problem*. London: Oxford University Press.

Den Boer, P. J. (1986). "The present status of the competitive exclusion principle." *Trends in Ecology and Evolution* 1: 25–28.

Denniston, C. (1978). "An incorrect definition of fitness revisited." *Annals of Human Genetics, London* 42: 77–85.

Desmond, Adrian (1984). "Robert E. Grant: The social predicament of a pre-Darwinian evolutionist." *Journal of the History of Biology* 17: 189–223.

——— (1989). *The Politics of Evolution: Medicine, Morphology, and Reform in Radical London*. Chicago: University of Chicago Press.

De Vries, Hugo (1889). *Intracellular Pangenesis*, trans. C. Stuart Gager, 1910. Chicago: Open Court.

——— (1900). "The law of segregation of hybrids." In *The Origin of Genetics, A Mendel Sourcebook*, ed. Kurt Stern and Eva Sherwood, trans. Eva Stern, 1966, 107–117. San Francisco: W. H. Freeman.

——— (1905). *Species and Varieties: Their Origin by Mutation.* Chicago: Open Court.

——— (1910). *The Mutation Theory,* vols. I and II, trans. J. C. Farmer and A. D. Darbeshire. Chicago: Open Court.

Diamond, J. (1978). "Niche shifts and the rediscovery of interspecific competition." *American Scientist* 66: 322–331.

Diamond, J., and Case, T. J., eds. (1986). *Community Ecology.* New York: Harper and Row.

Diara, A. (1987). "Les espèces sont-ils filles de la nature ou du naturaliste?" In *histoire du concept de l'espèce dans les sciences de la vie,* ed. S. Atran et al., 269–283. Paris: Fondation Singer-Polignac.

Dice, L. (1952). *Natural Communities.* Ann Arbor: University of Michigan Press.

Diels, L. (1932). "Die Methoden der Phytographie und der Systematik der Pflanzen." In *Handbuch der biologischen Arbeitsmethoden,* ed. E. Aberhalden, 67–190. Berlin: Urban and Schwarzenberg.

Dirzo, R., and Sarukhan, J., eds. (1984). *Perspectives on Plant Population Ecology.* Sunderland, Mass.: Sinauer.

Diver, C. (1940). "The problem of closely related snails living in the same area." In *The New Systematics,* ed. J. S. Huxley, 303–328. Oxford: Clarendon.

Dobzhansky, T. (1937). *Genetics and the Origin of Species.* New York: Columbia University Press.

——— (1943). "Genetics of natural populations. IX. Temporal changes in the composition of populations of *Drosophila pseudoobscura.*" *Genetics* 28: 162–186.

——— (1951). *Genetics and the Origin of the Species,* 3rd ed. New York: Columbia University Press.

——— (1955). "A review of some fundamental concepts and problems of population genetics." *Cold Spring Harbor Symposium in Quantitative Biology* 20: 1–15.

——— (1956). "What is an adaptive trait?" *American Naturalist* 90: 337–347.

——— (1959). "Discussion." Following Lamotte in *Cold Spring Harbor Symposia in Quantitative Biology.*

——— (1975). *Genetics of the Evolutionary Process.* New York: Columbia University Press.

Dobzhansky, T., and Queal, M. I. (1938). "Genic variation in populations of *Drosophila pseudoobscura* inhabiting isolated mountain ranges." *Genetics* 23: 463–484.

Donoghue, M. J. (1985). "A critique of the biological species concept and recommendations for a biological alternative." *Bryologist* 88: 172–181.

——— (1989). "Phylogenies and the analysis of evolutionary sequences, with examples from seed plants." *Evolution* 43: 1137–56.

Doolittle, R. F. (1981). "Similar amino acid sequences: Chance of common ancestry?" *Science* 214: 149–159.

Downhower, J. F. (1976). "Darwin's finches and the evolution of sexual dimorphism in body size." *Nature* 263(5578): 558–563.

Dubinin, M. P., and Romaschoff, D. D. (1932). "The genetic structure of species and their evolution." (In Russian.) *Biologiches Zhurnal* 1: 52–95.

Dunbar, R. I. M. (1982). "Adaptation, fitness and the evolutionary tautology." In *Current Problems in Sociobiology,* ed. King's College Sociobiology Group, Cambridge, 9-28. Cambridge: Cambridge University Press.

Dupré, J. (1981). "Natural kinds and biological taxa." *The Philosophical Review* 90: 66–90.

——— (1986). "Sex, gender, and essence." *Midwest Studies in Philosophy* 11: 441–457.

Dyson, F. (1985). *Origins of Life.* Cambridge: Cambridge University Press.

East, E. M. (1909). "The distinction between heredity and development in inbreeding." *American Naturalist* 43: 173–181.

——— (1936). "Heterosis." *Genetics* 21: 375–397.

Eberhard, W. G. (1979). "The function of horns in *Podischnus agenor* (Dynastinae) and other beetles." In *Sexual Selection and Reproductive Competition,* ed. M. S. Blum and N. A. Blum, 231–258. New York: Academic Press.

——— (1980). "Horned beetles." *Scientific American* 242(3): 166–181.

Edie, James M. (1976). *Speaking and Meaning.* Bloomington: Indiana University Press.

Edwards, A., and Cavalli-Sforza, L. (1963). "The reconstruction of evolution." *Annals of Human Genetics* 27: 105.

Egerton, F. (1977). "A bibliographical guide to the history of general ecology and population ecology." *History of Science* 15: 189–215.

——— (1983). "The history of ecology: Achievements and opportunities, part one." *Journal of the History of Biology* 16: 259–310.

Ehrlich, P. R., and Roughgarden, J. (1986). *Ecology.* New York: Macmillan.

Eibl-Eibesfeldt, I. (1970). *Ethology: The Biology of Behavior.* New York: Holt, Rinehart and Winston.

Eldredge, N. (1985). *Unfinished Synthesis: Biological Hierarchies and Modern Evolutionary Thought.* New York: Oxford University Press.

Eldredge, N., and Cracraft, J. (1980). *Phylogenetic Patterns and the Evolutionary Process.* New York: Columbia University Press.

Eldredge, N., and Gould, S. J. (1972). "Punctuated equilibria: An alternative to phyletic gradualism." In *Models in Paleobiology,* ed. T. J. M. Schopf, 82-115. San Francisco: Freeman, Cooper.

Elton, C. (1924). "Periodic fluctuations in the numbers of animals: Their causes and effects." *British Journal of Experimental Biology* 2: 119–163.

——— (1927). *Animal Ecology.* New York: Macmillan.

——— (1946). "Competition and the structure of ecological communities." *Journal of Animal Ecology* 15: 54–68.

——— (1958). *The Ecology of Invasions by Animals and Plants.* London: Methuen.

Emerson, A. E. (1939). "Social coordination and the superorganism." *American Midland Naturalist* 21: 182–209.

——— (1960). "The evolution of adaptation in population systems." In *Evolution after Darwin,* ed. S. Tax, 307–348. Chicago: University of Chicago Press.

Emlen, J. M. (1984). *Population Biology: The Coevolution of Population Dynamics and Behavior.* New York: Macmillan.

Emmel, T. (1973). *An Introduction to Ecology and Population Biology.* New York: Norton.
Endler, J. A. (1986). *Natural Selection in the Wild.* Princeton: Princeton University Press.
Endler, J. A., and McLellan, T. (1988). "The processes of evolution: Toward a newer synthesis." *Annual Review of Ecology and Systematics* 19: 395–421.
Eucken, Rudolf (1879). *Geschichte der Philosophischen Terminologie.* Reprint (1964), Hildesheim: Georg Olms.
Falconer, D. S. (1981). *Introduction to Quantitative Genetics,* 2nd ed. London: Longmans.
——— (1989). *Introduction to Quantitative Genetics,* 3rd ed. London: Longmans.
Farrall, L. (1979). "The history of eugenics: A bibliographical review." *Annals of Science* 36 (March): 111–123.
Farris, J. (1983). "The logical basis of phylogenetic analysis." In *Advances in Cladistics,* vol. 2, ed. N. Platnick and V. Funk, 7–36. New York: Columbia University Press.
Feldman, M. W., and Cavalli-Sforza, L. L. (1979). "Aspects of variance and covariance analysis with cultural inheritance." *Theoretical Population Biology* 15: 276–307.
Feldman, M. W., and Lewontin, R. C. (1975). "The heritability hang-up." *Science* 190: 1163–68.
Feldman, M. W., and Thomas, W. A. C. (1987). "Behavior-dependent contexts for repeated plays of the prisoner's dilemma II: Dynanical aspects of the evolution of cooperation." *Journal of Theoretical Biology* 128: 297–315.
Felsenstein, J. (1983). "Parsimony in systematics: Biological and statistical issues." *Annual Review of Ecology and Systematics* 14: 313–333.
Fisher, R. A. (1915). "The evolution of sexual preference." *Eugenics Review* 7: 184–192.
——— (1918). "The correlation between relatives on the supposition of Mendelian inheritance." *Transactions of the Royal Society of Edinburgh* 52: 399–433.
——— (1922). "On the dominance ratio." *Proceedings of the Royal Society of Edinburgh* 42: 321–341.
——— (1930a). *The Genetical Theory of Natural Selection.* Oxford: Oxford University Press.
——— (1930b). *The Genetical Theory of Natural Selection,* 2nd rev. ed., 1958. New York: Dover.
——— (1937). *Annals of Eugenics* 11: 355–369.
Fisher, R. A., and Ford, E. B. (1947). "The spread of a gene in natural conditions in a colony of the moth *Panaxia dominula.*" *Heredity* 1: 43–74.
——— (1950). "The Sewall Wright effect." *Heredity* 4: 117–119.
Fitch, W. M. (1966). "An improved method of testing evolutionary homology." *Journal of Molecular Biology* 16: 9–16.
——— (1970). "Distinguishing homologous from analogous proteins." *Systematic Zoology* 19: 99–113.
Foucault, M. (1966). *Les mots et les choses: Une archéologie des sciences humaines.* Paris: Gallimard.

Franklin, D. (1989). "What a child is given." *New York Times Magazine* (Sept. 3): 36.

Franklin, I., and Lewontin, R. C. (1970). "Is the gene the unit of selection?" *Genetics* 65: 707–734.

Furley, David (1987). *The Greek Cosmologists*. Cambridge: Cambridge University Press.

Futuyma D. J. (1986). *Evolutionary Biology*, 2nd ed. Sunderland, Mass.: Sinauer.

——— (1987). "On the role of species in anagenesis." *American Naturalist* 130: 465–473.

Futuyma, D. J., and Slatkin, M. (1983). *Coevolution*. Sunderland, Mass.: Sinauer.

Gaffney, P. (1975). "Roots of the niche concept." *American Naturalist* 109: 490.

Gallop, David (1975). *Plato, Phaedo*. Oxford: Clarendon Press.

Galton, F. (1883). *Inquiries into the Human Faculty*. London: Macmillan.

Gasman, D. (1971). *The Scientific Origins of National Socialism: Social Darwinism in Ernst Haeckel and the Monist League*. New York: Elsevier.

Gause, G. F. (1934). *The Struggle for Existence*. Baltimore: Williams and Wilkins.

——— (1947). "Problems of evolution." *Transactions of the Connecticut Academy of Science* 37: 17–68.

Gause, G. F., and Witt, A. A. (1935). "Behavior of mixed populations and the problem of natural selection." *American Naturalist* 69: 596–609.

Gee, J. H. R., and Giller, P. S. (1987). *Organization of Communities: Past and Present*. Oxford: Blackwell.

Geoffroy Saint-Hilaire, Etienne (1833). "Le degré d'influence du monde ambiant pour modifier les formes animales." *Mémoires de l'Academie Royale des Sciences de l'Institut de France*, 2nd ser., 12: 63–92.

Ghiselin, M. (1966). "An application of the theory of definitions to taxonomic principles." *Systematic Zoology* 15: 127–130.

——— (1969a). *The Triumph of the Darwinian Method*. Berkeley: University of California Press.

——— (1969b). "The distinction between similarity and homology." *Systematic Zoology* 18: 148–149.

——— (1974a). *The Economy of Nature and the Evolution of Sex*. Berkeley: University of California Press.

——— (1974b). "A radical solution to the species problem." *Systematic Zoology* 23: 536–544.

——— (1976). "The nomenclature of correspondence: A new look at 'homology' and 'analogy.'" In *Evolution, Brain, and Behavior: Persistent Problems*, ed. R. B. Masterson, W. Hodos, and H. Jerison, 129–142. Hillsdale, N.J.: Lawrence Erlbaum.

——— (1984). "'Definition,' 'character' and other equivocal terms." *Systematic Zoology* 33: 104–110.

——— (1987). "Species concepts, individuality, and objectivity." *Biology and Philosophy* 2: 127–143.

Giddens, A. (1981). "Agency, institution, and time-space analysis." In *Advances in Social Theory and Methodology*, ed. K. Knorr-Cetina and A. V. Cicourel, 161–174. Boston: Routledge and Kegan Paul.

Gigerenzer, G., et al. (1989). *The Empire of Chance: How Probability Changed Science and Life.* Cambridge: Cambridge University Press.
Giller, P. (1984). *Community Structure and the Niche.* New York: Chapman and Hall.
Gillespie, J. H. (1972). "Natural selection with varying selection coefficients: A haploid model." *Genetical Research* 21: 115–120.
——— (1973). "Natural selection for within-generation variance in offspring number." *Genetics* 76: 601–606.
——— (1975). "Natural selection for within-generation variance in offspring number, II." *Genetics* 81: 403–413.
——— (1977). "Natural selection for variances in offspring numbers: A new evolutionary principle." *American Naturalist* 111: 1010–14.
Gilmour, J. S. L. (1940). "Taxonomy and philosophy." In *The New Systematics,* ed. J. Huxley, 461–474. Oxford: Oxford University Press.
——— (1958). "The species: Yesterday and tomorrow." *Nature* 181: 379–380.
Gilmour, J. S. L., and Gregor, J. W. (1939). "Demes: A suggested new terminology." *Nature* 144: 333–334.
Gilpin, M. E. (1975). *Group Selection in Predator-Prey Communities.* Princeton: Princeton University Press.
Givnish, T. J. (1986). *On the Economy of Plant Form and Function.* Cambridge: Cambridge University Press.
——— (1988). "Theory and observations in plant community ecology: A review of recent approaches." Address to the ISEM Symposium, University of California, Davis, August 15, 1988.
Glass, Bentley, ed. (1959). *Forerunners of Darwin: 1745–1859.* Baltimore: Johns Hopkins University Press.
Gleason, H. (1926). "The individualistic concept of the plant association." *Bulletin of the Torrey Botanical Club* 53: 1–20.
——— (1927). "Further views on the succession concept." *Ecology* 8: 299–326.
Goldschmidt, R. B. (1928). "The gene." *Quarterly Review of Biology* 3: 307–356.
——— (1933). "Some aspects of evolution." *Science* 78: 539–547.
——— (1938a). *Physiological Genetics.* New York: McGraw-Hill.
——— (1938b). "The theory of the gene." *Science Monthly* 46: 268–273.
——— (1940). *The Material Basis of Evolution.* Seattle: University of Washington Press.
——— (1946). "Position effect and the theory of the corpuscular gene." *Experientia* 2: 197–232, 250–256.
——— (1952). "Evolution as viewed by one geneticist." *American Scientist* 40: 84–98.
Goodwin, B. C. (1982). "Development and evolution." *Journal of Theoretical Biology* 97: 43–55.
Goodwin, B. C., and Trainor, L. E. H. (1983). "The ontogeny and phylogeny of the pentadactyl limb." In *Development and Evolution,* ed. B. C. Goodwin, N. Holder, and C. C. Wylie, 353–379. Cambridge: Cambridge University Press.
Gordon, D. M. (1991). "Behavioral flexibility and the foraging ecology of seed-eating ants." American Naturalist 138: 411.

Gotthelf, Allan (1987). "Aristotle's conception of final causality." In *Philosophical Issues in Aristotle's Biology,* ed. Allan Gotthelf and James G. Lennox, 204–242. Cambridge: Cambridge University Press.

Goudge, T. A. (1961). *The Ascent of Life.* Toronto: University of Toronto Press.

Gould, S. J. (1971). "D'Arcy Thompson and the science of form." *New Literary History* 2: 229–258.

——— (1977a). *Ontogeny and Phylogeny.* Cambridge: Harvard University Press.

——— (1977b). *Ever Since Darwin.* New York: Norton.

——— (1981). "Hyena myths and realities." *Natural History* 90: 16–24.

——— (1982a). "The meaning of punctuated equilibrium and its role in validating a hierarchical approach to macroevolution." In *Perspectives on Evolution,* ed. Roger Milkman, 83–104. Sunderland, Mass.: Sinauer.

——— (1982b). "The uses of heresy: An introduction to Richard Goldschmidt's *The Material Basis of Evolution.*" In *The Material Basis of Evolution,* by Richard Goldschmidt, i–xlii. New Haven: Yale University Press.

——— (1983). "The hardening of the modern synthesis." In *Dimensions of Darwinism,* ed. M. Grene, 71–93. Cambridge: Cambridge University Press.

——— (1984a). "A most ingenious paradox." *Natural History* (December): 20–29.

——— (1984b). "Covariance sets and ordered geographic variation in *Cerion* from Aruba, Bonaire and Curacao: A way of studying nonadaptation." *Systematic Zoology* 33(2): 217–237.

——— (1985). "Taxonomy of death." *Nature* 313: 505–506.

Gould, S. J., and Lewontin, R. C. (1979). "The spandrels of San Marco and the Panglossian paradigm." *Proceedings of the Royal Society of London, B* 205: 581–598.

Gould, S. J., and Vrba, E. (1982). "Exaptation: A missing term in the science of form." *Paleobiology* 8: 4–15.

Grafen, A. (1982). "How not to measure inclusive fitness." *Nature* 298: 425–426.

——— (1984). "Natural selection, kin selection and group selection." In *Behavioral Ecology,* ed. J. R. Krebs and W. B. Davies, 62–91. Sunderland, Mass.: Sinauer.

Grant, V. (1957). "The plant species in theory and practice." *American Association for the Advancement of Science Publication* 50: 39–80.

——— (1963). *The Origin of Adaptations.* New York: Columbia University Press.

——— (1971). *Plant Speciation.* New York: Columbia University Press.

——— (1985). *The Evolutionary Process: A Critical Review of Evolutionary Theory.* New York: Columbia University Press.

Gray, A. (1850). *The Botanical Text Book,* 3rd ed. New York: Wiley and Putnam.

——— (1854). "Introductory essay, in Dr. Hooker's Flora of New Zealand." *American Journal of Science and Arts,* ser. 2, 17: 241–252, 334–350.

——— (1860). "Darwin's theory on the origin of species by means of natural selection." *American Journal of Science and Arts,* ser. 2, 29: 153–184.

——— (1863). "Species as to variation, geographic distribution, and succession." *American Journal of Science and Arts,* ser. 2, 35: 431–444.

——— (1880). *Natural Science and Religion.* New York: Charles Scribner's Sons.

Gray, A. J.; Crawley, M. J.; and Edwards, P. J. (1987). *Colonization, Succession and Stability.* Oxford: Blackwell.

Greene, J. C. (1977). "Darwin as a social evolutionist." *Journal of the History of Biology* 10: 1–27.

——— (1981). *Science, Ideology, and World View.* Berkeley: University of California Press.

Gregg, J. R. (1950). "Taxonomy, language and reality." *American Naturalist* 84: 421–433.

Griesemer, J. (1990). "Modeling in the museum: On the role of remnant models in the work of Joseph Grinnell." *Biology and Philosophy.* 5: 3–36.

Grinnell, J. (1904). "The origin and distribution of the chestnut-backed chickadee." *The Auk* 21: 364–379.

——— (1914). "An account of the mammals and birds of the Lower Colorado Valley with especial reference to the distributional problems presented." *University of California Publications in Zoology* 12: 51–294.

——— (1917a). "The niche-relationships of the California thrasher." *The Auk* 34: 427–433.

——— (1917b). "Field tests of theories concerning distributional control." *American Naturalist* 51: 115–128.

——— (1924). "Geography and evolution." *Ecology* 5: 225–229.

——— (1928). "Presence and absence of animals." *University of California Chronicle* 30: 429–450.

——— (1943). *Joseph Grinnell's Philosophy of Nature.* Berkeley: University of California Press.

Grinnell, J., and Swarth, H. (1913). "An account of the birds and mammals of the San Jacinto area of Southern California." *University of California Publications in Zoology* 10: 197–406.

Gruber, H. (1974). *Darwin on Man.* New York: Dutton.

Gulick, J. T. (1872). "On diversity of evolution under one set of external conditions." *Journal of the Linnean Society of Zoology* 11: 496–505.

——— (1889). "Intensive segregation, or divergence through independent transformation." *Journal of the Linnean Society of Zoology* 23: 312–380. Reprinted in Gulick (1905).

——— (1890). "Divergent evolution and the Darwinian Theory." *American Journal of Science* 34: 21–30.

——— (1905). *Evolution, Racial and Habitudinal.* Washington, D.C.: Carnegie Institution of Washington.

Gupta, A. P., and Lewontin, R. C. (1982). "A study of reaction norms in natural populations of *Drosophila pseudoobscura.*" *Evolution* 36(5): 934–948.

Haas, O., and Simpson, G. G. (1946). "Analysis of some phylogenetic terms, with attempts at redefinition." *Proceedings of the American Philosophical Society* 90: 319–349.

Haeckel, E. (1862–68). *Die Radiolarien,* 3 vols. Berlin: Georg Reimer.

——— (1866). *Generelle Morphologie der Organismen.* Berlin: Georg Reimer.

——— (1875). "Die Gastrula und die Eifurchung der Thiere." *Jenaische Zeitschrift fur Naturwissenschaft* 9: 402–508.

——— (1905). *The Evolution of Man.* In *Anthropogenie,* 5th ed., trans. J. McCabe. London: Watts.

Hagedoorn, A. L., and Hagedoorn, A. C. (1921). *On the Relative Value of the Processes Causing Evolution.* The Hague: Nijhoff.
Hagen, J. B. (1983). "The development of experimental methods in plant taxonomy, 1920–1950." *Taxon* 32: 406–416.
——— (1984). "Experimentalists and naturalists in twentieth century botany: Experimental taxonomy, 1920–1950." *Journal of the History of Biology* 17: 249–270.
Haigh, J., and Maynard Smith, J. (1972). "Can there be more predators than prey?" *Theoretical Population Biology* 3: 290–299.
Hailman, J. P. (1982). "Evolution and behavior: An iconoclastic view." In *Learning, Development, and Culture,* ed. H. C. Plotkin, 205–254. New York: Wiley.
Haldane, J. B. S. (1932). *The Causes of Evolution.* Ithaca, N.Y.: Cornell University Press; London: Longmans, Green.
——— (1937a). "The effect of variation on fitness." *American Naturalist* 71: 337–349.
——— (1937b). "Human biology and politics." In *Adventures of a Biologist.* New York: Harper and Row.
——— (1954). "The measurement of natural selection." *Proceedings of the Ninth International Congress of Genetics* (*Caryologica* supplement) 1: 480–487.
——— (1957). "The cost of natural selection." *Journal of Genetics* 55: 511–524.
Haller, M. (1963). *Eugenics: Hereditarian Attitudes in American Thought.* New Brunswick, N.J.: Rutgers University Press.
Halliday, T. R. (1983). "The study of mate choice." in *Mate Choice,* ed. P. Bateson, 3–32. Cambridge: Cambridge University Press.
Hamilton, W. D. (1964). "The genetical evolution of social behaviour I. and II." *Journal of Theoretical Biology* 7: 1–52.
——— (1975). "Innate social aptitudes in man: An approach from evolutionary genetics." In *Biosocial Anthropology,* ed. R. Fox, 133–155. New York: Wiley.
Hamilton, W. D., and Axelrod, R. (1981). "The evolution of cooperation." *Science* 211: 1390–96.
Hamilton, W. D., and Zuk, M. (1982). "Heritable true fitness and bright birds: A role for parasites?" *Science* 218: 384–87.
Hampe, M., and Morgan, S. R. (1988). "Two consequences of Richard Dawkins' view of genes and organisms." *Studies in History and Philosophy of Science* 19: 119–138.
Hardin, G. (1960). "The Competitive Exclusion Principle." *Science* 131: 1292–97.
——— (1968). "The tragedy of the commons." *Science* 162: 1243–48.
Harper, J. L. (1961). "Approaches to the study of plant competition." *Symposia of the Society for Experimental Biology* 15: 1–39.
——— (1977). *Population Biology of Plants.* London: Academic Press.
Harrison, J. W. H., and Garrett, F. C. (1925–1926). "The induction of melanism in the lepidoptera and its subsequent inheritance." *Proceedings of the Royal Society of London B* 101: 115–126.
Hartl, D. L., and Clark, A. G. (1989). *Principles of Population Genetics,* 2nd ed. Sunderland, Mass.: Sinauer.

Harvell, C. D. (1986). "The ecology and evolution of inducible defenses in a marine bryozoan: Cues, costs and consequences." *American Naturalist* 128: 810–823.

Harwood, J. (1989). "Editor's introduction: Genetics, eugenics, and evolution." *British Journal for the History of Science* 22: 257–265.

Hay, D. A. (1985). *Essentials of Behavior Genetics.* Melbourne: Blackwell.

Hazen, W., ed. (1964). *Readings in Population and Community Ecology.* Philadelphia: W. B. Saunders.

Hecht, M. K. (1976). "Phylogenetic inference and methodology, as applied to the vertebrate record." In *Evolutionary Biology,* ed. M. K. Hecht, W. Steere, and B. Wallace, 335–363. New York: Plenum Press.

Hecht, M. K., and Edwards, J. L. (1977). "The methodology of phylogenetic inference above the species level." In *Major Patterns in Vertebrate Evolution,* ed. M. K. Hecht, P. C. Goody, and B. M. Hecht, 3–51. New York: Plenum.

Hedrick, P. W. (1986). "Genetic polymorphism in heterogeneous environments: A decade later." *Annual Review of Ecology and Systematics* 17: 535–566.

Heisler, I. L., and Damuth, J. (1987). "A method of analyzing selection in hierarchically structured populations." *American Naturalist* 130: 582–602.

Hennig, W. (1966). *Phylogenetic Systematics.* Urbana: University of Illinois Press.

Herbert, S. (1980). "Introduction." In *The Red Notebook of Charles Darwin.* Ithaca, N.Y.: Cornell University Press.

Herbert, W. (1837). *Amaryllidaceae.* London: James Ridgway.

Herbold, B., and Moyle, P. B. (1986). "Introduced species and vacant niches." *American Naturalist* 128: 751–760.

Herrera, C. M. (1986). "Vertebrate-dispersed plants: Why they don't behave the way they should." In *Frugivores and Seed Dispersal,* ed. A. Estrada and T. H. Fleming, 5–18. Dordrecht: Dr. W. Junk.

Hesse, Mary (1966). *Models and Analogies in Science.* South Bend, Ind.: Notre Dame University Press.

——— (1980). *Revolutions and Reconstructions in the Philosophy of Science.* Bloomington: Indiana University Press.

——— (1985). "Texts without types and lumps without laws." *New Literary History* 17: 31–48.

Hillis, D. M. (1988). "Systematics of the *Rana pipiens* complex: Puzzle and paradigm." *Annual Review of Ecology and Systematics* 19: 39–63.

Hodge, M. J. S. (1971). "Lamarck's science of living bodies." *British Journal for the History of Science* 5: 323–352.

——— (1983). "Darwin and the laws of the animate part of the terrestrial system (1835–1837): On the Lyellian origins of his zoonomical explanatory program." *Studies in the History of Biology* 6: 1–106.

——— (1985). "Darwin as a lifelong generation theorist." In *The Darwinian Heritage,* ed. David Kohn, 207–243. Princeton: Princeton University Press.

——— (1987a). "Darwin, species, and the theory of natural selection." In *Histoire du concept de l'espèce dans les sciences de la vie,* ed. S. Atran et al., 227–252. Paris: Fondation Singer-Polignac.

——— (1987b). "Natural selection as a causal, empirical, and probabilistic theory." In *The Probabilistic Revolution*, ed. L. Krüger et al., 2: 233–270. Cambridge, Mass.: MIT Press.

——— (1989). "Darwin's theory and Darwin's argument." In *What the Philosophy of Biology Is*, ed. M. Ruse, 163–182. Dordrecht: Kluwer.

Hodge, M. J. S., and Kohn, D. (1986). "The immediate origins of natural selection." In *The Darwinian Heritage*, 185–206. Princeton: Princeton University Press.

Hoffstader, A. (1941). "Objective teleology." *Journal of Philosophy* 38: 29–39.

Holmes, E. B. (1980). "Reconsideration of some systematic concepts and terms." *Evolutionary Theory* 5: 35–87.

Holsinger, K. E. (1984). "The nature of biological species." *Philosophy of Science* 51: 293–307.

Holt, R. D. (1977). "Predation, apparent competition and the structure of prey communities." *Theoretical Population Biology* 12: 197–229.

Hooker, J. D. (1853). "Flora Novae-Zelandicae." Part I: *Flowering Plants.* London: Lovell Reeve.

——— (1861). "Outlines of the distribution of Arctic plants." *Transactions of the Linnean Society* 23: 251–348.

Hooker, J. D., and Thomson, T. (1855). *Flora Indica.* London: Pamplin.

Hubbs, C. L. (1944). "Concepts of homology and analogy." *American Naturalist* 78: 289–307.

Hull, David (1965). "The effect of essentialism on taxonomy: 2000 years of stasis." *British Journal of Philosophy of Science* 15: 314–326; 16: 1–18.

——— (1967). "Certainty and circularity in evolutionary taxonomy." *Evolution* 21: 174–189.

——— (1968). "The operational imperative: Sense and nonsense in operationism." *Systematic Zoology* 17: 438–457.

——— (1973). *Darwin and His Critics.* Chicago: University of Chicago Press.

——— (1974). *The Philosophy of Biological Science.* Englewood Cliffs, N.J.: Prentice-Hall.

——— (1976). "Are species really individuals?" *Systematic Zoology* 25: 174–191.

——— (1978). "A matter of individuality." *Philosophy of Science* 44: 335–360.

——— (1979). "The limits of cladism." *Systematic Zoology* 28: 416–440.

——— (1980). "Individuality and selection." *Annual Review of Ecology and Systematics* 11: 311–332.

——— (1984). "A matter of individuality." In *Conceptual Issues in Evolutionary Biology*, ed. E. Sober, 623–645. Cambridge, Mass.: MIT Press.

——— (1988). *Science as a Process.* Chicago: University of Chicago Press.

Huston, M.; DeAngelis, D. L.; and Post, W. M. (1988). "From individuals to ecosystems: A new approach to ecological theory." *Bioscience* 38: 682–691.

Hutchinson, G. E. (1948). "Circular causal systems in ecology." *Annals of the New York Academy of Science* 50: 221–246.

——— (1957). "Concluding remarks." *Cold Spring Harbor Symposium on Quantitative Biology* 22: 415–427.

——— (1959). "Homage to Santa Rosalia, or why are there so many kinds of animals?" *American Naturalist* 93: 145–159.

——— (1961). "The paradox of the plankton." *American Naturalist* 95: 137–145.

——— (1965). *The Ecological Theater and the Evolutionary Play.* New Haven: Yale University Press.

——— (1975). "Variations on a theme by Robert MacArthur." In *Ecology and Evolution of Communities,* ed. M. Cody and J. Diamond, 492–521. Cambridge, Mass.: Harvard University Press.

——— (1978). *An Introduction to Population Ecology.* New Haven: Yale University Press.

Hutchinson, G. E., and Deevey, E. (1949). "Ecological studies on populations." *Survey of Biological Progress* 1: 325–359.

Huxley, J. S. (1912). *The Individual in the Animal Kingdom.* Cambridge: Cambridge University Press.

——— (1938a). "The present standing of the theory of sexual selection." In *Evolution: Essays on Aspect of Evolutionary Biology, Presented to Professor E. S. Goodrich on His Seventieth Birthday,* ed. G. R. de Beer, 11–42. Oxford: Clarendon Press.

——— (1938b). "Darwin's theory of sexual selection and the data subsumed by it, in the light of recent research." *American Naturalist* 72: 416–433.

——— (1942). *Evolution: The Modern Synthesis.* London: Allen and Unwin; New York: Harper.

——— (1960). "The openbill's open bill: A teleonomic enquiry." *Zool. Jb. Syst.* 88: 9–30.

Huxley, L. (1918). *Life and Letters of Sir Joseph Dalton Hooker,* 2 vols. New York: D. Appleton.

Huxley, T. H. (1852). "Upon animal individuality." *Proceedings of the Royal Institution,* New Series, vol. 1 (1851–1854). Reprinted in *The Scientific Memoirs of Thomas Henry Huxley,* ed. M. Foster and W. R. Lankester (1892), 1: 146–151. London: Macmillan.

——— (1878). "Evolution." In *Encyclopaedia Britannica,* vol. 8, 9th ed. New York: Charles Scribner's Sons.

——— (1893). *Darwiniana: Collected Essays,* 2. London: Macmillan.

Hyman, L. H. (1940). *The Invertebrates: Protozoa through Ctenophora.* New York: McGraw-Hill.

Inglis, W. G. (1970). "The observational basis of homology." *Systematic Zoology* 15: 219–228.

Jablonski, D. (1986a). "Causes and consequences of mass extinctions: A comparative approach." In *Dynamics of Extinction,* ed. D. K. Elliot, 183–229. New York: Wiley.

——— (1986b). "Background and mass extinctions: The alteration of macroevolutionary regimes." *Science* 231: 129–133.

Jackson, J. B. C. (1985). "Clonality: A preface." In *Population Biology and Evolution of Clonal Organisms,* ed. J. B. C. Jackson, L. W. Buss, and R. E. Cook, ix–xi. New Haven: Yale University Press.

Jackson, J. B. C.; Buss, L. W.; and Cook, R. E., eds. (1985). *Population Biology and Evolution of Clonal Organisms.* New Haven: Yale University Press.

Jacob, F. (1989). *The Logic of Life: The Possible and the Actual.* London: Penguin Group.

James, F.; Johnston, R.; Wamer, N.; Niemi, G.; and Boecklen, W. (1984). "The Grinnellian niche of the wood thrush." *American Naturalist* 124: 17–47.

Janzen, D. H. (1977). "What are dandelions and aphids?" *American Naturalist* 111: 586–589.

Jardine, N. (1967). "The concept of homology in biology." *British Journal for the Philosophy of Science* 18: 125–139.

Jardine, N., and Sibson, R. (1971). *Mathematical Taxonomy.* London: John Wiley and Sons.

Jensen, A. B. (1989). "Raising IQ without increasing intelligence?" *Developmental Review* 9: 234–258.

Jeuken, M. (1952). "The concept of 'individual' in biology." *Acta Biotheoretica* 10: 57–86.

Johannsen, W. (1909). *Elemente der Exakten Erblichkeitslehre.* Jena: Gustav Fischer.

Johnson, M. E., and Colville, V. R. (1982). "Regional integration of evidence for evolution in the Silurian *Pentamerus-Pentameroides* lineage." *Lethaia* 15: 41–54.

Jones, Donald F. (1917). "Dominance of linked factors as a means of accounting for heterosis." *Genetics* 2: 466–479.

Jones, G. (1980). *Social Darwinism and English Thought.* Brighton, Sussex: Harvester.

Joravsky, D. (1970). *The Lysenko Affair.* Cambridge, Mass.: Harvard University Press.

Jordan, D. S. (1905). "The origin of species through isolation." *Science* n.s. 22: 545–562.

——— (1911). *The Heredity of Richard Rowe.* Boston: American Universalist Association.

Jordan, K. (1896). "On mechanical selection and other problems." *Novitates in Zoology* 3: 426–525.

Jordanova, L. (1984). *Lamarck.* Oxford: Oxford University Press.

Kammerer, P. (1924). *The Inheritance of Acquired Characteristics.* New York: Boni and Liveright.

Kamin, L. J. (1974). *The Science and Politics of IQ.* Potomac, Md.: Erlbaum.

Kant, I. (1973). *Critique of Judgement,* trans. J. C. Meredith. New York: Oxford University Press.

Kaplan, D. R. (1984). "The concept of homology and its central role in the elucidation of plant systematic relationships." In *Cladistics: Perspectives on the Reconstruction of Evolutionary History,* ed. T. Duncan and T. Stuessy, 51–69. New York: Columbia University Press.

Karlin, S. (1978). "Comparisons of positive assortative mating and sexual selection models." *Theoretical Population Biology* 14: 281–312.

Karlin, S.; Ghandour, G.; Ost, F.; Tavare, S.; and Korn, L. J. (1983). "New approaches for computer analysis of nucleic acid sequences." *Proceedings of the National Academy of Sciences, USA* 80: 5660–64.

Karlin, S., and Matessi, C. (1983). "Kin selection and altruism." *Proceedings of the Royal Society of London B* 219: 327–353.

Keibel, F. (1895). "Normentafeln zur Entwickelungsgeschichte der Wirbeltiere." *Ant. Anz.* 11: 225–234.

——— (1898). "Das biogentische Grundegesetz und die Cenogenese." *Ergebnisse der Anatomie und Entwicklungsgeschichte* 7: 722–792.

Keller, E. F. (1987). "Reproduction and the central project of evolutionary theory." *Biology and Philosophy* 2: 383–396.

——— (1988). "Demarcating public from private values in evolutionary discourse." *Journal of the History of Biology* 21(2): 195–211.

——— (1991). "Language and ideology in evolutionary biology: Reading cultural norms into natural law." In *The Boundaries of Humanity: Humans, Animals, and Machines,* ed. J. J. Sheehan and M. Sosna. Berkeley: University of California Press.

Kellogg, V. L. (1907). *Darwinism To-Day.* New York: Holt.

Kelly, A. (1981). *The Descent of Darwin: The Popularization of Darwinism in Germany, 1860–1914.* Chapel Hill: University of North Carolina Press.

Kerner von Marilaun, A. (1891). *Pflanzenleben,* vol. 2. Leipzig and Vienna: Geschichte der Pflanzen, Bibliographisches Institut.

Kevles, D. J. (1985). *In the Name of Eugenics: Genetics and the Uses of Human Heredity.* New York: Knopf. 2nd ed. (1986), Berkeley: University of California Press.

Key, K. H. L. (1967). "Operational homology." *Systematic Zoology* 16: 275–276.

Kielmeyer, Karl Friedrich (1793). *Ueber die Verhältnis der organischen Kräfte. Sudhoff's Archiv für die Geschichte der Medizin und der Naturwissenschaften* 23 (1930): 247–265.

Kimler, W. C. (1983). "Mimicry: Views of naturalists and ecologists before the modern synthesis." In *Dimensions of Darwinism,* ed. M. Grene, 97–128. Cambridge: Cambridge University Press.

——— (1986). "Advantage, adaptiveness and evolutionary ecology." *Journal of the History of Biology* 19: 215–234.

Kimura, M. (1968). "Evolutionary rate at the molecular level." *Nature* 217: 624–626.

——— (1969). "The rate of molecular evolution considered from the standpoint of population genetics." *Proceedings of the National Academy of Sciences, USA* 63: 1181–88.

——— (1977). "Preponderance of synonymous changes as evidence for the Neutral Theory of Molecular Evolution." *Nature* 267: 275–276.

——— (1983). *The Neutral Theory of Molecular Evolution.* Cambridge: Cambridge University Press.

——— (1989). "The Neutral Theory of Molecular Evolution and the world view of the naturalists." *Proceedings of the 16th International Congress of Genetics: Génome.*

Kimura, M., and Crow, J. F. (1964). "The number of alleles that can be maintained in a finite population." *Genetics* 49: 725–738.

Kimura, M., and Ohta, T. (1969). "The average number of generations until fixation of a mutant gene in a finite population." *Genetics* 61: 763–771.

——— (1974). "On some principles governing molecular evolution." *Proceedings of the National Academy of Sciences, USA* 71: 2848–52.

King, J. L. (1967). "The gene interaction component of the genetic load." *Genetics* 53: 403–413.

King, J. L., and Jukes, T. H. (1969). "Non-Darwinian Evolution." *Science* 164: 788–798.

Kingsland, S. E. (1985). *Modelling Nature: Episodes in the History of Population Ecology.* Chicago: University of Chicago Press.

Kingsolver, J. G., and Koehl, M. A. R. (1985). "Aerodynamics, thermoregulation, and the evolution of insect wings: Differential scaling and evolutionary change." *Evolution* 39: 488–504.

Kirkpatrick, M. (1982). "Sexual selection and the evolution of female choice." *Evolution* 36: 1–12.

——— (1985). "Evolution of female choice and male parental investment in polygynous species: The demise of the 'sexy son.'" *American Naturalist* 125: 788–810.

——— (1986). "The handicap mechanism of sexual selection does not work." *American Naturalist* 127: 222–240.

——— (1987). "Sexual selection by female choice in polygynous animals." *Annual Review of Ecology and Systematics* 187: 43–70.

Kitcher, P. (1982). "Genes." *British Journal for the Philosophy of Science* 33: 337–359.

——— (1984a). "Species." *Philosophy of Science* 51: 308–333.

——— (1984b). "Against the monism of the moment." *Philosophy of Science* 51: 616–630.

Kitts, D. B. (1983). "Can baptism alone save a species?" *Systematic Zoology* 32: 27–33.

Klaauw, C. J. van der (1966). "Introduction to the philosophic background and prospects of the supraspecific comparative anatomy of conservative characters in the adult stages of conservative elements of Vertebrata with enumeration of many examples." *Nerh. Kon. Ned. Akad. Wetensch., Afd. Natuurk, Tweede Sect.* 57: 1–196.

Koestler, A. (1967). *The Ghost in the Machine.* New York: Macmillan.

——— (1971). *The Case of the Midwife Toad.* London: Hutchinson.

Kohn, David (1980). "Theories to work by: Rejected theories, reproduction, and Darwin's path to natural selection." *Studies in the History of Biology* 4: 67–170.

Kondrashov, A. S. (1988). "Deleterious mutations and the evolution of sexual reproduction." *Nature* 336: 435–440.

Kondrashov, A. S., and Crow, J. F. (1988). "King's formula for the mutation load with epistasis." *Genetics* 120: 853–856.

Krebs, C. (1978). *Ecology, The Experimental Analysis of Distribution and Abundance.* New York: Harper and Row.

Krebs, J. R., and Harvey, P. (1988). "Lekking in Florence." *Nature* 333: 12–13.

Kropotkin, P. (1902). *Mutual Aid: A Factor in Evolution.* London: Heinemann.

——— (1908). *Mutual Aid,* 3rd ed. London: Heinemann.

Kuhn, T. S. (1979). "Metaphor in Science." In *Metaphor and Thought,* ed. Andrew Ortony. Cambridge: Cambridge University Press.

Lack, D. (1947a). *Darwin's Finches.* Cambridge: Cambridge University Press.

——— (1947b). "The significance of clutch size." *Ibis* 89: 302–352.
——— (1954). *The Natural Regulation of Animal Numbers.* Oxford: Oxford University Press.
——— (1966). *Population Studies of Birds.* Oxford: Oxford Univesity Press.
——— (1968). *Ecological Adaptations for Breeding in Birds.* London: Methuen.
Lamarck, Jean Baptiste de (1809). *Philosophie zoologique,* 2 vols. Paris: Dentu.
——— (1815–1822). *Histoire naturelle des animaux sans vertèbres,* 7 vols. Paris: Verdiere.
——— (1914). *Zoological Philosophy,* trans. Hugh Eliot. London. Reprint (1963), New York: Haffner.
Lambert, D. M.; Kingett, P. D.; and Slooten, E. (1982). "Intersexual selection: The problem and a discussion of the evidence." *Evolutionary Theory* 6: 67–78.
Lamotte, M. (1959). "Polymorphism of natural populations of *Cepaea nemoralis.*" *Cold Spring Harbor Symposia in Quantitative Biology* 24: 65–84.
Lande, R. (1976). "The maintenance of genetic variability by mutation in a polygenic character with linked loci." *Genetic Research* 26: 221–235.
——— (1980). "Sexual dimorphism, sexual selection, and adaptation in polygenic characters." *Evolution* 37: 1201–27.
——— (1981). "Models of speciation by sexual selection on polygenic traits." *Proceedings of the National Academy of Sciences, USA* 78: 3721–25.
——— (1987). "Genetic correlations between the sexes in the evolution of sexual dimorphism and mating preferences." In *Dahlem Workshop on Sexual Selection: Testing the Alternatives,* ed. J. W. Bradbury and M. B. Andersson, 83–94. Chichester: Wiley.
Lande, R., and Arnold, S. J. (1983). "The measurement of selection on correlated characters." *Evolution* 37: 1210–26.
Lankester, E. R. (1870). "On the use of the term homology in modern zoology." *Annual Magazine of Natural History,* ser. 4, 6: 34–43.
Larwood, G. P., ed. (1988). *Extinction and Survival in the Fossil Record.* Oxford: Clarendon Press.
Lauder, G. V. (1986). "Homology, analogy, and the evolution of behavior." In *Evolution of Animal Behavior,* ed. M. H. Nitecki and J. A. Kitchell, 9–40. New York: Oxford University Press.
——— (1990). "Functional morphology and systematics: Studying functional patterns in an historical context." *Annual Review of Ecology and Systematics* 21: 317–340.
La Vergata, A. L. (1987). "Au nom de l'espèce: Classification et nomenclature au XIX siècle." In *Histoire du concept de l'espèce dans les sciences de la vie.* Paris: Fondation Singer-Polignac.
Lawton, J. H. (1982). "Vacant niches and unsaturated communities: A comparison of bracken herbivores at sites on two continents." *Journal of Animal Ecology* 51: 573–595.
——— (1984). "Non-competitive populations, non-convergent communities, and vacant niches: The herbivores of bracken." In *Ecological Communities: Conceptual Issues and the Evidence,* ed. D. R. Strong, Jr., D. Simberloff, L. G. Abele, and A. B. Thistle, 67–100. Princeton: Princeton University Press.

Le Boeuf, B. J. (1974). "Male-male competition and reproductive success in elephant seals." *American Zoologist* 14: 163–176.
Lehninger, Albert L. (1971). *Bioenergetics*, 2nd ed. Menlo Park: W. A. Benjamin.
Leibniz, G. W. (1768). *Opera Omnia.* Geneva: Louis Dutens.
Lennox, James G. (1981). "Enç on Harvey and consequence etiologies." *Philosophy of Science* 48: 323–326.
——— (1983). "Robert Boyle's defense of teleological inference in experimental science." *Isis* 74: 38–52.
——— (1985). "Plato's unnatural teleology." In *Platonic Investigations,* ed. Dominic J. O'Meara, 195–218. Washington, D.C.: Catholic University Press.
Lenoir, Timothy (1982). *The Strategy of Life: Teleology and Mechanics in Nineteenth Century German Biology.* Dordrecht: D. Riedel.
Lerner, I. M. (1954). *Genetic Homeostasis.* New York: Dover Publications.
Lesch, J. E. (1975). "The role of isolation in evolution: George J. Romanes and John T. Gulick." *Isis* 66: 483–503.
Levins, R. (1966). "The strategy of model building in population biology." *American Scientist* 54: 421–431.
——— (1968). *Evolution in Changing Environments.* Princeton: Princeton University Press.
Levins, R., and Lewontin, R. C. (1980). "Dialectics and reductionism in ecology." *Synthese* 43.
——— (1985). *The Dialectical Biologist.* Cambridge, Mass.: Harvard University Press.
Lewontin, R. C. (1970). "The units of selection." *Annual Review of Ecology and Systematics* 1: 1–18.
——— (1974a). *The Genetic Basis of Evolutionary Change.* New York: Columbia University Press.
——— (1974b). "Darwin and Mendel: The materialist revolution." In *The Heritage of Copernicus,* ed. Jerzy Neyman, 166–183. Cambridge, Mass.: MIT Press.
——— (1974c). "The analysis of variance and the analysis of causes." *American Journal of Human Genetics* 26: 400–411.
——— (1977a). "Adattamento." In *Enciclopedia Einaudi,* vol. 1. Turin. Translated as "Adaptation," in *The Dialectical Biologist,* ed. R. Levins and R. C. Lewontin, 65–84. Cambridge, Mass.: Harvard University Press.
——— (1977b). "Caricature of Darwinism." *Nature* 266: 283–284.
——— (1978). "Adaptation." *Scientific American* 239(3): 212–230.
——— (1982). "Organism and environment." In *Learning, Development, and Culture,* ed. E. H. C. Plotkin. New York: John Wiley and Sons.
Lewontin, R. C., and Hubby, J. L. (1966). "A molecular approach to the study of genic heterozygosity in natural populations. II. Amount of variation and degree of heterozygosity in natural populations of *Drosophila pseudoobscura.*" *Genetics* 54: 595–609.
Li, C. C. (1967). "Fundamental theorem of natural selection." *Nature* 214: 505–506.
——— (1975). *Path Analysis: A Primer.* Pacific Grove, Calif.: Boxwood Press.

Limoges, C. (1970). *La Selection Naturelle.* Paris: Presses Universitaires de France.
Lincoln, R. J.; Boxshall, G. A.; and Clark, P. F. (1982). *A Dictionary of Ecology, Evolution, and Systematics.* Cambridge: Cambridge University Press.
Linnaeus, C. (1751). *Philosophia botanica.* Stockholm: Kiesewetter.
Lloyd, E. A. (1986). "Evaluation of evidence in group selection debates." *Proceedings of the Philosophy of Science Association 1986* 1: 483–493.
——— (1988). *The Structure and Confirmation of Evolutionary Theory.* Westport, Conn.: Greenwood Press.
——— (1989). "A structural approach to defining units of selection." *Philosophy of Science* 56: 395–418.
——— (1990). "Different questions, different answers: An anatomy of the units of selection debates." Manuscript.
Lloyd, E. A., and Gould, S. J. (1992). "Species selection on variability." *Proceedings of the National Academy of Sciences, USA.*
Lorenz, K. (1966). *On Aggression.* London: Methuen.
Ludmerer, K. L. (1972). *Genetics and American Society: A Historical Appraisal.* Baltimore: Johns Hopkins University Press.
Lush, J. L. (1937). *Animal Breeding Plans.* 3rd ed., 1945. Ames, Iowa: Iowa State University Press.
Lyell, Charles (1830–1833). *Principles of Geology,* 3 vols. London: Murray.
Lynch, M. (1984). "The limits to life history evolution in *Daphnia.*" *Evolution* 38(3): 465–482.
MacArthur, R. (1958). "Population ecology of some warblers of northern coniferous forests." *Ecology* 39: 599–619.
——— (1968). "The theory of the niche." In *Population Biology and Evolution,* ed. R. C. Lewontin, 159–176. Syracuse: Syracuse University Press.
——— (1972a). "Coexistence of species." In *Challenging Biological Problems,* ed. J. Behnke, 253–259. New York: Oxford University Press.
——— (1972b). *Geographical Ecology: Patterns in the Distribution of Species.* New York: Harper and Row.
MacArthur, R., and Levins, R. (1967). "The limiting similarity, convergence and divergence of coexisting species." *American Naturalist* 101: 377–385.
MacArthur, R., and Wilson, E. O. (1967). *The Theory of Island Biogeography.* Princeton: Princeton University Press.
MacDougall, William (1927). "An experiment for the testing of the hypothesis of Lamarck." *British Journal of Psychology* 17: 267–304.
MacPhail, E. M. (1982). *Brain and Intelligence in Vertebrates.* Oxford: Clarendon Press.
Maienschein, J.; Collins, J.; and Beatty, J., eds. (1986). "Reflections on ecology and evolution." *Journal of the History of Biology* 19: 167–312.
Margalef, R. (1968). *Perspectives in Ecological Theory.* Chicago: University of Chicago Press.
Margulis, L. (1982). *Symbiosis in Cell Evolution.* San Francisco: W. H. Freeman.
Marshall, D. R., and Jain, S. K. (1968). "Phenotypic plasticity of *Avena fatua* and *A. barbata.*" *American Naturalist* 102: 457–466.
Martin, P. S., and Klein, R. G., eds. (1984). *Quaternary Extinctions: A Prehistoric Revolution.* Tucson: University of Arizona Press.

Matessi, C., and Karlin, S. (1984). "On the evolution of altruism by kin selection." *Proceedings of the National Academy of Sciences, USA* 81: 1754–58.
——— (1986). "Altruistic behavior in sibling groups with unrelated intruders." In *Evolutionary Processes and Theory,* ed. S. Karlin and E. Nevo, 689–724. Orlando, Fla.: Academic Press.
Matthew, P. (1831). *On Naval Timber and Arboriculture.* London: Longmans.
May, R. M. (1973). *Stability and Complexity in Model Ecosystems.* Princeton: Princeton University Press.
Maynard Smith, J. (1964). "Group selection and kin selection." *Nature* 201: 1145–47.
——— (1974). *Models in ecology.* Cambridge: Cambridge University Press.
——— (1976a). "Sexual selection and the handicap principle." *Journal of Theoretical Biology* 57: 239–242.
——— (1976b). "Group selection." *Quarterly Review of Biology* 51: 277-283.
——— (1978). The Evolution of Sex. Cambridge: Cambridge University Press.
——— (1982). *Evolution and the Theory of Games.* Cambridge: Cambridge University Press.
——— (1985). "Mini review. Sexual selection, handicaps and true fitness." *Journal of Theoretical Biology* 115: 1–8.
——— (1987a). "How to model evolution." In *The Latest on the Best: Essays on Evolution and Optimality,* ed. J. Dupré, 119–131. Cambridge, Mass.: MIT Press.
——— (1987b). "Sexual selection: A classification of models." In *Dahlem Workshop on Sexual Selection: Testing the Alternatives,* ed. J. W. Bradbury and M. B. Andersson, 9–20. Chichester: Wiley.
——— (1988). "Can a mixed strategy be stable in a finite population?" *Journal of Theoretical Biology* 130: 247–251.
Mayr, Ernst (1941). *Systematics and the Origin of Species.* New York: Dover Publications.
——— (1942). *Systematics and the Origin of Species.* New York: Columbia University Press.
——— (1949). "Speciation and selection." *Proceedings of the American Philosophical Society* 93: 514–519.
——— (1959). "Agassiz, Darwin and evolution." *Harvard Library Bulletin* 13: 165–194.
——— (1960). "The emergence of evolutionary novelties." In *Evolution after Darwin,* vol. 1, ed. Sol Tax, 349–380. Chicago: University of Chicago Press.
——— (1962). "Accident or design: The paradox of evolution." In *The Evolution of Living Organisms.* Proceedings of the Darwin Centenary Symposium, Royal Society of Victoria, 1–14. Melbourne: Melbourne University Press.
——— (1963). *Animal Species and Evolution.* Cambridge, Mass.: Harvard University Press.
——— (1965a). "Cause and effect in biology." In *Cause and Effect,* ed. Daniel Lerner, 33–50. Cambridge, Mass.: MIT Press. Reprinted in *Man and Nature,* ed. R. Munson, 101–118. New York: Dell Publishing, 1971.
——— (1965b). "Selection and gerichtete Evolution." *Naturwissenschaften* 52: 173–180. Translated as "Selection and directional evolution," in *Evolution*

and the Diversity of Life, ed. E. Mayr, 44–52. Cambridge, Mass.: Harvard University Press, 1976.

——— (1969). *Principles of Systematic Zoology.* New York: McGraw-Hill.

——— (1970). *Populations, Species, and Evolution.* Cambridge, Mass.: Harvard University Press.

——— (1972a). "Lamarck revisited." *Journal of the History of Biology* 5: 55–94. Reprinted in *Evolution and the Diversity of Life,* ed. E. Mayr, 222–250. Cambridge, Mass.: Harvard University Press.

———(1972b). "Sexual selection and natural selection." In *Sexual Selection and the Descent of Man, 1871–1971,* ed. B. Campbell, 87–104. London: Heinemann.

——— (1973). "Alden Holmes Miller." In *Biographical Memoirs of the National Academy of Sciences.* New York: Columbia University Press, 176–214.

——— (1974). "Cladistic analysis of cladistic classification?" *Zeitschrift für Zoologische Systematik und Evolutionsforschung* 12: 94–128.

——— (1976). "Typological versus population thinking." In *Evolution and the Diversity of Life,* ed. E. Mayr, 26–29. Cambridge, Mass.: Harvard University Press.

——— (1982a). *The Growth of Biological Thought.* Cambridge, Mass.: Harvard University Press.

———(1982b). "Adaptation and selection." *Biologisches Zentralblatt* 102: 161–174. Abridged reprint in E. Mayr, *Toward a New Philosophy of Biology,* 133–148. Cambridge, Mass.: Harvard University Press.

——— (1983). "How to carry out the adaptationist program." *American Naturalist* 121: 324–334.

——— (1986). "Natural selection: The philosopher and the biologist." *Paleobiology* 12(2): 233–239.

——— (1987). "The ontological status of species: Scientific progress and philosophical terminology." *Biology and Philosophy* 2: 145–166.

——— (1988). *Toward a New Philosophy of Biology.* Cambridge, Mass.: Harvard University Press.

Mayr, Ernst, and Provine, William, eds. (1980). *The Evolutionary Synthesis.* Cambridge, Mass.: Harvard University Press.

McIntosh, R. P. (1967). "A continuum concept of vegetation." *Botanical Review* 33: 130–187.

McIntosh, R. (1985). *The Background of Ecology, Concept and Theory.* New York: Columbia University Press.

McKinney, M. L. (1988). *Heterochrony in Evolution: A Multidisciplinary Approach.* New York: Plenum Press.

Meckel, Johann (1821). *System der vergleichenden Anatomie,* vol. 1. Halle: Renger.

Medvedev, Z. (1969). *The Rise and Fall of T. D. Lysenko.* New York: Columbia University Press.

Mehnert, E. (1895). "Die individuelle Variation des Wirbelthierembryo." *Morphologische Arbeiten* 5: 386–444.

——— (1897). "Kainogenese." *Morphologische Arbeiten* 7: 1–156.

——— (1898). *Biomechanik.* Jena: Gustav Fischer.

Mendel, Gregor (1865). "Experiments on plant hybrids." In *The Origin of Genetics, A Mendel Sourcebook,* ed. Curt Stern and Eva Sherwood (1966), 1–48. San Francisco: W. H. Freeman.
Merrell, D. (1981). *Ecological Genetics.* Minneapolis: University of Minnesota Press.
Meyen, S. V. (1973). "Plant morphology in its nomothetic aspects." *The Botanical Review* 39: 205–260.
Meyer, A. (1987). "Phenotypic plasticity and heterochrony in *Cichlasoma managuense* (Pisces, Cichlidae) and their implications for speciation in cichlid fishes." *Evolution* 41(6): 1357–69.
Michod, R. E. (1982). "The theory of kin selection." *Annual Review of Ecology and Systematics* 13: 23–55.
Michod, R. E., and Abugov, R. (1980). "Adaptive topographies in family structured models of kin selection." *Science* 210: 667–669.
Michod, R. E., and Hamilton, W. D. (1980). "Coefficients of relatedness in sociobiology." *Nature* 288: 694–697.
Michod, R. E., and Hasson, O. (1990). "On the evolution of reliable indicators of fitness." *American Naturalist* 135: 788–808.
Mills, S., and Beatty, J. (1979). "The propensity interpretation of fitness." *Philosophy of Science* 46: 263–286.
Milne, A. (1961). "Definition of competition among animals." *Symposia of the Society for Experimental Biology* 15: 40–61.
Milthorpe, F. L. (1961). "Preface." *Symposia of the Society for Experimental Biology* 15: vii.
Mishler, B. D., and Brandon, R. N. (1987). "Individuality, pluralism, and the phylogenetic species concept." *Biology and Philosophy* 2: 397–414.
Mishler, B. D., and Donoghue, M. J. (1982). "Species concepts: A case for pluralism." *Systematic Zoology* 31: 491–503.
Moment, G. B. (1945). "The relationship between serial and special homology and organic similarities." *American Naturalist* 79: 445–455.
Moore, James R. (1979). *The Post-Darwinian Controversies.* Cambridge: Cambridge University Press.
Moran, P. A. P. (1962). "The statistical processes of evolutionary theory." *Heredity* 12: 145–167.
Morgan, T. H. (1917). "The theory of the gene." *American Naturalist* 51: 513–544.
——— (1926). *The Theory of the Gene.* New Haven: Yale University Press.
——— (1932). *The Scientific Basis of Evolution.* New York: Norton.
Morgan, T. H., et al. (1915). *The Mechanism of Mendelian Heredity.* New York: Henry Holt.
Morris, H. M. (1974). *Scientific Creationism.* San Diego: Creation-Life Publishers.
Morton, N. E.; Crow, J. F.; and Muller, H. J. (1956). "An estimate of the mutational damage in man from data on consanguineous marriages." *Proceedings of the National Academy of Sciences, USA* 42: 855–863.
Moulton, M. P., and Pimm, S. L. (1983). "An experimental test of community-wide character displacement." *Evolution* 37.
Moyle, P. B. (1986). "Fish introductions into North America: Patterns and ecolog-

ical impact." In *Ecology of Biological Invasions of North America and Hawaii,* ed. H. Mooney and J. Drake, 27–43. New York: Springer-Verlag.

Muller, H. J. (1918). "Genetic variability, twin hybrids and constant hybrids, in a case of balanced lethal factors." *Genetics* 3: 422–429.

——— (1922). "Variation due to change in the individual gene." *American Naturalist* 56: 32–50.

——— (1927). "Artificial transmutation of the gene." *Science* 66: 84–87.

——— (1940). "Bearings of the 'Drosophila' work on systematics." In *The New Systematics,* ed. Julian Huxley. Oxford: Oxford University Press.

——— (1950). "Our load of mutations." *American Journal of Human Genetics* 2: 111–176.

Müller-Hill, Benno. (1988). *Murderous Science: Elimination by Scientific Selection of Jews, Gypsies.* Oxford: Oxford University Press.

Müntzing, A.; Tedin, O.; and Turesson, G. (1931). "Field studies and experimental methods in taxonomy." *Hereditas* 15: 1–12.

Myers, N. (1988). "Tropical forests and their species: Going, going . . . ?" In *Biodiversity,* ed. E. O. Wilson, 28–35. Washington, D.C.: National Academy Press.

Nagel, Ernst (1961). *The Structure of Science.* New York: Harcourt, Brace and World.

Nakatsuru, K., and Kramer, D. L. (1982). "Is sperm cheap? Limited male fertility in the lemon tetra (Pisces, Characidae)." *Science* 216: 753–755.

Neff, N. A. (1986). "A rational basis for *a priori* character weighting." *Systematic Zoology* 35: 110–123.

Nelson, G. (1970). "Outline of a theory of comparative biology." *Systematic Zoology* 19: 373–384.

——— (1989). "Cladistics and evolutionary models." *Cladistics* 5: 275–289.

Nelson, G., and Platnick, N. (1981). *Systematics and Biogeography: Cladistics and Vicariance.* New York: Columbia University Press.

Nevo, E.; Beiles, A.; and Ben-Schlomo, R. (1984). "The evolutionary significance of genetic diversity: Ecological, demographic, and life history correlates." In *Evolutionary Dynamics of Genetic Diversity,* ed. S. S. Mani, 13–213. Berlin: Springer-Verlag.

Neyman, J.; Park, T.; and Scott, E. (1956). "The struggle for existence, the *Tribolium* model: Biological and statistical aspects." In *Proceedings of the Third Berkeley Symposium on Mathematical Statistics and Probability,* vol. 4, ed. J. Neyman, 41–79.

Nicholson, A. J. (1933). "The balance of animal populations." *Journal of Animal Ecology* 2: 132–178.

——— (1954). "An outline of the dynamics of animal populations." *Australian Journal of Zoology* 2: 9–65.

Nitecki, M. H., ed. (1984). *Extinction.* Chicago: University of Chicago Press.

——— ed. (1988). *Evolutionary Progress.* Chicago: University of Chicago Press.

Norton, B. J. (1973). "The biometric defense of Darwin." *Journal of the History of Biology* 6: 283–316.

Nunney, L. (1985). "Group selection, altruism, and structured deme models." *American Naturalist* 126: 212–230.

Nuovo, V. (1992). "The concept of adaptation from Paley through Darwin." *Synthese.*

O'Donald, P. (1962). "The theory of sexual selection." *Heredity* 17: 541–552.

——— (1980). *Genetic Models of Sexual Selection.* Cambridge: Cambridge University Press.

——— (1983). "Sexual selection by female choice." In *Mate Choice,* ed. P. Bateson, 53–66. Cambridge: Cambridge University Press.

Odum, E. P. (1953). *Fundamentals of Ecology,* 1st ed. (2nd ed., 1959; 3rd ed., 1971). Philadelphia: Saunders.

——— (1971). *Fundamentals of Ecology,* 3rd ed. Philadelphia: Saunders.

Ohta, D. (1983). "Hierarchical theory of selection: The covariance formula of selection and its application." *Bulletin of the Biometrical Society of Japan* 4: 25–33.

Olby, R. (1974). *The Path to the Double Helix.* Seattle: University of Washington Press.

——— (1985). *Origins of Mendelism,* 2nd ed. Chicago: University of Chicago Press.

O'Malley, C. D.; Poynter, F. N. L.; and Russell, K. F. (1961). *William Harvey's Lectures on the Whole of Anatomy.* Berkeley: University of California Press.

O'Neill, R. V.; DeAngelis, D. L.; Waide, J. B.; and Allen, T. F. H. (1986). *A Hierarchical Concept of the Ecosystem.* Princeton: Princeton University Press.

Oppel, A. (1891). *Vergleichung des Entwicklungsgrades der Organe zu verscheidenen Entwicklungszeiten bei Werbeltieren.* Jena: Gustav Fischer.

Ospovat, D. (1981). *The Development of Darwin's Theory.* Cambridge: Cambridge University Press.

Owen, R. (1843). *Lectures on the Comparative Anatomy and Physiology of the Invertebrate Animals.* London: Longman, Brown, Greene, and Longmans.

——— (1848). *On the Archetype and Homologies of the Vertebrate Skeleton.* London: R. and J. E. Taylor.

——— (1849). *On the Nature of Limbs.* London: Voorst.

Oxford English Dictionary: Compact Edition (1971). Oxford: Oxford University Press.

Oyama, S. (1985). *The Ontogeny of Information.* Cambridge: Cambridge University Press.

Packard, A. (1888). "The cave fauna of North America." *Memoirs of the National Academy of Science* 4, part 1: 1–156.

Pagel, Walter (1967). *William Harvey's Biological Ideas.* Basel: Karger.

——— (1976). *New Light on William Harvey.* Basel: Karger.

Palumbi, S. (1984). "Tactics of acclimation: Morphological changes of sponges in an unpredictable environment." *Science* 225: 1478–80.

Park, T. (1939). "Analytical population studies in relation to general ecology." *American Midland Naturalist* 21: 235–255.

——— (1948). "Experimental studies of interspecies competition. I. Competition between populations of the flour beetles, *Tribolium confusum* Duval and *Tribolium castaneum* Herbst." *Ecological Monographs* 18: 265–307.

——— (1954a). "Competition." *Physiological Zoology* 27: 177–238.

——— (1954b). "Experimental studies of interspecies competition. II. Temperature, humidity, and competition in two species of *tribolium*." *Physiological Zoology* 27: 177–238.

——— (1962). "Beetles, competition, and populations." *Science* 138: 1369–75.

Parker, G. A. (1982). "Phenotype-limited evolutionarily stable strategies." In *Current Problems in Sociobiology,* ed. King's College Sociobiology Group, 173–201. Cambridge: Cambridge University Press.

Partridge, L., and Halliday, T. (1984). "Mating patterns and mate choice." In *Behavioral Ecology: An Evolutionary Approach,* 2nd ed., ed. J. R. Krebs and N. B. Davies, 222–250. Oxford: Blackwell.

Pateman, C. (1988). *The Sexual Contract.* Stanford: Stanford University Press.

Paterson, H. E. H. (1978). "More evidence against speciation by reinforcement." *South African Journal of Science* 74: 369–371.

——— (1982). "Perspective on speciation by reinforcement." *South African Journal of Science* 78: 53–57.

——— (1985). "The recognition concept of species." *Species and Speciation,* Transvaal Museum Monograph No. 4, ed. E. Vrba, 21–29. Pretoria: Transvaal Museum.

——— (1989). "A view of species." In *Dynamic Structures in Biology,* ed. B. Goodwin, A. Sibatani, and G. Webster, 77–88. Edinburgh: Edinburgh University Press.

Patterson, C. (1982). "Morphological characters and homology." In *Problems of Phylogenetic Reconstruction,* ed. K. A. Joysey and A. E. Friday, 21–74. London: Academic Press.

——— (1987). "Introduction." In *Molecules and Morphology in Evolution: Conflict or Compromise?,* ed. C. Patterson, 1–22. Cambridge: Cambridge University Press.

——— (1988). "Homology in classical and molecular biology." *Molecular Biology and Evolution* 5: 603–625.

Paul, D. B. (1988). "The selection of the 'Survival of the Fittest.'" *Journal of the History of Biology* 21: 411–425.

Pearl, R., and Reed, L. J. (1920). "On the rate of growth of the population of the United States since 1870 and its mathematical representation." *Proceedings of the National Academy of Sciences* 6: 275–288.

Peck, J., and Feldman, M. W. (1986). "The evolution of helping behavior in large randomly-mixed populations." *American Naturalist* 127: 209.

Perry, R. B. (1921). "Purpose." *Journal of Philosophy* 18: 85–105.

Peters, J. A., ed. (1959). *Classic Papers in Genetics.* Englewood Cliffs, N.J.: Prentice-Hall.

Peters, R. (1976). "Tautology in evolution and ecology." *American Naturalist* 110: 1–12.

Pianka, E. (1974). *Evolutionary Ecology.* New York: Harper and Row.

——— (1976). "Competition and niche theory." In *Theoretical Ecology: Principles and Applications,* ed. R. M. May, 114–141. Philadelphia: W. B. Saunders.

——— (1981). "Competition and niche theory." In *Theoretical Ecology: Principles and Applications,* ed. R. M. May, 167–196. Oxford: Blackwell Scientific.

——— (1983). *Evolutionary Ecology,* 3rd ed. New York: Harper and Row.

Pickett, S. T. A., and White, P. S. (1985). *The Ecology of Natural Disturbance and Patch Dynamics.* Orlando Fla.: Academic Press.
Pielou, E. C. (1975). *Ecological Diversity.* New York: Wiley.
Pimentel, R. A., and Riggins, R. (1987). "The nature of cladistic data." *Cladistics* 3: 201–209.
Pimm, S. L.; Jones, H. L.; and Diamond, J. (1988). "On the risk of extinction." *American Naturalist* 132: 757–785.
Platnick, N. I. (1979). "Philosophy and the transformation of cladistics." *Systematic Zoology* 128: 537–547.
Playfair, J. (1802). *Illustrations of the Huttonian Theory of the Earth.* London: Caddell and Davies; Edinburgh: William Creech.
Pollack, E. (1978). "With selection for fecundity the mean fitness does not necessarily increase." *Genetics* 90: 383–389.
Pollack, E., and Kempthorne, O. (1971). "Malthusian parameters in genetics populations: II. Random mating populations in infinite habitats." *Theoretical Population Biology* 2: 357–390.
Pollock, G. B. (1988). "Suspending disbelief: Of Wynne-Edwards and his critics." *Journal of Evolutionary Biology* 2: 205–221.
Pomiankowski, A. (1987a). "The costs of choice in sexual selection." *Journal of Theoretical Biology* 128: 195–218.
——— (1987b). "Sexual selection: The handicap principle does work—sometimes." *Proceedings of the Royal Society of London, B* 231: 123–145.
Popper, K. R. (1972). *Objective Knowledge.* Oxford: Oxford University Press.
Poulton, E. B. (1896). *Charles Darwin and the Theory of Natural Selection.* London: Cassell.
——— (1908). *Essays on Evolution, 1889–1907.* Oxford: Clarendon Press.
Pound, R. (1893). "Symbiosis and mutualism." *American Naturalist* 27: 509–520.
Price, G. R. (1970). "Selection and covariance." *Nature* 227: 520–521.
——— (1972). "Extension of covariance selection mathematics." *Annals of Human Genetics* 35: 485–490.
Price, P. W. (1980). *Evolutionary Biology of Parasites.* Princeton: Princeton University Press.
——— (1984). "Communities of specialists: Vacant niches in ecological and evolutionary time." In *Ecological Communities: Conceptual Issues and the Evidence,* ed. D. R. Strong, Jr., D. Simberloff, L. G. Abele, and A. B. Thistle, 510–524. Princeton: Princeton University Press.
Pringle, J. W. S. (1951). "On the parallel between learning and evolution." *Behaviour* 3: 90–110.
Proctor, R. (1988). *Racial Hygiene: Medicine under the Nazis.* Cambridge, Mass.: Harvard University Press.
Provine, W. B. (1985). "Adaptation and mechanisms of evolution after Darwin." In *The Darwinian Heritage,* ed. D. Kohn, 825–866. Princeton: Princeton University Press.
———, ed. (1986a). *Sewall Wright: Evolution.* Chicago: University of Chicago Press.

——— (1986b). *Sewall Wright and Evolutionary Biology.* Chicago: University of Chicago Press.

Pusey, J. R. (1983). *China and Charles Darwin.* Cambridge, Mass.: Harvard University Press.

Quamman, D. (1987). "Aliens." *The Pan American Review* (October): 27–30. Reprinted from *Outdoors.*

Queiroz, K. de (1985). "The ontogenetic method for determining character polarity and its relevance to phylogenetic systematics." *Systematic Zoology* 34: 280–299.

Queiroz, K. de, and Donoghue, M. J. (1990). "Phylogenetic systematics of Nelson's version of cladistics?" *Cladistics* 6: 61–76.

Randall, J. H., Jr. (1961). *The School of Padua and the Rise of Modern Science.* Padua: Antenore.

Rao, D. C.; Morton, N. E.; Lalouel, J. M.; and Lew, R. (1982). "Path analysis under generalized assortative mating, II. American I.Q." *Genetical Research Cambridge* 39: 187–198.

Rao, D. C.; Morton, N. E.; and Yee, S. (1976). "Resolution of cultural and biological inheritance by path analysis." *American Journal of Human Genetics* 28: 228–242.

Raup, D. M. (1966). "Geometric analysis of shell coiling: General problems." *Journal of Paleontology* 40: 1178–90.

——— (1967). "Geometric analysis of shell coiling: Coiling in ammonoids." *Journal of Paleontology* 41: 43–65.

——— (1984). "Death of Species." In *Extinction,* ed. M. H. Nitecki, 1–19. Chicago: University of Chicago Press.

——— (1986). "Biological extinction in earth history." *Science* 231: 1528–33.

——— (1988). "Diversity crises in the geological past." In *Biodiversity,* ed. E. O. Wilson, 51–57. Washington, D.C.: National Academy Press.

Raup, D. M., and Boyajian, G. E. (1988). "Patterns of generic extinction in the fossil record." *Paleobiology* 14: 109–125.

Raup, D. M., and Sepkoski, J. J., Jr. (1984). "Periodicity of extinctions in the geologic past." *Proceedings of the National Academy of Sciences, U.S.A.* 81: 801–805.

Raup, D. M., and Stanley, S. M. (1971). *Principles of Paleontology,* 1st ed. San Francisco: W. H. Freeman.

——— (1978). *Principles of Paleontology,* 2nd ed. San Francisco: W. H. Freeman.

Reanney, D. (1976). "Extrachromosomal elements as possible agents of adaptation and development." *Bacteriological Reviews* 40(3): 552–590.

——— (1977). "Genetic engineering as an adaptive strategy." In *Brookhaven Symposium in Biology: Genetic Interaction and Gene Transfer* 29: 248–271.

Reeck, G. R.; de Haen, C.; Teller, D. C.; Doolittle, R. F.; Fitch, W. M.; Dickerson, R. E.; Chambon, P.; McLachlan, A. D.; Margoliash, E.; Jukes, T. H.; and Zuckerkandl, E. (1987). "Homology." In *Proteins and Nucleic Acids: A Terminological Muddle and a Way Out of It. Cell* 50: 667.

Reed, E. S. (1981). "The lawfulness of natural selection." *American Naturalist* 118: 61–71.

Regan, C. T. (1926). "Organic evolution." *Report of the British Association for the Advancement of Science* (1925): 75–86.

Remane, A. (1952). *Die Grundlagen des naturlichen Systems der vergleichenden Anatomie und der Phylogenetik*. Leipzig: Geest and Portig.

Rensch, B. (1959). *Evolution above the Species Level.* New York: Columbia University Press.

Rice, J.; Cloninger, C. R.; and Reich, T. (1978). "Multifactor inheritance with cultural transmission and assortative mating. I. Description and basic properties of the unitary models." *American Journal of Human Genetics* 30: 618–643.

Richards, O. W. (1927). "Sexual selection and allied problems in the insects." *Biological Reviews* 2: 298–360.

Richards, R. J. (1980). "Christian Wolff's prolegomena to empirical and rational psychology: Translation and commentary." *Proceedings of the American Philosophical Society* 124: 227–239.

——— (1987). *Darwin and the Emergence of Evolutionary Theories of Mind and Behavior.* Chicago: University of Chicago Press.

——— (1988). "The moral foundations of the idea of evolutionary progress: Darwin, Spencer, and the Neo-Darwinians." In *Evolutionary Progress,* ed. Matther Nitecki, 129–148. Chicago: University of Chicago Press.

——— (1992). *The Meaning of Evolution: The Morphological Construction and Ideological Reconstruction of Darwin's Theory.* Chicago: University of Chicago Press.

Ricklefs, R. (1979). *Ecology,* 2nd ed. New York: Chiron Press.

Riedl, R. J. (1978). *Order in Living Organisms.* New York: John Wiley and Sons.

Rieger, R., and Tyler, S. (1979). "The homology theorem in ultrastructural research." *American Zoologist* 19: 655–664.

Rieppel, O. (1980). "Homology, a deductive concept?" *Zeitshrift für Zoologische Systematik und Evolutionsforschung* 18: 315–319.

——— (1988). *Fundamentals of Comparative Biology.* Basel: Birkhauser Verlag.

Rinard, R. G. (1981). "The problem of the organic individual: Ernst Haeckel and the development of the Biogenetic Law." *Journal of the History of Biology* 14: 249–275.

Rodrigues, P. D. (1986). "On the term character." *Systematic Zoology* 35: 140–141.

Roe, K. E., and Frederick, R. G. (1981). *Dictionary of Theoretical Concepts in Biology.* Metuchen, N.J.: Scarecrow Press.

Roe, S. (1981). *Matter, Life, and Generation: 18th-Century Embryology and the Haller-Wolff Debate.* Cambridge: Cambridge University Press.

Rolfe, R. A. (1900). "Hybridisation viewed from the standpoint of systematic botany." *Journal of the Royal Horticultural Society* 24: 181–202.

Roll-Hansen, N. (1985). "A new perspective on Lysenko?" *Annals of Science* 42: 261–276.

——— (1988). "The progress of eugenics: Growth of knowledge and change in ideology." *History of Science* 26: 295–331.

Romer, A. S. (1949). "Time series and trends in animal evolution." In *Genetics,*

Paleontology, and Evolution, ed. G. L. Jepsen, E. Mayr, and G. G. Simpson, 103–120. Princeton: Princeton University Press.

Rorty, Richard (1985). "Texts and lumps." *New Literary History* 17(1): 1–17.

Rosenberg, A. (1985). *The Structure of Biological Science*. Cambridge: Cambridge University Press.

Rosenbueth, A.; Wiener, N.; and Bigelow, J. (1943). "Behavior, purpose and teleology." *Philosophy of Science* 10: 18–19.

Ross, H. H. (1962). *A Synthesis of Evolutionary Theory*. Englewood Cliffs, N.J.: Prentice-Hall.

Roth, V. L. (1984). "On homology." *Biological Journal of the Linnean Society* 22: 13–29.

——— (1988). "The biological basis of homology." In *Ontogeny and Systematics*, ed. C. J. Humphries, 1–26. New York: Columbia University Press.

——— (1989). "Fabricational noise in elephant dentitions." *Paleobiology* 15(2): 165–179.

Rothschild, W. (1895). "A revision of the *Papilios* of the eastern hemisphere exclusive of Africa." *Novitates Zoologicae* 2: 166–463, pl. 6.

Roughgarden, J. (1972). "Evolution of niche width." *American Naturalist* 106: 683–718.

——— (1976). "Resource partitioning among competing species: A coevolutionary approach." *Theoretical Population Biology* 9: 388–424.

Rudwick, M. J. S. (1976). *The Meaning of Fossils*, 2nd ed. Chicago: University of Chicago Press.

Ruse, M. (1973). *The Philosophy of Biology*. London: Hutchinson University Library.

——— (1979). *The Darwinian Revolution: Science Red in Tooth and Claw*. Chicago: University of Chicago Press.

——— (1982). *Darwinism Defended: A Guide to the Evolution Controversies*. London: Addison-Wesley.

——— (1988a). *Philosophy of Biology Today*. Albany: SUNY Press.

——— (1988b). *But Is It Science? The Philosophical Question in the Creation/Evolution Controversy*. Buffalo: Prometheus.

——— (1989). *The Darwinian Paradigm: Essays on Its History, Philosophy and Religious Implications*. London: Routledge.

Russel, D. A. (1984). "Terminal cretaceous extinctions of large reptiles." In *Catastrophes and Earth History: The New Uniformitarianism*, ed. W. A. Berggren and J. A. Van Couvering, 373–384. Princeton: Princeton University Press.

Russell, E. S. (1945). *The Directiveness of Organic Activities*. Cambridge: Cambridge University Press.

Russett, C. E. (1976). *Darwin in America: The Intellectual Response, 1865–1912*. San Francisco: Freeman.

Ryder, R. (1975). *Victims of Science*. London: Davis-Poynter.

Ryle, G. (1935). "Internal relations." *Proceedings of the Aristotelian Society* (suppl.), 14: 154–172.

Sanderson, M. J. (1989). "Patterns of homoplasy in North American *Astragalus* L. (Fabaceae)." Ph.D. dissertation, University of Arizona, Tucson.

Sanderson, M. J., and Donoghue, M. J. (1989). "Patterns of variation in levels of homoplasy." *Evolution* 43: 1781–95.
Sapp, J. (1987). *Beyond the Gene: Cytoplasmic Inheritance and the Struggle for Authority in Genetics*. New York: Oxford University Press.
Sattler, R. (1966). "Towards a more adequate approach to comparative morphology." *Phytomorphology* 16: 417–429.
——— (1984). "Homology: A continuing challenge." *Systematic Botany* 9: 382–394.
——— (1988). "Homeosis in plants." *American Journal of Botany* 75: 1606–17.
Schankler, D. M. (1981). "Local extinction and ecological re-entry of early Eocene mammals." *Nature* 293: 135–138.
Scheman, N. (1983). "Individualism and the objects of psychology." In *Discovering Reality*, ed. S. Harding and M. Hintikka, 225–244. Boston: D. Reidel.
Schindewolf, Otto. (1936). *Palaeontologie, Entwicklungslehre und Genetik*. Berlin: Borntraeger.
——— (1950). *Grundfragen der Palaeontologie*. Stuttgart: Schweizerbart.
Schlichting, C. D., and Levin, D. A. (1986). "Phenotypic plasticity: An evolving plant character." *Biological Journal of the Linnean Society* 29: 37–47.
Schmalhausen, I. I. (1949). *Factors of Evolution: The Theory of Stabilizing Selection*. Philadelphia: Blakiston.
Schmidt-Nielsen, K. (1983). *Animal Physiology*, 3rd ed. New York: Cambridge University Press.
Schmitt, Charles (1983). *Aristotle and the Renaissance*. Cambridge, Mass.: Harvard University Press.
Schneider, W. (1982). "Toward the improvement of the human race: The history of eugenics in France." *Journal of Modern History* 54: 268–291.
Schoener, T. W. (1983). "Field experiments on competition." *American Naturalist* 122: 241–285.
——— (1986). "Resource partitioning." In *Community Ecology: Pattern and Process*, ed. J. Kikkawa and D. Anderson, 91–126. Oxford: Blackwell Scientific Publications.
——— (1987). "Axes of controversy in community ecology." In *Community and Evolutionary Ecology of North American Stream Fishes*, ed. W. J. Matthews and D. C. Heins, 8–16. Norman: University of Oklahoma Press.
——— (1989). "The ecological niche." *Ecological Concepts*, ed. J. M. Cherrett. London: Blackwell Scientific Publications.
Scriven, M. (1959). "Explanation and prediction in evolutionary theory." *Science* 130: 477–482.
Scudder, G. E., and Reveal, J. L. (1981). *Evolution Today*. Proceedings of the Second International Congress of Systematic and Evolutionary Biology, University of British Columbia, 1980. Pittsburgh: Hunt Institute for Botanical Documentation, Carnegie-Mellon University.
Searcy, W. A. (1982). "The evolutionary effects of mate selection." *Annual Review of Ecology and Systematics* 13: 57–85.
Sebright, J. (1809). *The Art of Improving the Breeds of Domestic Animals*. London: J. Harding. Reprinted with Darwin's annotations in *Artificial Selection*

and the Development of Evolutionary Theory, ed. C. J. Bajema. Stroudsburg, Pa.: Hutchinson Ross, 1982.
Seger, J. (1981). "Kinship and covariance." *Journal of Theoretical Biology* 91: 191–213.
——— (1985). "Unifying genetic models for the evolution of female choice." *Evolution* 39: 1185–93.
Seger, J., and Brockmann, J. (1987). "What is bet-hedging?" *Oxford Studies in Evolutionary Biology* 4: 182–211.
Seilacher, A. (1973). "Fabricational noise in adaptive morphology." *Systematic Zoology* 22: 451–465.
Selander, R. K. (1985). "Protein polymorphism and the genetic structure of natural populations of bacteria." In *Population Genetics and Molecular Evolution,* ed. T. Ohta and K. Aoki, 85–106. Tokyo: Japanese Scientific Societies Press.
Sepkoski, J. J., Jr. (1987). "Reply to Patterson and Smith." *Nature* 330: 251–252.
Serres, E. R. (1824–26). *Anatomie comparée du cerveau,* vols. 1 and 2. (Printed in 1827). Paris: Gabon.
——— (1827). "Théorie des formations organiques." *Annales des Sciences Naturelles* 12: 82–143.
——— (1837). "Zoologie: Anatomie des mollusques." *L'Institut, Section des Sciences Mathématiques, Physiques et Naturelles* no. 191 (4 Jan.): 370–371.
Shaw, G. B. (1974). *The Bodley Head Bernard Shaw.* London: Bodley Head.
Sheppard, P. M. (1951). "Fluctuations in the selective value of certain phenotypes in the polymorphic land snail *Cepaea nemoralis (L).*" *Heredity* 5: 125–134.
——— (1952). "Natural selection in two colonies of the polymorphic land snail *Cepaea nemoralis.*" *Heredity* 6: 233–238.
Shull, G. H. (1909). "A pure-line method of corn breeding." *Proceedings of the American Breeding Association* 5: 51–59.
——— (1914). "Duplicate genes for which capsule form in Brusa bursa-pastoris." *Zeitschrift für Induktive abstammungs- und Vererbungslehre* 12: 97–149.
——— (1948). "What is heterosis?" *Genetics* 33: 439–446.
Silvertown, J. W. (1984). "Phenotypic variety in seed germination behavior: The ontogeny and evolution of somatic polymorphism in seeds." *American Naturalist* 124: 1–16.
Simberloff, D. (1981). "Community effects of introduced species." In *Biotic Crises in Ecological and Evolutionary Time,* ed. T. H. Nitecki, 53–81. New York: Academic Press.
——— (1982). "The status of competition theory in ecology." *Annales Zoologici Fennici* 19: 241–253.
——— (1984). "The great god of competition." *The Sciences* 24(4): 16–22.
Simberloff, D., and Colwell, R. K. (1984). "Release of engineered organisms: A call for ecological and evolutionary assessment of risks." *Genetic Engineering News* 4(7): 4.
Simpson, G. G. (1944). *Tempo and Mode in Evolution.* New York: Columbia University Press.
——— (1949). *The Meaning of Evolution.* New Haven: Yale University Press.
——— (1950). *The Meaning of Evolution: A Study of the History of Life and of Its Significance for Man.* London: Oxford University Press.

——— (1953). *The Major Features of Evolution.* New York: Simon and Schuster.
——— (1961). *Principles of Animal Taxonomy.* New York: Columbia University Press.
——— (1964). *This View of Life.* New York: Harcourt, Brace and World.
Singer, P. (1976). *Animal Liberation.* London: Jonathan Cape.
Sinnott, E., et al. (1958). *Principles of Genetics,* 5th ed. New York: McGraw-Hill.
Slatkin, M. (1981). "A diffusion model of species selection." *Paleobiology* 7(4): 421–425.
Sloan, P. R. (1979). "Buffon, German biology, and the historical interpretation of biological species." *British Journal for the History of Science* 12: 109–153.
——— (1986). "Darwin, vital matter, and the transformation of species." *Journal of the History of Biology* 19: 369–445.
——— (1987). "Buffon's idea of biological species." In *Histoire du concept d'espèce dans les sciences de la vie,* ed. S. Atran et al., 101–140. Paris: Fondation Singer-Polignac.
——— (1992). "Introduction." In *Richard Owen's Hunterian Lectures, May–June 1836,* ed. P. R. Sloan. Chicago: University of Chicago Press.
Slobodkin, L. B., and Rapoport, A. (1974). "An optimal strategy of evolution." *Quarterly Review of Biology* 49: 181–200.
Smith-Gill, S. J. (1983). "Developmental plasticity: Developmental conversion versus phenotypic modulation." *American Zoology* 23: 47–55.
Sneath, P., and Sokal, R. R. (1973). *Numerical Taxonomy.* San Francisco: W. H. Freeman.
Sober, E. (1981). "Holism, individualism, and the units of selection." *Proceedings of the PSA 1980,* vol. 2: 93–121.
——— (1984a). *The Nature of Selection: Evolutionary Theory in Philosophical Focus.* Cambridge, Mass.: MIT Press.
——— (1984b). "Sets, species, and evolution: Comments on Philip Kitcher's 'Species.'" *Philosophy of Science* 51: 334–341.
——— ed. (1984c). *Conceptual Issues in Evolutionary Biology.* Cambridge, Mass.: MIT Press.
——— (1988). *Reconstructing the Past: Parsimony, Evolution and Inference.* Cambridge, Mass.: MIT Press.
——— (1990). "The Poverty of pluralism: A reply to Sterelny and Kitcher." *Journal of Philosophy* 87: 151–158.
Sober, E., and Lewontin, R. C. (1982). "Artifact, cause, and genic selection." *Philosophy of Science* 49: 157–180.
Sokal, R. R., and Rohlf, F. J. (1981). *Biometry.* San Francisco: W. H. Freeman.
Sokal, R. R., and Sneath, P. H. A. (1963). *Principles of Numerical Taxonomy.* San Francisco: W. H. Freeman.
Soloway, Richard A. (1990). *Demography and Degeneration: Eugenics and the Declining Birthrate in Twentieth-Century Britain.* Chapel Hill: University of North Carolina Press.
Sommerhoff, George (1950). *Analytical Biology.* Oxford: Oxford University Press.
Soskice, J. M. (1985). *Metaphor and Religious Language.* Oxford: Clarendon Press.

Spencer, H. (1852). "A theory of population, deduced from the general law of animal fertility." *Westminister Review*, n.s. 1, 468–501.
——— (1864). *Principles of Biology*, 2 vols. London: Williams and Norgate.
——— (1893). *Principles of Ethics*. London: Williams and Norgate.
——— (1898). *Principles of Biology*. New York: Appleton.
Stanley, S. (1975). "A theory of evolution above the species level." *Proceedings of the National Academy of Sciences, USA* 72: 646–650.
——— (1981). *The New Evolutionary Timetable: Fossils, Genes, and the Origin of Species*. New York: Basic Books.
——— (1987). *Extinction*. New York: Scientific American Books.
Star, S., and Griesemer, J. (1989). "Institutional ecology translations, and boundary objects: Amateurs and professionals in Berkeley's Museum of Vertebrate Zoology, 1907–1939." *Social Studies of Science* 19(3): 387–420.
Stauffer, R. C. (1975). *Charles Darwin's Natural Selection*. New York: Cambridge University Press.
Stearns, S. C. (1982). "The role of development in the evolution of life histories." In *Evolution and Development*, ed. J. T. Bonner, 237–258. Dahlem Konferenzen, Berlin: Springer-Verlag.
——— (1989). "The evolutionary significance of phenotypic plasticity." *Bioscience* 39(7): 436–445.
Stebbins, G. L. (1950). *Variation and Evolution in Plants*. New York: Columbia University Press.
——— (1966). *Processes of Organic Evolution*. Englewood Cliffs, N.J.: Prentice-Hall.
Steele, E. J. (1979). *Somatic Selection and Adaptive Evolution*. Toronto. Reprint, Chicago: University of Chicago Press.
Steenis, C. G. G. J. van (1948). "General considerations." In *Flora Malensiana*, ser. 1, ed. C. G. G. J. van Steenis. 4(1): xiii–lxix.
Sterelny, K., and Kitcher, P. (1988). "The return of the gene." *Journal of Philosophy* 85: 339–361.
Stevens, P. F. (1984). "Homology and phylogeny: Morphology and systematics." *Systematic Botany* 9: 395–409.
——— (1986). "Evolutionary classifications in botany, 1960–1985. *Journal of the Arnold Arboretum* 67: 313–339.
Stevens, P. F., and Cullen, S. (1990). "Linnaeus and the cortex-medulla theory, the key to his understanding of plant form and appreciation of natural relationships." *Journal of the Arnold Arboretum* 71: 179–220.
Stocking, George W., Jr. (1962). "Lamarckianism in American social science." *Journal of the History of Ideas* 23: 239–256.
Stonehouse, B., and Perrins, C., eds. (1977). *Evolutionary Ecology*. London: Macmillan.
Strawson, P. F. (1959). *Individuals*. New York: Doubleday.
Stresemann, E. (1975). *Ornithology from Aristotle to the Present*. Cambridge, Mass.: Harvard University Press.
Strong, D. R.; Lawton, J. H.; and Southwood, R. (1984). *Insects on Plants*. Cambridge, Mass.: Harvard University Press.
Strong, D. R.; Simberloff, D.; Abele, L. G.; and Thistle, A. B. (1984). *Ecological*

Communities: Conceptual Issues and the Evidence. Princeton: Princeton University Press.

Sulloway, Frank (1982). "Darwin and his finches: The evolution of a legend." *Journal of the History of Biology* 15: 1–52.

Sultan, S. E. (1987). "Evolutionary implications of phenotypic plasticity in plants." *Evolutionary Biology* 21: 127–178.

Suzuki, D. T.; Griffiths, A. J. F.; Miller, J. H.; and Lewontin, R. C. (1989). *An Introduction to Genetic Analysis,* 4th ed. New York: W. H. Freeman.

Swammerdam, Jan (1685). *Historia Insectorum Generalis.* Translated from Dutch by H. Henninius. Batavorum: Jordanum Luchtmans.

Taylor, Charles (1964). *The Explanation of Behaviour.* London: Routledge and Kegan Paul.

Taylor, P. J. (1988). "Technocratic optimism, H. T. Odum, and the partial transformation of ecological metaphor after World War II." *Journal of the History of Biology* 21: 213–244.

——— (1989). "Developmental versus morphological approaches to modeling ecological complexity." *Oikos* 54: 121–126.

Taylor, W. P. (1916). "The status of the beavers of Western North America, with a consideration of the factors in their speciation." *University of California Publications in Zoology* 12: 413–495.

Teilhard de Chardin, P. (1955). *The Phenomenon of Man.* New York: Harper.

Templeton, A. R. (1989). "The meaning of species and speciation: A genetic perspective." In *Speciation and Its Consequences,* ed. D. Otte and J. A. Endler, 3–27. Sunderland, Mass.: Sinauer.

Templeton, A. R., and Gilbert, L. (1985). "Population genetics and the coevolution of mutualism." In *The Biology of Mutualism: Ecology and Evolution,* ed. D. H. Boucher, 128–144. London: Croom Helm.

Templeton, A. R.; Sing, C. F.; and Brokaw, B. (1976). "The unit of selection in Drosophila mercatorium. I. The interaction of selection and meiosis in parthenogenesis." *Genetics* 82: 349–376.

Tennyson, A. (1851). *In Memoriam.* Reprinted in *The Works of Tennyson,* ed. H. Tennyson, 247–286. London: Macmillan, 1913.

Thoday, J. M. (1953). "Components of fitness." *Symposia of the Society for Experimental Biology* 7: 96–113.

Thompson, A. (1839). "Generation." In *The Cyclopaedia of Anatomy and Physiology,* ed. R. B. Todd, 5 vols. London: 1836–1859.

Thornhill, R. (1980a). "Sexual selection in the black-tipped hangingfly." *Scientific American* 242(6): 138–145.

——— (1980b). "Competitive, charming males and choosy females: Was Darwin correct?" *Florida Entomologist* 63: 5–30.

Thornhill, R., and Alcock, J. (1983). *The Evolution of Insect Mating Systems.* Cambridge, Mass.: Harvard University Press.

Tiedje, J. M.; Colwell, R. K.; Grossman, Y. L.; Hodson, R. E.; Lenski R. E.; Mack, R. N.; and Regal, P. J. (1989). "The planned introduction of genetically engineered organisms: Ecological considerations and recommendations." *Ecology* 70: 298–315.

Tilman, G. D. (1982). *Resource Competition and Community Structure.* Princeton: Princeton University Press.

Tinbergen, N. (1967). "Adaptive features of the black-headed gull *Larus ridibandus.*" *Proceedings of the Fourteenth International Ornthological Congress*: 43–59.

Tolman, E. C. (1932). *Purposeful Behavior in Animals and Man.* Berkeley and Los Angeles: University of California Press. Reprint (1967), New York: Appleton-Century Crofts.

Trivers, R. L. (1971). "The evolution of reciprocal altruism." *Quarterly Review of Biology* 46: 35–57.

——— (1972). "Parental investment and sexual selection." In *Sexual Selection and the Descent of Man,* ed. B. Campbell, 136–179. London: Heinemann.

——— (1985). *Social Evolution.* Menlo Park, Calif.: Benjamin Cummings.

Turelli, M. (1986). "Gaussian versus non-Gaussian genetic analyses of polygenic mutation-selection balance." In *Evolutionary Processes and Theory,* ed. S. Karlin and E. Nevo, 607–628. Orlando, Fla.: Academic Press.

Turesson, G. (1922). "The genotypical response of the plant species to the habitat." *Hereditas* 3: 211–350.

Turkington, R.; Cahn, M. A.; Vardy, A.; and Harper, J. L. (1979). "The growth, distribution and neighbour relationships of *Trifolium repens* in a permanent pasture. III. The establishment and growth of *Trifolium repens* in natural and disturbed sites." *Journal of Ecology* 67: 231–243.

Turkington, R., and Harper, J. L. (1979). The growth, distribution and neighbor relationships of *Trifolium repens* in a permanent pasture. IV. Fine Scale biotic differentiation." *Journal of Ecology* 67: 245–264.

Turner, J. R. G. (1983). "'The hypothesis that explains mimetic resemblance explains evolution': The gradualist-saltationist schism." In *Dimensions of Darwinism,* ed. M. Grene, 129–169. Cambridge: Cambridge University Press.

Udvardy, M. (1959). "Notes on the ecological concepts of habitat, biotope and niche." *Ecology* 40: 725–728.

Ulanowicz, R. E. (1986). *Growth and Development: Ecosystems Phenomenology.* New York: Springer-Verlag.

Uyenoyama, M. K., and Feldman, M. W. (1980). "Theories of kin and group selection: A population genetics perspective." *Theoretical Population Biology* 17: 380–414.

——— (1981). "On relatedness and adaptive topography in kin selection." *Theoretical Population Biology* 19: 89–123.

Valentine, J. W., ed. (1985). *Phanerozoic Diversity Patterns.* Princeton: Princeton University Press.

Van Beneden, P. J. (1873). "Un mot sur la vie sociale des animaux inferieurs." *Bulletin de l'Academie Royale de Belgique,* ser. 2, 28: 621–648.

Vandermeer, J. (1972). "Niche theory." *Annual Review of Ecology and Systematics* 3: 107–132.

Van Valen, L. M. (1973). "A new evolutionary law." *Evolutionary Theory* 1: 1–30.

——— (1976). "Ecological species, multispecies, and oaks." *Taxon* 25: 233–239.

——— (1982). "Homology and causes." *Journal of Morphology* 173: 305–312.
——— (1985). "How constant is extinction?" *Evolutionary Theory* 7: 93–106.
Vavilov, N. (1922). "The law of homologous series in variation." *Journal of Genetics* 12: 47–49.
Vehrencamp, S. L., and Bradbury, J. W. (1984). "Mating systems and ecology." In *Behavioural Ecology: An Evolutionary Approach,* ed. J. R. Krebs, 251–278. Sunderland, Mass.: Sinauer.
Via, S., and Lande, R. (1985). "Genotype-environment interaction and the evolution of phenotypic plasticity." *Evolution* 39: 505–522.
Volterra, V. (1928). "Variations and fluctuations of the number of individuals in an animal species living together." *Journal du Conseil. Conseil International pour l'Exploration de la Mer* 3: 3–51.
Vrba, E. (1984). "What is species selection?" *Systematic Zoology* 33: 318–328.
Vrba, E., and Eldredge, N. (1984). "Individuals, hierarchies, and processes: Towards a more complete evolutionary theory." *Paleobiology* 10: 146–172.
Vrba, E., and Gould, S. J. (1986). "The hierarchical expansion of sorting and selection: Sorting and selection cannot be equated." *Paleobiology* 12: 217–228.
Waddington, C. H. (1942). "Canalization of development and the inheritance of acquired characters." *Nature* 150: 563–565.
——— (1956). "Genetic assimilation of the Bithorax phenotype." *Evolution* 10: 1–13.
——— (1957a). "Selection by, of, and for." In *The Strategy of the Genes,* 59–108. London: Allen and Unwin.
——— (1957b). *The Strategy of the Genes.* London: Allen and Unwin.
Wade, M. J. (1977). "An experimental study of group selection." *Evolution* 31: 134–153.
——— (1978). "A critical review of models of group selection." *Quarterly Review of Biology* 53: 101-114.
——— (1980). "Kin selection: Its components." *Science* 210: 665–666.
——— (1985). "Soft selection, hard selection, kin selection, and group selection." *American Naturalist* 125: 61–75.
Wade, M. J., and Arnold, S. J. (1980). "The intensity of sexual selection in relation to male sexual behaviour, female choice, and sperm precedence." *Animal Behavior* 28: 446–461.
Wagner, G. P. (1989a). "The origin of morphological characters and the biological basis of homology." *Evolution* 43: 1157–71.
——— (1989b). "The biological homology concept." *Annual Review of Ecology and Systematics* 20: 51–69.
Wainwright, S. A.; Biggs, W. D.; Currey, J. D.; and Gosline, J. M. (1976). *Mechanical Design in Organisms.* Princeton: Princeton University Press.
Walker, T. D., and Valentine, J. W. (1984). "Equilibrium models of evolutionary species diversity and the number of empty niches." *American Naturalist* 124: 887–899.
Wallace, A. R. (1866). "On the phenomena of variation and geographical distribution as illustrated by the Papilionidae of the Malayan region." *Transactions of the Linnean Society London* 25: 1–72.

——— (1889). *Darwinism.* 3rd ed., 1901. London: Macmillan.
Wallace, B. (1970). *Genetic Load: Its Biological and Conceptual Aspects.* Englewood Cliffs, N.J.: Prentice Hall.
Walton, D. (1990). "The units of selection and the bases of selection." *Philosophy of Science* 58(3): 417–435.
Waters, C. Kenneth (1991). "Tempered realism about the forces of selection." *Philosophy of Science* 58(4): 553–573.
Watson, James D. (1988). *Molecular Biology of the Gene,* 4th ed. Menlo Park, Calif.: Benjamin.
Weatherhead, P. J., and Robertson, R. J. (1979). "Offspring quality and the polygyny threshold: 'The sexy son hypothesis.'" *American Naturalist* 113: 201–208.
Weismann, A. (1881). "The origin of the markings of caterpillars." In *Studies in the Theory of Descent,* ed. and trans. R. Meldola, 161–389. London: Sampson Low, Marston, Searle and Rivington.
Weiss, S. (1987). *Race Hygiene and National Efficiency: The Eugenics of Wilhelm Schallmayer.* Berkeley: University of California Press.
West-Eberhard, M. J. (1975). "The evolution of social behavior by kin selection." *Quarterly Review of Biology* 50: 1–35.
——— (1989). "Phenotypic plasticity and the origins of diversity." *Annual Review of Ecology and Systematics* 20: 249–278.
Whaley, W. G. (1944). "Heterosis." *The Botanical Review* 10: 461–498.
Whewell, William (1833). *Astronomy and General Physics Considered with Reference to Natural Theology,* 2nd ed. London: W. Pickering.
White, M. J. D. (1978). *Modes of Speciation.* San Francisco: Freeman.
Whittaker, R., and Levin, S., eds. (1975). *Niche: Theory and Application.* Stroudsburg: Dowden Hutchinson and Ross.
Whittaker, R.; Levin, S.; and Root, R. (1973). "Niche, habitat, and ecotope." *American Naturalist* 107: 321–338.
Wickler, W. (1966). "Ursprung und biologische Deutung des Genitalprasentierens mannlicher Primaten." *Tierpsychologiogiee* 23: 422–437.
Wiens, J. A. (1984). "Resource systems, populations and communities." In *A New Ecology,* ed. P. W. Price, C. N. Slobodchikoff, and W. S. Gaud, 397–436. New York: John Wiley.
Wiley, E. O. (1975). "Karl Popper, systematics, and classification: A reply to Walter Bock and other evolutionary taxonomists." *Systematic Zoology* 24: 233–243.
——— (1979). "An annotated Linnean hierarchy, with comments on natural taxa and competing systems." *Systematic Zoology* 28: 308–337.
——— (1981). *Phylogenetics: The Theory and Practice of Phylogenetic Systematics.* New York: John Wiley.
Williams, G. C. (1966). *Adaptation and Natural Selection.* Princeton: Princeton University Press.
——— (1975). *Sex and Evolution.* Princeton: Princeton University Press.
——— (1985). "A defense of reductionism in evolutionary biology." *Oxford Surveys in Evolutionary Biology* 2: 1–17.
——— (1986). "Comments." *Biology and Philosophy* 1(1): 114–122.

Williams, G. C., and Williams, D. C. (1957). "Natural selection of individually harmful social adaptations among sibs with special reference to the social insects." *Evolution* 11: 32–39.

Williams, M. B. (1970). "Deducing the consequences of evolution: A mathematical model." *Journal of Theoretical Biology* 29: 343–385.

——— (1973). "Falsifiable predictions of evolutionary theory." *Philosophy of Science* 40(4): 535–536.

——— (1989). "Evolvers are individuals: Extension of the species as individuals claim." In *What the Philosophy of Biology Is: Essays for David Hull,* ed. M. Ruse, 305–312. Amsterdam: Kluwer.

Williams, Raymond (1976). *Keywords.* 2nd ed., 1983. Oxford: Oxford University Press.

Williamson, M. H. (1957). "An elementary theory of interspecific competition." *Nature* 180: 422–425.

——— (1972). *An Analysis of Biological Populations.* London: Edward Arnold.

Willis, J. C. (1922). *Age and Area.* Cambridge: Cambridge University Press.

Wilson, D. S. (1980). *The Natural Selection of Populations and Communities.* Menlo Park, Calif.: Benjamin Cummings.

——— (1983a). "The group selection controversy: History and current status." *Annual Review of Ecology and Systematics* 14: 159–187.

——— (1983b). "The effect of population structure on the evolution of mutualism: A field test involving burying beetles and their phoretic mites." *American Naturalist* 121: 851–870.

Wilson, D. S., and Colwell, R. K. (1981). "Evolution of sex ratio in structured demes." *Evolution* 35: 882–897.

Wilson, D. S., and Sober, E. (1989). "Reviving the superorganism." *Journal of Theoretical Biology* 136: 337–356.

Wilson, E. O. (1975). *Sociobiology: The New Synthesis.* Cambridge, Mass.: Harvard University Press.

——— (1978). *On Human Nature.* Cambridge, Mass.: Harvard University Press.

——— (1984). *Biophilia.* Cambridge, Mass.: Harvard University Press.

Wilson, L. G. (1970). *Sir Charles Lyell's Scientific Journals on the Species Question.* New Haven: Yale University Press.

Wimsatt, W. C. (1972). "Teleology and the logical structure of function statements." *Studies in History and Philosophy of Science* 3: 1–80.

——— (1980). "Reductionist research strategies and their biases in the units of selection controversy." In *Scientific Discovery,* ed. T. Nickles, 213–259. Dordrecht: Reidel.

——— (1981). "Units of selection and the structure of the multi-level genome." *Proceedings of the Philosophy of Science Association 1980* 2: 122–183.

——— (1986). "Developmental constraints, generative entrenchment, and the innate acquired distinction." In *Integrating Scientific Disciplines,* ed. W. Bechtel, 185–208. Dordrecht: Martinus-Nijhoff.

Winsor, M. P. (1976). *Starfish, Jellyfish and the Order of Life.* New Haven: Yale University Press.

——— (1979). "Louis Agassiz and the species question." *Studies in the History of Biology* 13: 89–117.

Winter, W. P.; Walsh, K. A.; and Neurath, H. (1968). "Homology as applied to proteins." *Science* 162: 1433.
Wittgenstein, L. (1923). *Tractatus Logico-Philosophicus.* London: Routledge and Kegan Paul.
Wolf, E. (1982). *Europe and the People without History.* Berkeley: University of California Press.
Wolff, C. F. (1759). *Theoria Generationis.* Halae: Hendelianis.
Wollaston, T. V. (1856). *On the Variation of Species.* London: John van Voorst.
Woodfield, Andrew (1976). *Teleology.* Cambridge: Cambridge University Press.
——— (1979). "Teleology." In *Dictionary of the History of Science,* ed. W. F. Bynam, E. J. Browne, and Roy Porter, 415–416. Princeton: Princeton University Press.
Woodger, S. (1945). "On biological transformations." In *Essays on Growth and Form,* ed. E. LeGros Clark and P. B. Medawar, 95–120. Oxford: Clarendon Press.
Worster, D. (1977). *Nature's Economy: A History of Ecological Ideas.* (Reprinted in 1988.) Cambridge: Cambridge University Press.
Wright, L. (1973). "The case against teleological reductionism." *British Journal for the Philosophy of Science* 19: 211–223.
——— (1976). *Teleological Explanations.* Berkeley: University of California Press.
Wright, S. (1921). "Systems of mating. I–V." *Genetics* 6: 111-178.
——— (1930). "Review of the genetical theory of natural selection." *Journal of Heredity* 21: 349–356.
——— (1931). "Evolution in Mendelian populations." *Genetics* 16: 97–159.
——— (1932). "The roles of mutation, inbreeding, crossbreeding and selection in evolution." *Proceedings of the Sixth International Congress of Genetics* 1: 356–366.
——— (1940). "The statistical consequences of Mendelian heredity in relation to speciation." In *The New Systematics,* ed. J. S. Huxley, 161–184. Oxford: Oxford University Press.
——— (1945). "Tempo and mode in evolution: A critical review." *Ecology* 26: 415–419.
——— (1948). "On the roles of directed and random changes in gene frequency in the genetics of populations." *Evolution* 2: 279–294.
——— (1949a). "Adaptation and selection." In *Genetics, Paleontology, and Evolution,* ed. G. L. Jepson, G. G. Simpson, and E. Mayr, 365–389. Princeton: Princeton University Press.
——— (1949b). Letter to Richard Goldschmidt, August 12, 1949, Goldschmidt Papers, Bancroft Library, University of California, Berkeley, Box 4.
——— (1949c). "Population structure in evolution." *Proceedings of the American Philosophical Society* 93: 471–478.
———(1951). "Fisher and Ford on the 'Sewall Wright Effect.'" *American Scientist* 39: 452–458, 479.
——— (1955). "Classification of the factors of evolution." *Cold Spring Harbor Symposia in Quantitative Biology* 20: 16–24.
——— (1959). Discussion following Lamotte. *Cold Spring Harbor Symposia in Quantitative Biology* 24: 84.

——— (1968–1978). *Evolution and the Genetics of Populations,* 4 vols. Chicago: University of Chicago Press.

——— (1980). "Genic and organismic selection." *Evolution* 34: 825–843.

Wright, S., and Dobzhansky, T. (1946). "Genetics of natural populations. XII. Experimental reproduction of some of the changes caused by natural selection in certain populations of *Drosophila pseudoobscura.*" *Genetics* 31: 125–150.

Wynne-Edwards, V. C. (1962). *Animal Dispersion in Relation to Social Behaviour.* Edinburgh: Oliver and Boyd.

——— (1963). "Intergroup selection in the evolution of social systems." *Nature* 200: 623–626.

Young, R. M. (1985). *Darwin's Metaphor: Nature's Place in Victorian Culture.* Cambridge: Cambridge University Press.

Zahavi, A. (1975). "Mate selection: A selection for a handicap." *Journal of Theoretical Biology* 53: 205–214.

——— (1977). "The cost of honesty (further remarks on the handicap principle)." *Journal of Theoretical Biology* 67: 603–605.

——— (1981). "Natural selection, sexual selection and the selection of signals." In *Evolution Today.* Proceedings of the Second International Congress of Systematic and Evolutionary Biology, University of British Columbia, Vancouver, Canada, 1980, ed. G. E. Scudder and J. L. Reveal, 133–138. Pittsburgh: Hunt Institute for Botanical Documentation, Carnegie-Mellon University.

Zirkle, C. (1946). "The early history of the idea of the inheritance of acquired characters and pangenesis." *Transactions of the American Philosophical Society* 35: 91–151.

INDEX

Page numbers in italics indicate keyword entries.